# Oxidative Stress in Applied Basic Research and Clinical Practice



**Editor-in-Chief**
Donald Armstrong

For further volumes:
http://www.springer.com/series/8145

## Note from the Editor-in-Chief

All books in this series illustrate point-of-care testing and critically evaluate the potential of antioxidant supplementation in various medical disorders associated with oxidative stress. Future volumes will be updated as warranted by emerging new technology, or from studies reporting clinical trials.

Donald Armstrong
Editor-in-Chief

Domenico Praticò • Patrizia Mecocci
Editors

# Studies on Alzheimer's Disease

Humana Press

*Editors*
Domenico Praticò
Department of Pharmacology
Center for Translational Medicine
Temple University School of Medicine
Philadelphia, PA, USA

Patrizia Mecocci
Department of Clinical and Experimental Medicine
Institute of Gerontology and Geriatrics
Ospedale S. Maria della Misericordia Perugia
University of Perugia, Perugia, Italy

ISBN 978-1-4939-6034-7 ISBN 978-1-62703-598-9 (eBook)
DOI 10.1007/978-1-62703-598-9
Springer New York Heidelberg Dordrecht London



Printed on acid-free paper

Humana Press is a brand of Springer
Springer is part of Springer Science+Business Media (www.springer.com)

# Preface

Alzheimer's disease (AD) has no racial, socioeconomic, or geographic boundaries, since it is the most common cause of dementia in elderly individuals throughout the world.

As people worldwide live to an older age, AD has become a serious ever-growing public health problem. The importance and the scientific challenges of this disease have been well recognized by the neuroscience research community. Approximately 5 % of AD is caused by missense mutations in the gene for either the amyloid β precursor protein (APP) or some of the enzymes (i.e., presenilin-1) involved in its metabolism. However, in its sporadic form, which is the most frequent, the cause(s) remains unknown.

Among different potential mechanisms, the combination of genetic risk factors with abnormal metabolic oxidative reactions in the central nervous system has been consistently implicated in the pathogenesis of the disease, and represents the biological basis for the "oxidative stress hypothesis of AD" which was originally formulated by Markesbery in his article of 1997.

Since then, many studies, ranging from basic research to clinical and epidemiological observations, strongly support what moved from a hypothesis to a certainty: the core role of the slight unbalance between oxidants and antioxidants (redox status) that represents an input for life and, in the meantime, a risk for dysfunction and disease when uncontrolled (oxidative stress). So, after several years from its proposal, it seems appropriate to look and consider all the remarkable advances achieved in understanding the biological complexity and mechanistic relationship between oxidative stress and AD.

In constructing this book, we have tried to draw upon our own experiences and those of our colleagues in order to present the current status of both basic and clinical science research in this specific area.

The authors are expert scientists who are directly involved in the laboratory and clinical research in the field of oxidative metabolism and neurodegenerative diseases with special focus on AD.

The book is divided into three main sections. The first section focuses on new working hypotheses implicating oxidative stress and AD. Its chapters will provide the reader with insights into the impact that the oxidative stress hypothesis had in neuroscience research (Chap. 1), an overview on the general phenomenon of oxidative stress (Chap. 2), its intricate connection with inflammation (Chap. 3), new experimental evidence involving an old enzyme such as GADPH (Chap. 4), the relationship between mitochondria and Aβ (Chap. 5), and the new role that plasma membrane plays in controlling the cellular redox system (Chap. 6).

The second section reviews some of the cellular and metabolic oxidative mechanisms that have been involved in AD pathogenesis. After focusing on metal dyshomeostasis (Chap. 7) and oxidative posttranslational modification of proteins (Chap. 8), it continues with the analysis of cholesterol oxidative metabolism (Chap. 9), and the provocative theme of brain hypometabolism, oxidative stress, and maternal transmission of AD (Chap. 10), finishing with the role on HPA dysfunction and AD (Chap. 11).

The third section describes the clinical aspects and implications of the oxidative stress hypothesis in the AD neurobiology. In these final chapters the attention is shifted towards the translational aspects of the oxidative stress hypothesis, with the main goal to apply the knowledge accumulated so far into a clinical scenario by looking at preventative as well as therapeutic opportunities derived from it.

The role that diabetes plays in AD pathophysiology will be considered (Chap. 12), and the clinical topics will expand on the current state of the art on peripheral biological measure of oxidative stress in AD (Chap. 13). These two chapters will be followed by the analysis of modifiable environmental risk factors such as nutrition and lifestyle and the risk of AD (Chap. 14), and the intriguing question of the usefulness of nutrients as an important source for antioxidants in preventing cognitive decline (Chap. 15). This section will conclude our editorial effort with an analytical review of the clinical trials that have used antioxidants in mild cognitive impairments and AD (Chap. 16).

No introduction would be complete without a grateful acknowledgement of many friends and colleagues who did the work and wrote the chapters. They are not just experts in their fields but also extraordinary individuals. Thus, without any shadow of doubt this endeavor was enabled by the contributions of many friends, collaborators, and colleagues, without whose enthusiasm and engagement this work could have never been born. An extraordinary link of science and friendship that was really supportive in constructing this book.

While we largely underestimated the devotion and effort necessary on our side at the beginning, we strongly believe that it ultimately yielded an excellent product which fully achieves our original plans. Despite the fact that we confront a continuously evolving topic, where frequent updates would be desirable, if not necessary, we believe in the value of a book like ours that attempts to organize in a snapshot-picture type of work, the enormous amount of literature available in a comprehensive and fulfilling fashion. We are aware of possible gaps, some emerging materials not included, and the ever-rapid evolution of some of the themes highlighted in the book. It is our hope that our colleagues, neurobiologists, neurologists, internists, geriatricians, and other physicians alike will find this compendium a useful guide to this most exciting time in the neurobiology of AD.

Philadelphia, PA, USA
Perugia, Italy

Domenico Praticò
Patrizia Mecocci

# Contents

# Contributors

**Paul A. Adlard, Ph.D.** Synaptic Neurobiology Laboratory, The Florey Institute of Neuroscience and Mental Health, Parkville, VIC, Australia

**Maria Ankarcrona, Ph.D.** Department of Neurobiology, Care Sciences and Society, KI-Alzheimer's Disease Center (KI-ADCR), Karolinska Institutet, Stockholm, Sweden

**Pascale Barberger-Gateau, Ph.D.** ISPED, Centre INSERM U897 – Epidemiologie-Biostatistique, University of Bordeaux, Bordeaux, France

INSERM, ISPED, Centre INSERM U897–Epidemiologie-Biostatistique, Bordeaux, France

**Ashley I. Bush, Ph.D.** Oxidation Biology Laboratory, The Florey Institute of Neuroscience and Mental Health, Parkville, VIC, Australia

**D. Allan Butterfield, Ph.D.** Department of Chemistry, Center of Membrane Sciences, The Sanders-Brown Center on Aging, University of Kentucky, Lexington, KY, USA

**Rudy J. Castellani, Ph.D.** Division of Neuropathology, Department of Pathology, University of Maryland, Baltimore, MD, USA

**Fabio Di Domenico, Ph.D.** Department of Biochemical Sciences, Sapienza University of Rome, Rome, Italy

**Amna Farshori, Ph.D.** Department of Pathology, University of Maryland, Baltimore, MD, USA

**Catherine Feart, Ph.D.** ISPED, Centre INSERM U897 – Epidemiologie-Biostatistique, University of Bordeaux, Bordeaux, France

INSERM, ISPED, Centre INSERM U897 – Epidemiologie-Biostatistique, Bordeaux, France

**David I. Finkelstein, Ph.D.** Parkinson's Disease Laboratory, The Florey Institute of Neuroscience and Mental Health, Parkville, VIC, Australia

**Sara M. Hancock, Ph.D.** Synaptic Neurobiology Laboratory, The Florey Institute of Neuroscience and Mental Health, Parkville, VIC, Australia

**Kenneth Hensley, Ph.D.** Department of Pathology and Neurosciences, University of Toledo Medical Center, Toledo, OH, USA

**Luigi Iuliano, M.D.** Laboratory of Vascular Medicine and Mass Spectrometry, Department of Medico-Surgical Sciences and Biotechnology, Sapienza University of Rome, Rome, Italy

**Yash B. Joshi, B.Sc.** Department of Pharmacology, Center for Translational Medicine, Temple University School of Medicine, Philadelphia, PA, USA

**Mony de Leon, D.Ed.** Department of Psychiatry, Center for Brain Health, NYU School of Medicine, New York, NY, USA

**Valerio Leoni, M.D., Ph.D.** Laboratory of Clinical Pathology and Medical Genetics, Foundation IRCCS Institute of Neurology Carlo Besta, Milan, Italy

**Luc Letenneur, Ph.D.** ISPED, Centre INSERM U897 – Epidemiologie-Biostatistique, University of Bordeaux, France

INSERM, ISPED, Centre INSERM U897–Epidemiologie-Biostatistique, Bordeaux, France

**Bei-Xu Li, Ph.D.** Department of Pathology, University of Maryland, Baltimore, MD, USA

**Pauline McHugh, M.D.** Department of Psychiatry, Center for Brain Health, NYU School of Medicine, New York, NY, USA

**Patrizia Mecocci, M.D., Ph.D.** Department of Clinical and Experimental Medicine, Institute of Gerontology and Geriatrics, Ospedale S. Maria della Misericordia Perugia, University of Perugia, Perugia, Italy

**Lisa Mosconi, Ph.D.** Department of Psychiatry, Center for Brain Health, NYU School of Medicine, New York, NY, USA

**John Murray, M.S.** Department of Psychiatry, Center for Brain Health, NYU School of Medicine, New York, NY, USA

**Georgy Perry, Ph.D.** College of Sciences, University of Texas, San Antonio, TX, USA

**Ludger Pientka** Department of Geriatrics, Marienhospital Herne, Ruhr University, Bochum, Germany

**Marzia Perluigi, Ph.D.** Department of Biochemistry Sciences, Sapienza University of Rome, Rome, Italy

**M. Cristina Polidori, Ph.D.** Institute of Geriatrics, University of Cologne, Cologne, Germany

**Renato Polimanti, Ph.D.** Department of Biology, "Tor Vergata" University, Rome, Italy

**Domenico Praticò, M.D.** Department of Pharmacology, Center for Translational Medicine, Temple University School of Medicine, Philadelphia, PA, USA

**Carlo Salustri, Ph.D.** Institute of Cognitive Sciences and Technologies (CNR), MEG Unit, Fatebenefratelli Hospital, Rome, Italy

**Helmut Sies, M.D., Ph.D.(hon)** Institute of Biochemistry and Molecular Biology I, Heinrich-Heine-University Düsseldorf, Düsseldorf, Germany

**Mariacristina Siotto, Ph.D.** Don Carlo Gnocchi Foundation, ONLUS, Milan, Italy

**Michael A. Sirover, Ph.D.** Department of Pharmacology, Temple University School of Medicine, Philadelphia, PA, USA

**Rosanna Squitti, Ph.D.** Department of Neuroscience, AFaR—Fatebenefratelli Hospital, Rome, Italy

Department of Neurology, "Campus Bio-medico" University, Rome, Italy

**Rukhsana Sultana, Ph.D.** Department of Chemistry, Center for Membrane Sciences, The Sanders-Brown Center on Aging, University of Kentucky, Lexington, KY, USA

**Aaron M. Swomley, Ph.D.** Department of Chemistry, Center for Membrane Sciences, The Sanders-Brown Center on Aging, University of Kentucky, Lexington, KY, USA

**Weili Xu, M.D., Ph.D.** Aging Research Center, Karolinska Institutet-Stockholm University, Stockholm, Sweden

Department of Epidemiology and Statistics, Tianjin Medical University, Tianjin, People's Republic of China

# Part I
# Hypotheses

# Chapter 1
# Oxidative Stress: Impact in Neuroscience Research

**Helmut Sies**

**Abstract** "Oxidative Stress" denotes an imbalance between oxidants and antioxidants in favor of the oxidants, leading to a disruption of redox signaling and control and/or molecular damage. It is fascinating to note the development of the general concept of "oxidative stress" from basic chemistry of biomolecules to detailed ramifications in neurodegenerative diseases. This brief introduction to the book, "Oxidative Stress and Alzheimer's Disease," intends to delineate this development which extends over more than a quarter-century. One exemplary field of research in neuroscience, namely that of selenium, selenocompounds and selenoproteins, will be presented in some detail.

## 1.1 Definition of Oxidative Stress

In 1985, the term "oxidative stress" was initially defined to denote a disturbance in the prooxidant/antioxidant balance in favor of the prooxidants, resulting in potential damage [1], and knowledge on the biochemistry of oxidative stress was described in a review [2]. Since then, an enormous amount of research has expanded our grasp of free radicals and reactive oxygen and nitrogen species in biology and medicine. In particular, the field has advanced to appreciate that fundamental processes in biology and medicine involve redox processes (apart from oxidative phosphorylation), leading to exciting discoveries in "redox regulation" and "redox signaling." Accommodating these new insights, the definition was updated: "'Oxidative Stress' denotes an imbalance between oxidants and antioxidants in favor of the oxidants,

H. Sies, M.D., Ph.D.(hon) (✉)
Institute of Biochemistry and Molecular Biology I, Heinrich-Heine-University Düsseldorf, Düsseldorf, Germany
e-mail: sies@uni-duesseldorf.de

D. Praticò and P. Mecocci (eds.), *Studies on Alzheimer's Disease*, Oxidative Stress in Applied Basic Research and Clinical Practice, DOI 10.1007/978-1-62703-598-9_1, 

leading to a disruption of redox signaling and control and/or molecular damage" [3]. This definition is intentionally global, to cover a large scope, and it needs to be filled with detail as specific biological or medical problems are being considered.

## 1.2 Relation to Neurodegenerative Disorders

A highly cited article from 1993, entitled "Oxidative stress, glutamate, and neurodegenerative disorders," states that "there is an increasing amount of experimental evidence that oxidative stress is a causal, or at least ancillary, factor in the neuropathology of several adult neurodegenerative disorders, as well as in stroke, trauma, and seizures" [4]. This statement can probably rightly be upheld at present in view of much new information on the underlying biochemistry [5]. Hydrogen peroxide was recognized early on as being a mediator of amyloid-beta protein toxicity [6]. However, results from intervention studies attempting to translate such knowledge into clinical benefit were less than promising, e.g., results from a recent randomized clinical trial with cerebrospinal fluid biomarker measures employing antioxidants for Alzheimer's disease [7].

So, where are the research issues regarding oxidative stress in this context? One major issue relates to the molecular targets and their accessibility. The important prooxidant enzymes which generate superoxide anions and hydrogen peroxide are being controlled in their gene expression, subcellular localization and activity and, in disease processes, may be dysregulated/overstimulated. One major group of such enzymes is the group of NADPH oxidases, which are being addressed as therapeutic targets for neurodegenerative diseases [8]. Other prooxidant enzymes such as lipoxygenases of different type are important to consider as well. Similarly, mechanisms keeping in check the amounts of unliganded iron have again come into focus [9, 10].

Particularly sensitive molecular targets and subcellular sites in oxidation reactions deserve continuing attention. This refers, for example, to the aspects of nuclear and mitochondrial DNA oxidation in Alzheimer's disease [11]. Interest currently seems to focus to a great extent on the role of mitochondria, our energy powerhouse [11–17]. There are strategies to target molecules to mitochondria, which potentially could counteract excessive prooxidative activity [18]. However, intervention studies in humans have been performed with micronutrients, not with such targeted compounds yet [19]. Treatment with the latter may be confronted with considerable obstacles, particularly in long-term studies.

## 1.3 Research Lines in Oxidative Stress: Selenium, Selenocompounds and Selenoproteins

These introductory remarks do not attempt to cover the many advances which have emerged in recent times, e.g., the redox control of ion channels and its fundamental implications to neurobiology. Coming from the author's ongoing research interest in micronutrients, one aspect will be briefly mentioned here as an example: the role of selenium, selenocompounds and selenoproteins. Following the finding that a selenoorganic compound, ebselen (then called PZ 51), is capable of serving as an enzyme mimic which catalyzes the glutathione peroxidase reaction [20], it was noted that there could be interest in prospective therapy for cerebral ischemia [21] and that ebselen lowers plasma interleukin-6 levels and glial heme oxygenase-1 expression after focal photothrombotic brain ischemia [22]. We turned our attention to selenoprotein P, which not only is considered as a selenium transporter protein in plasma [23] but also as being capable of protecting against oxidation and nitration reactions [24]. Selenoprotein P was also found in the cerebrospinal fluid [25]. Recent work in our group addressed the protection afforded by selenoprotein P in astrocytes as well as in a broader context [26–28].

A novel level in selenoprotein research was opened when a multisystem selenoprotein deficiency disorder was described in humans [29]. The spectrum of diseases related to inborn defects of selenium biology is expanding [30]. There are relationships between selenium and cognitive impairment [31]. Regarding neuroscience research, deletion of selenoprotein P results in impaired function of parvalbumin (PV) interneurons and alterations in fear learning and sensorimotor gating in mice [32]. Furthermore, the absence of selenoprotein P but not selenocysteine lyase results in neurological dysfunction [33]. Neuronal and axonal degeneration was detected in selenoprotein P-deleted mice [34]. These latter publications unravel further avenues of more refined research for better understanding of one micronutrient's potential role in distinctive neurobiological functions. Similar breakthrough studies in oxidative stress research may be anticipated as additional components of the multifaceted redox realm are being examined using the newly available methods in cell biology and genetics [35]. The implications of selenium homeostasis and antioxidant selenoproteins in brains for disorders in the central nervous system have been discussed recently [36].

## 1.4 Outlook

Thus, a refined analysis of genetic and cell-biological control of prooxidative reactions at specific cellular and subcellular targets can well be expected to shed new light on the biological impact of oxidative stress in long-term neurodegenerative and other diseases. The interest is shifting from externally supplied small-molecule antioxidants in the form of dietary supplements at high doses towards a focus on highly distinct roles of redox reactions in cellular functions and their potential pharmacological control.

## References

1. Sies H. Oxidative stress: introductory remarks. In: Sies H, editor. Oxidative stress. London: Academic; 1985. p. 1–8.
2. Sies H. Biochemistry of oxidative stress. Angew Chem Int Ed Engl. 1986;25:1058–71.

3. Sies H, Jones D. Oxidative Stress. In: Fink G, editor. Encyclopedia of stress. Amsterdam: Elsevier; 2007. p. 45–8.
4. Coyle JT, Puttfarcken P. Oxidative stress, glutamate, and neurodegenerative disorders. Science. 1993;262:689–95.
5. Zhu X, Castellani RJ, Moreira PI, Aliev G, Shenk JC, Siedlak SL, Harris PL, Fujioka H, Sayre LM, et al. Hydroxynonenal-generated crosslinking fluorophore accumulation in Alzheimer disease reveals a dichotomy of protein turnover. Free Radic Biol Med. 2012;52:699–704.
6. Behl C, Davis JB, Lesley R, Schubert D. Hydrogen-peroxide mediates amyloid-beta protein toxicity. Cell Jun 17 1994;77:817–827.
7. Galasko DR, Peskind E, Clark CM, Quinn JF, Ringman JM, Jicha GA, Cotman C, Cottrell B, Montine TJ, et al. Antioxidants for Alzheimer disease: a randomized clinical trial with cerebrospinal fluid biomarker measures. Arch Neurol. 2012;69(7):836–41.
8. Gao HM, Zhou H, Hong JS. NADPH oxidases: novel therapeutic targets for neurodegenerative diseases. Trends Pharmacol Sci. 2012;33(6):295–303.
9. Castellani RJ, Moreira PI, Perry G, Zhu X. The role of iron as a mediator of oxidative stress in Alzheimer disease. Biofactors. 2012;38:133–8.
10. Kell DB. Towards a unifying, systems biology understanding of large-scale cellular death and destruction caused by poorly liganded iron: Parkinson's, Huntington's, Alzheimer's, prions, bactericides, chemical toxicology and others as examples. Arch Toxicol. 2010 Nov;84: 825–89.
11. Santos RX, Correia SC, Zhu X, Lee HG, Petersen RB, Nunomura A, Smith MA, Perry G, Moreira PI. Nuclear and mitochondrial DNA oxidation in Alzheimer's disease. Free Radic Res. 2012;46:565–76.
12. Du H, Guo L, Yan SS. Synaptic mitochondrial pathology in Alzheimer's disease. Antioxid Redox Signal. 2012 Jun 15;16:1467–75.
13. Hedskog L, Zhang S, Ankarcrona M. Strategic role for mitochondria in Alzheimer's disease and cancer. Antioxid Redox Signal. 2012 Jun 15;16:1476–91.
14. Leuner K, Schutt T, Kurz C, Eckert SH, Schiller C, Occhipinti A, Mai S, Jendrach M, Eckert GP, et al. Mitochondrion-derived reactive oxygen species lead to enhanced amyloid beta formation. Antioxid Redox Signal. 2012;16:1421–33.
15. Schmitt K, Grimm A, Kazmierczak A, Strosznajder JB, Gotz J, Eckert A. Insights into mitochondrial dysfunction: aging, amyloid-beta, and tau-A deleterious trio. Antioxid Redox Signal. 2012;16:1456–66.
16. Swerdlow RH. Mitochondria and cell bioenergetics: increasingly recognized components and a possible etiologic cause of Alzheimer's disease. Antioxid Redox Signal. 2012;16:1434–55.
17. Atamna H, Mackey J, Dhahbi JM. Mitochondrial pharmacology: electron transport chain bypass as strategies to treat mitochondrial dysfunction. Biofactors. 2012 Mar;38:158–66.
18. Smith RA, Hartley RC, Cocheme HM, Murphy MP. Mitochondrial pharmacology. Trends Pharmacol Sci. 2012;33(6):341–52.
19. Packer L, Sies H, Eggersdorfer M, Cadenas E. Micronutrients and brain health. Boca Raton: CRC Press; 2010.
20. Muller A, Cadenas E, Graf P, Sies H. A novel biologically active seleno-organic compound–I. Glutathione peroxidase-like activity in vitro and antioxidant capacity of PZ 51 (Ebselen). Biochem Pharmacol. 1984;33:3235–9.
21. Parnham M, Sies H. Ebselen: prospective therapy for cerebral ischaemia. Expert Opin Investig Drugs. 2000;9:607–19.
22. Gladilin S, Bidmon HJ, Divanach A, Arteel GE, Witte OW, Zilles K, Sies H. Ebselen lowers plasma interleukin-6 levels and glial heme oxygenase-1 expression after focal photothrombotic brain ischemia. Arch Biochem Biophys. 2000;380:237–42.
23. Burk RF, Hill KE. Selenoprotein P-expression, functions, and roles in mammals. Biochim Biophys Acta. 2009 Nov;1790:1441–7.
24. Arteel GE, Mostert V, Oubrahim H, Briviba K, Abel J, Sies H. Protection by selenoprotein P in human plasma against peroxynitrite-mediated oxidation and nitration. Biol Chem. 1998 Aug;379:1201–5.

25. Scharpf M, Schweizer U, Arzberger T, Roggendorf W, Schomburg L, Kohrle J. Neuronal and ependymal expression of selenoprotein P in the human brain. J Neural Transm. 2007 Jul;114: 877–84.
26. Steinbrenner H, Alili L, Bilgic E, Sies H, Brenneisen P. Involvement of selenoprotein P in protection of human astrocytes from oxidative damage. Free Radic Biol Med. 2006 May 1;40: 1513–23.
27. Steinbrenner H, Bilgic E, Alili L, Sies H, Brenneisen P. Selenoprotein P protects endothelial cells from oxidative damage by stimulation of glutathione peroxidase expression and activity. Free Radic Res. 2006 Sep;40:936–43.
28. Steinbrenner H, Sies H. Protection against reactive oxygen species by selenoproteins. Biochim Biophys Acta. 2009 Nov;1790:1478–85.
29. Schoenmakers E, Agostini M, Mitchell C, Schoenmakers N, Papp L, Rajanayagam O, Padidela R, Ceron-Gutierrez L, Doffinger R, et al. Mutations in the selenocysteine insertion sequence-binding protein 2 gene lead to a multisystem selenoprotein deficiency disorder in humans. J Clin Invest. 2010 Dec;120:4220–35.
30. Schweizer U, Dehina N, Schomburg L. Disorders of selenium metabolism and selenoprotein function. Curr Opin Pediatr. 2011 Aug;23:429–35.
31. Berr C, Arnaud J, Akbaraly TN. Selenium and cognitive impairment: a brief-review based on results from the EVA study. Biofactors. 2012 Mar;38:139–44.
32. Pitts MW, Raman AV, Hashimoto AC, Todorovic C, Nichols RA, Berry MJ. Deletion of selenoprotein P results in impaired function of parvalbumin interneurons and alterations in fear learning and sensomotor gating. Neuroscience. 2012;208:58–68.
33. Raman AV, Pitts MW, Seyedali A, Hashimoto AC, Seale LA, Bellinger FP, Berry MJ. Absence of selenoprotein P but not selenocysteine lyase results in severe neurological dysfunction. Genes Brain Behav. 2012;6:10–183X.
34. Caito SW, Milatovic D, Hill KE, Aschner M, Burk RF, Valentine WM. Progression of neurodegeneration and morphologic changes in the brains of juvenile mice with selenoprotein P deleted. Brain Res. 2011;1398:1–12. Epub@2011 May 5:1–12.
35. Murphy MP, Holmgren A, Larsson NG, Halliwell B, Chang CJ, Kalyanaraman B, Rhee SG, Thornalley PJ, Partridge L, et al. Unraveling the biological roles of reactive oxygen species. Cell Metab. 2011;13:361–6.
36. Steinbrenner H, Sies H. Selenium homeostasis and antioxidant selenoproteins in brain: implications for disorders in the central nervous system. Arch Biochem Biophys. 2013 Mar 13. doi:pii: S0003–9861(13)00081–7. 10.1016/j.abb.2013.02.021.

# Chapter 2
# Oxidative Stress and Alzheimer's Disease

**Rudy J. Castellani, Bei-Xu Li, Amna Farshori, and Georgy Perry**

**Abstract** Oxidative stress is the inevitable result of life's requirement to reduce molecular oxygen to water for cellular respiration and energy metabolism. For a number of reasons, the human brain appears particularly vulnerable to oxidative stress, which has necessitated elaboration of complex antioxidant defenses in order to maintain oxidative balance. With advanced age, oxidative balance wanes in favor of oxidative stress, which sometimes results in disease, in particularly age associated sporadic or environmentally driven diseases such as Alzheimer's disease, cardiovascular disease, and cancer. Over the last 20 years, our laboratory has investigated oxidative stress by numerous in situ techniques and have identified oxidative stress-associated adducts, redox active transition metals, and metal associated proteins, not only within pathological lesions of the AD brain, but also unaffected brain prior to the onset of overt structural pathology. We have further demonstrated that oxidative stress decreases with increasing pathology, especially amyloid, suggesting that hallmark lesions in AD are more likely a productive response than a deleterious event. These and other findings continue to indicate the need to examine oxidative stress in greater detail, as well as expand the universe of antioxidant therapies, particularly as classical lesion-based therapies continue to fail.

R.J. Castellani, Ph.D. (✉)
Division of Neuropathology, Department of Pathology, University of Maryland, 22 South Greene Street, Baltimore, MD 21201, USA
e-mail: rcastellani@som.umaryland.edu

B.-X. Li, Ph.D. • A. Farshori, Ph.D.
Department of Pathology, University of Maryland, Baltimore, MD, USA

G. Perry, Ph.D.
College of Sciences, University of Texas, San Antonio, TX, USA

D. Praticò and P. Mecocci (eds.), *Studies on Alzheimer's Disease*, Oxidative Stress in Applied Basic Research and Clinical Practice, DOI 10.1007/978-1-62703-598-9_2,

## 2.1 Oxidative Stress

The reduction of molecular oxygen to water during the process of cellular respiration drives both the ATP synthesis necessary to sustain life, as well as the production of free radicals which can potentially destroy life [1]. The latter no doubt precipitated the evolution of elaborate antioxidant defenses, such that the pro-oxidant tendency of molecular oxygen in the intact organisms is balanced somewhere near the organism's ability to neutralize free radicals with near 100 % efficiency. Such is the nature of life for much of the lifespan of the organism.

Commensurate with the aging process, however, is a slow diminution in antioxidant defenses, and the "slow burn" of oxidative stress as a gradual but steady destructive force. Among the antioxidant mechanisms are vitamins A, C, and E, glutathione, and a number of enzymes that facilitate electron transfer to a nontoxic species, such as catalase, superoxide dismutase, and glutathione peroxidase, and each has been shown to decrease with age. It is therefore not surprising that virtually all sporadic or environmentally driven chronic diseases (cancer, cardiovascular disease, neurodegeneration) are both age-related and associated with oxidative stress [2].

The brain appears to be vulnerable to oxidative stress via a number of mechanisms. Glutathione and vitamin E have been shown to be in limited supply in the brain, in the face of an inherently high metabolic activity, dominated by oxidative metabolism [3]. Iron content is increased in some brain regions (e.g., substantia nigra), which is a potent source of free radicals via the Fenton reaction (ferrous iron plus hydrogen peroxide yielding ferric iron and an hydroxyl radical, the latter being a potent, nonselective oxidant species).

## 2.2 Involvement of Oxidative Stress in Alzheimer's Disease

*Heme oxygenase*. Once the steady state is breached in favor of oxidative stress, the deleterious effects are far-reaching. Indeed, every category of biomacromolecule, e.g., carbohydrates, protein, lipid, nucleic acids, is a potential target [2]. Among the earliest in vivo studies indicating the presence of oxidative stress in brain tissue was the pioneering study by Mark Smith in the 1990s on the role of heme oxygenase in the AD brain [4]. Heme oxygenase is an endogenous antioxidant that exists in multiple isoforms. Most notable is Heme oxygenase-1 (HO-1), which is inducible in the presence of oxidative stress, and processes heme to biliverdin, while generating redox-active iron and carbon monoxide. Carbon monoxide then may have a neurotransmitter function, and it is further of note that recent studies have shown a neuroprotective effect of carbon monoxide. Heme oxygenase was shown to be markedly upregulated in the AD brain within lesional as well as unaffected tissue, which laid the groundwork for subsequent studies and an expansion of our knowledge of oxidative stress in human brain in vivo, prior to which oxidative stress was only studied in highly artificial experimental constructs.

*Advanced glycation and lipid peroxidation.* In previous studies, we have since shown that protein adducts pentosidine and pyralline, formed via the Maillard reaction and advanced glycation, a process accelerated by molecular oxygen, are abundant in AD brains, not only within pathological lesions, but within pre-pathological vulnerable neurons [5]. We have also demonstrated lipid peroxidation end products hydroxynonenal and malondialdehyde in AD, both within lesions and within pathologically normal tissue. Both sets of adducts, advanced glycation end products and advanced lipid peroxidation end products, lead to the formation of intramolecular and intermolecular cross-links which render otherwise soluble proteins insoluble and resistant to degradation [6]. Carboxymethyllysine, an adduct associated both with advanced glycation and lipid peroxidation, has similarly been shown by us to be intimately associated with the AD brain [7]. In a recent study, we have also shown that hydroxynonenal in particular may accumulate less within lesions as they do within neurofilaments of axons [8]. The lysine-rich nature of neurofilaments, and lysine-lysine adducts may therefore contribute to the aging and disease process by disrupted slow axonal transport.

*Nitration.* In another study, we examined the brain for nitration of tyrosine residues, as peroxynitrite is a source of hydroxyl radical-like reactivity, and it directly oxidizes proteins and other macromolecules with resultant carbonyl formation from side-chain and peptide-bond cleavage [9]. Indeed, we found increased protein nitration in neurons, including but not restricted to neurofibrillary pathology, while control brains showed no such involvement. These data not only supported oxidative damage in AD, but that it was not necessarily limited to poorly soluble fibrils such as neurofibrillary tangles, but much like other indicators of oxidative stress such as AGE and lipid peroxidation adducts, reflects a broad and early event.

*Damage to nucleic acids.* DNA modification intuitively speaking appears more relevant to cancer biology than neurodegeneration, given the importance of acquired genetic alterations in cancer and the fact that the vast majority of neurons, including vulnerable neurons in AD, are post mitotic. Nevertheless, 8-hydroxyguanosine modification has been shown to be increased in AD brain tissue. Additionally, and perhaps more importantly, oxidation of RNA has been linked to AD in recent studies. There is of interest given the single stranded nature of RNA. It may be that oxidative stress induced damage to RNA results in sublethal cellular injury, altering protein translation, and disrupted cellular metabolism in favor of neurodegeneration [10].

*Heavy metals.* Transition metals, and in particular copper, iron, and zinc, are potent catalysts of free radicals [11–14]. Evidence suggests that metals may accumulate in the brain with age and disease, accelerating oxidative damage and possibly contributing to amyloidosis [15]. Using a modification of the Prussian blue reaction to select for redox state, we have demonstrated specifically redox-active iron in the AD as well as Parkinson disease. Moreover, the accumulation within lesions suggests that insoluble proteinaceous accumulations may be a "sink" for free radicals within the brain, suggesting that lesions themselves are more in line with Darwinian theory, or adaptation to the environment, than a priori indicators of toxicity [3, 16–18].

*Are hallmark lesions a response to oxidative stress?* The selective nature of the neurodegenerative process in the AD brain is difficult to explain in the face of a relatively selective disease process, affecting certain brain regions and certain neuronal subsets during the course of the disease. Whether the overall level of free radical production differs in vulnerable areas, or whether antioxidant defenses are deficient in those same areas for one reason or another, is difficult to discern precisely, although evidence for both mechanisms has been offered in the literature [19, 20]. It may further be worth noting that the major pathogenic hypothesis, or the putative amyloid cascade, has shown little region selectivity aside from "cortex" and is in fact less selective than phospho-tau accumulation, which is manifestly a downstream event [16]. It is also of some interest that oxidative stress precedes all types of pathological lesions, and that in one study, the magnitude of oxidative stress *decreased* with increasing amyloid, suggestive of a neuroprotective effect of amyloid production [21]. Clinical trials investigating the efficacy of antioxidant therapy in AD have been disappointing (although no less disappointing that all of the trials targeting amyloid-β (Aβ)), and it may be that the complexity of redox biology and approaching and altering brain chemistry in this regard, exceeds both understanding and therapeutic strategies available to date. Nevertheless, as the scientific community comes to realize more and more that lesion targeting is based on rare Mendelian amyloidosis and not strict age-related sporadic conditions, continued study of oxidative stress in vitro, in vivo, and as a therapeutic approach, is certainly warranted.

## 2.3 Conclusion

Oxidative stress is intimately intertwined with chronic, age-related disease processes, including AD. The specific vulnerability of the brain to oxidative stress may in part explain the abundance of evidence for its existence within both lesional and nonlesion tissue. Moreover, the existence of specific adducts of oxidative stress has, since the 1990s facilitated the production of antisera against those adducts, which has in turn allowed in-situ examination for oxidative stress hallmarks in humans, rather than through highly artificial experimental constructs prior to this time. The data have generally furthered the evidence for oxidative stress and provided a temporal dynamic that precedes traditional hallmark lesions. As lesion targeting continues to fail, a more substantial investment in oxidative biology as it pertains to AD is warranted.

## References

1. Castellani R, Hirai K, Aliev G, et al. Role of mitochondrial dysfunction in Alzheimer's disease. J Neurosci Res. 2002;70(3):357–60.
2. Perry G, Nunomura A, Hirai K, et al. Is oxidative damage the fundamental pathogenic mechanism of Alzheimer's and other neurodegenerative diseases? Free Radic Biol Med. 2002; 33(11):1475–9.

3. Castellani RJ, Siedlak SL, Perry G, Smith MA. Sequestration of iron by Lewy bodies in Parkinson's disease. Acta Neuropathol. 2000;100(2):111–4.
4. Smith MA, Kutty RK, Richey PL, et al. Heme oxygenase-1 is associated with the neurofibrillary pathology of Alzheimer's disease. Am J Pathol. 1994;145(1):42–7.
5. Smith MA, Sayre LM, Monnier VM, Perry G. Radical AGEing in Alzheimer's disease. Trends Neurosci. 1995;18(4):172–6.
6. Sayre LM, Zelasko DA, Harris PL, Perry G, Salomon RG, Smith MA. 4-Hydroxynonenal-derived advanced lipid peroxidation end products are increased in Alzheimer's disease. J Neurochem. 1997;68(5):2092–7.
7. Castellani RJ, Harris PL, Sayre LM, et al. Active glycation in neurofibrillary pathology of Alzheimer disease: N(epsilon)-(carboxymethyl) lysine and hexitol-lysine. Free Radic Biol Med. 2001;31(2):175–80.
8. Zhu X, Castellani RJ, Moreira PI, et al. Hydroxynonenal-generated crosslinking fluorophore accumulation in Alzheimer disease reveals a dichotomy of protein turnover. Free Radic Biol Med. 2012;52(3):699–704.
9. Smith MA, Richey Harris PL, Sayre LM, Beckman JS, Perry G. Widespread peroxynitrite-mediated damage in Alzheimer's disease. J Neurosci. 1997;17(8):2653–7.
10. Castellani RJ, Nunomura A, Rolston RK, et al. Sublethal RNA oxidation as a mechanism for neurodegenerative disease. Int J Mol Sci. 2008;9(5):789–806.
11. Perry G, Taddeo MA, Petersen RB, et al. Adventiously-bound redox active iron and copper are at the center of oxidative damage in Alzheimer disease. Biometals. 2003;16(1):77–81.
12. Perry G, Sayre LM, Atwood CS, et al. The role of iron and copper in the aetiology of neurodegenerative disorders: therapeutic implications. CNS Drugs. 2002;16(5):339–52.
13. Castellani RJ, Moreira PI, Liu G, et al. Iron: the Redox-active center of oxidative stress in Alzheimer disease. Neurochem Res. 2007;32(10):1640–5.
14. Castellani RJ, Moreira PI, Perry G, Zhu X. The role of iron as a mediator of oxidative stress in Alzheimer disease. Biofactors. 2012;38(2):133–8.
15. Bush AI. The metal theory of Alzheimer's disease. J Alzheimers Dis. 2013;33 Suppl 1: S277–81.
16. Castellani RJ, Perry G. Pathogenesis and disease-modifying therapy in Alzheimer's disease: the flat line of progress. Arch Med Res. 2012;43(8):694–8.
17. Castellani RJ, Lee HG, Zhu X, Nunomura A, Perry G, Smith MA. Neuropathology of Alzheimer disease: pathognomonic but not pathogenic. Acta Neuropathol. 2006;111(6):503–9.
18. Castellani RJ, Lee HG, Zhu X, Perry G, Smith MA. Alzheimer disease pathology as a host response. J Neuropathol Exp Neurol. 2008;67(6):523–31.
19. Smith MA, Perry G, Richey PL, et al. Oxidative damage in Alzheimer's. Nature. 1996; 382(6587):120–1.
20. Beal MF. Less stress, longer life. Nat Med. 2005;11(6):598–9.
21. Nunomura A, Castellani RJ, Zhu X, Moreira PI, Perry G, Smith MA. Involvement of oxidative stress in Alzheimer disease. J Neuropathol Exp Neurol. 2006;65(7):631–41.

Castellani RJ, Smith MA, [illegible]
[illegible] 2001;[illegible]

# Chapter 3
# Inflammation and Oxidation: A Link in Alzheimer's Disease Pathogenesis

**Kenneth Hensley**

**Abstract** Over the past three decades, two paradigms have been developing that largely describe the difference between normal and pathological brain aging: Namely, the twin paradigms of neural oxidative stress and neuroinflammation. These two concepts both emerged amidst an intellectual environment of either casual neglect or antipathy but have come to permeate much basic and applied neuroscience research. Current science has come to understand that these two phenomena are mutually but subtly linked to such an extent as to be practically inseparable. In Alzheimer's disease (AD) both neural oxidative stress and neuroinflammatory events are crucial components of the neurodegenerative landscape. Both appear to arise, in part, as consequences of the classic features of AD pathology namely amyloid peptide (Aβ)-containing senile plaques and neurofibrillary tangles (NFTs). Viewed from a different perspective, however, oxidative stress and neuroinflammation also contribute fundamentally to cell biological changes that predispose the AD brain to these classic histopathologies. This review will attempt to summarize the current state of knowledge regarding neural oxidative stress in AD, with emphasis on mechanistic connections between the two phenomena, and their joint relationship to both plaques and tangles.

## 3.1 Introduction

Alzheimer's disease (AD) is an exigent medical, humanitarian, and economic problem that affects almost five million Americans [1], a number that certainly will rise in the developed world due to its aging demography. In 2010, 5.1 % of the US

K. Hensley, Ph.D. (✉)
Department of Pathology and Neurosciences,
University of Toledo Medical Center, Toledo, OH 43614, USA
e-mail: Kenneth.hensley@utoledo.edu

D. Praticò and P. Mecocci (eds.), *Studies on Alzheimer's Disease*, Oxidative Stress in Applied Basic Research and Clinical Practice, DOI 10.1007/978-1-62703-598-9_3,

population was older than 65 years and 454,000 new AD cases were diagnosed; this was roughly 10 % increase over 2,000 numbers [1]. Worldwide, AD and similar dementias may afflict 35 million people (0.5 % of the world population) [1]. There are no cures, effective treatments, or robust preventive measures for AD. For some people in early stage AD, the drugs tacrine (Cognex), donepezil (Aricept), rivastigmine (Exelon), or galantamine (Razadyne) are prescribed to improve the cholinergic system [2]. The strategy of attempting to boost the cholinergic system is predicated on scientific understanding that was essentially established prior to 1980, namely, that AD impact on cognition is largely manifest through loss of cholinergic neurons. Current cholinergic agonists do provide partial, limited relief of cognitive symptoms but carry significant adverse effects such as gastrointestinal irritation. Another drug, Memantine (Namenda), an NMDA antagonist thought to combat excitotoxic neuronal stress, has been approved to treat moderate to severe AD with limited success [2]. Thus, there is an urgent but chronically unmet need to better understand AD etiology in order to develop more effective treatments or preventive measures for the disease.

This review will systematically examine AD from four distinct but complementary perspectives that have been each intensively explored in the past but which, perhaps, have never been completely unified into a cohesive intellectual framework. First, we will briefly recapitulate the cardinal histological description of AD as a manifestation of brain region-specific deposition of amyloid β-peptide (Aβ) rich plaques and neurofibrillary tangles (NFTs). Second, we will focus upon the "degeneration" part of AD neurodegeneration by considering AD primarily as a disease of neural atrophy, which logically forces one to consider AD as a fundamental pathology of neuronal cytoskeletons. Third, we will consider why AD neurodegeneration can be viewed as an exacerbation of normal, age-associated oxidative stress. Fourth, we will consider how the preceding components of AD biology can be described within the emerging rubric of neuroinflammation theory. The goal of this exercise is to find common, unifying links amongst disparate sub-specialties of AD research that might suggest new therapeutic approaches to slow AD clinical progression.

## 3.2 A Histological Consideration of AD: Plaques and Tangles Are Only Part of the Story

AD is formally defined and diagnosed by the presence of both abundant dense (neuritic) amyloid beta peptide (Aβ)-rich plaques and abundant neurofibrillary tangles (NFTs) in areas including the entorhinal cortex, hippocampus and isocortex; combined with synapse loss and clinical dementia [3]. Aβ plaques have been intensively studied for almost three decades and have been thoroughly reviewed elsewhere [4, 5]. For the purposes of the present manuscript, it is sufficient to summarize that 40–42 amino acid Aβ peptides derived from a classic single-pass type 1 transmembrane protein (amyloid precursor protein, APP) form neurotoxic oligomers and ultimately collect in very hydrophobic extracellular deposits called plaques. Initially diffuse, these plaques condense over time to become dense-core or neuritic structures in more advanced AD [4, 5]. These plaques are typically surrounded and infiltrated by reactive microglia, and by hypertrophic astrocytes that can be recognized by their upregulation of homotypic substances such as glial fibrillary acidic protein (GFAP) and S100β [4, 5]. In recent years the dogma of Alzheimer's disease research is that pathological alterations of amyloid precursor protein processing result in runaway accumulation of Aβ in various forms, which precipitates downstream processes including neuroinflammatory activation of microglia, neuritic pathology (e.g., tangles), and cell death. This "Amyloid Cascade Hypothesis," which places Aβ first and foremost as a trigger for the initiation of other AD pathophenomena, has resulted in a number of clinical trials of anti-amyloidogenic therapies, none of which has succeeded and several of which have done harm [5]. Thus, the preeminence of amyloid peptide biology as a guiding light for therapy development has been called sharply into question recently [5].

NFTs are located throughout the neocortical grey matter parenchyma, but dystrophic neurites (axons) form in a distinctive corona about the plaques [3]. NFTs are composed of hyperphosphorylated microtubule associated proteins (MAPs) and many other components which have been identified by immunohistochemistry, immunoprecipitation, and laser capture microdissection-mass spectrometry studies [6]. The principle and most-studied NFT component is the MAP tau, which adopts a characteristic paired helical filament conformation in the tangles (PHF-tau) [7]. Tau functions to bind and stabilize microtubules (MTs), which give length and rigidity to axons [7]. Phosphorylation of tau by cyclin dependent kinase 5 (CDK5) and glycogen synthase kinase-3β (GSK3β) contributes to MT destabilization [7].

Notable amongst other cytoskeletal proteins that co-aggregate with PHF-tau in NFTs is collapsin response protein-2 (CRMP2) [6, 8, 9]. CRMP2 is a member of a separate class of microtubule-associated proteins that is phylogenetically distinct from tau and which serve additional adaptor functions not directly related to MT stabilization [10, 11]. CRMP2 is hyperphosphorylated in NFTs by the same CDK5 and GSK3β enzymes implicated in tau hyperphosphorylation [10, 11]. Thus, CRMPs are likely to share importance with tau in causing NFT deposition or contributing to NFT-related neuron damage. This is a rapidly evolving, relatively new area of research investigation that is likely to gain more attention in coming years [10, 11].

Although AD diagnosis formally requires a burden of plaques plus tangles, the mere presence of abundant plaques and tangles does not always coincide with neuron loss or clinical dementia status prior to death and autopsy [12, 13]. For instance, in a study of octa- and nonagenarians who did not present with clinical AD, postmortem analysis revealed that 20–40 % of the subjects met neuropathological criteria for AD depending on the exact scoring criteria employed [13]. Such clinico-histological discrepancies may be because some individuals have more functional cognitive reserve capacity than others, hence, can endure a greater neuropathological burden (plaques and tangles) before suffering downstream consequences including neuron death and clinically evident dementia [12]. In any event, NFTs are better correlate than amyloid plaques to synapse loss and clinical dementia in formally demented AD patients [14].

## 3.3 Alzheimer's Disease Is, Fundamentally, a Disease of Neural Atrophy and Hence, Cytoskeletal Failure

While plaques and tangles are histologically dramatic, and are a necessary feature for confirmation of an AD diagnosis, these phenomena are not the only (or, necessarily, the best) correlates of cognitive decline. Structural neural atrophy and accompanying synapse loss also have been correlated significantly and consistently to clinical dementia, generally exceeding plaque or tangle counts as a predictor of cognitive dysfunction [15–19]. Neurons in AD are frankly atrophic with loss of dendritic arborization in human AD: For instance, manifesting as decreased average length per cell of Golgi-stainable branches and numbers of branches in particular amygdaloid and hippocampal neurons [15, 16] and by measurable synapse loss [17, 18]. Synapse loss is evident even in mild cognitive impairment (MCI), a prodromal state of AD [18]. Ultrastructural stereology indicates a 20 % loss of synapses in hippocampal CA1 of MCI patients progressing to 55 % loss in mild-moderate AD [18], with a generally consistent correlation to local NFT density [17]. The most striking synaptic changes occur in close apposition to Aβ plaques where loss of dendritic spines (orthogonal presynaptic termini) is remarkable [18]. While frank synaptic neuropathology is subtle in some AD models, synaptic deficits are a feature of the 3× Tg-AD mouse [19–21].

If plaques and NFTs are characteristic of AD, frank neuron loss and gross cortical atrophy is even more obvious even upon cursory examination. AD brain suffers severe loss of neurons with particular carnage amongst certain subsets, for instance, cholinergic circuits projecting from the basal forebrain into the cortex and hippocampus [22]. It is understandable therefore that much AD research has focused on the mechanisms of neuron death in efforts to protect the neurons against death. Unfortunately, focus on neuron death has largely led the AD community and especially potential therapy developers away from consideration of the processes that precede death, i.e., what pathways drive neural atrophy (degeneration)? Understanding the dynamic processes of synapse loss, neuritic de-aborization and axonal contraction might suggest new strategies for making "sick" neurons more healthful. This is a subtle but fundamentally different approach than trying to block end-stage death mechanisms, which thus far has been remarkably unsuccessful at producing new clinical tools.

Having logically turned the focus of AD neurodegeneration in the direction of understanding the processes leading to axonal dystrophy, neuritic dearborization and synaptic paring, it becomes apparent that one should focus upon the cytoskeleton in order to understand these features. A neuron has a very complex and dynamic structure which gives each neuron its own particular functional characteristics, and this structure is dependent upon a sophisticated cytoskeletal framework. As discussed in the remainder of this review, AD brain suffers oxidative stress and cytoskeletal proteins are particularly vulnerable. Neuroinflammatory processes engender much of this oxidative stress while redox-sensitive signal transduction also can lead to posttranslational alterations of cytoskeletal proteins or cytoskeletal regulatory pathways. Hence, a deeper understanding of cytoskeletal restructuring could help identify pathogenic pathways in AD that are amenable to therapeutic manipulation in order to slow the atrophic features of neurodegeneration.

## 3.4 Exacerbated Oxidative Stress Is a Confirmed Component of AD Pathology and a Plausible Factor Contributing to Cytoskeletal Degeneration

Protein oxidative lesions naturally occur in the brain and increase with age in a nonlinear fashion that appears exacerbated in AD and other age-associated neurodegenerative pathologies [23–26]. As early as 1991, Earl Stadtman and colleagues assessed an approximate 25 % increase in total protein carbonyls, indexing protein oxidative modifications, through the use of protein carbonyl derivatization techniques [23]. This work has been widely reproduced and extended upon by other measurement methods [24–26].

In recent years, immunostaining and instrumental analyses studies have measured specific protein oxidative modifications such as nitration. These studies generally find that specific oxidative modifications in AD brain can far exceed the relative disease-associated burden of general protein carbonyls. For instance, high performance liquid chromatography with electrochemical array detection (HPLC-ECD) has been used to demonstrate that 3-nitrotyrosine and 3,3′-dityrosine are increased at least twofold in affected regions of the AD brain relative to age-matched control brain [25]. Interestingly, lipid-phase nitration in the AD brain correlates well with protein nitration on a brain region-by-region pattern [26] but may be quantitatively more severe. For instance, the amount of protein tyrosine nitration in AD brain is typically on the order of 1:1,000 or less. Contrastingly, in an HPLC-ECD study of the lipid-phase nitration marker5-nitro-gamma-tocopherol (5-NO2-$\gamma$T) in AD brain, we found that the 5-NO2-$\gamma$T/$\gamma$T ratio may reach 30 %, and mean quantities present in AD brain exceeded those in control brain by twofold [26].

A variety of oxidation-sensitive enzyme systems are also significantly compromised in AD brain [24]. Of particular note in this regard, at least one study assessed cytoskeletal protein structural correlates of oxidative stress in AD brain using specialized techniques of electron paramagnetic resonance (EPR) spin-labeling applied to synaptosomal preparations from rapid postmortem-derived human brain tissue [24]. This study found a collapse of protein structure in AD synaptosomes, similar to that observed with in vitro synaptosomal oxidation [24]. Thus, there is ample empirical reason to postulate that oxidative stress occurs in AD and creates a measurable alteration in cytoskeletal structure, particularly within neurites and synaptic compartments.

When specific proteins are assessed on a case-by-case basis, the degree of oxidative modification difference between AD and age-matched normal brain can become even more apparent and, perhaps, may shed mechanistic light on fundamental pathways of cytoskeletal disruption. Recent proteomics studies by Sultana, Butterfield and colleagues [27–31] go far toward such elucidation. These researchers discovered that global protein adduction by the lipid oxidation product 4-hydroxynonenal (HNE) is not particularly severe in AD, although particular proteins including CRMP2 (discussed above) are HNE-adducted to a degree 2–3 times that of normal brain CRMP2 [30, 31]. Given the importance of CMRP2 in maintaining neurite and synaptic structure/function, and the physical association of phosphorylated CRMP2 with NFTs, this recent data suggests that focus of research upon consequences of specific protein oxidation events may be more important than looking at brain oxidative stress as a global and stochastic process.

The discussion about oxidative stress in AD has moved from asking "what and where" type questions to assessing the "why and how." Why does oxidative stress increase in AD brain, how might this stress contribute to neurodegeneration, and how might these pathological pathways be pharmacologically mitigated?

## 3.5 Relationship of AD Pathophenomena as Viewed Through the Lens of Neuroinflammation Theory

The topics of neuroinflammation as a driving force in AD pathogenesis and the history of neuroinflammation as a sub-discipline of neuroscience research have been thoroughly reviewed recently [32] and therefore will be discussed here in an abbreviated fashion. In the late 1980s-early 1990s several neuropathologists and epidemiologists began noting that (1) immune-signaling substances such as interleukins were elaborate in the AD brain, concomitant with proliferation of apparently reactive microglia and (2) long-term use of nonsteroidal anti-inflammatory drugs for arthritic conditions seemed to reduce the risk for onset of AD (reviewed in ref. 32). Other work in the 1990s revealed that activation of neuro-inflammatory pathways, such as microglial NADPH oxidase and iNOS recruitment, can cause a pronounced local increase in oxidizing species [33].

Moreover, contemporaneous studies revealed that reactive oxygen species such as $H_2O_2$ and •NO can activate redox-sensitive signal transduction machinery linked to microglial activation, cytokine elaboration, and a host of biochemical pathways with possible relevance to AD pathogenesis (reviewed in ref. 33). As a case in point, the redox-sensitive p38 mitogen-activated protein kinase (p38 MAPK) pathway was documented to be hyper-activated in AD brain, in neurons juxtaposed with amyloid plaques and in plaque-associated microglia [34]. This stress-activated kinase is exquisitely sensitive to redox activation, and works together with other similar pathways to induce nitric oxide synthase expression in inflammatory states [35, 36]. Mechanistically, these redox signaling pathways are probably triggered mostly by transient, reversible oxidation of thiol active sites in sensitive phosphatases, followed by glutathionylation of the oxidized thiol [33, 37, 38]. Stress-activated kinase systems could plausibly contribute to posttranslational modifications of microtubule-associated proteins such as lead to the deposition of paired-helical filament tau in

AD [36]. Thus, p38 MAPK and its relatives are still receiving significant attention as potential therapeutic targets in AD [36]. Combined with other forms of direct cytoskeletal protein oxidative lesioning, the net result likely would be to the detriment of the neuron's structure, function and ultimately viability [36, 37].

Thus, neuroinflammation provides a guiding and unifying perspective that unites the phenomena of oxidative stress with AD pathophenomena. In this perspective, neuro-immune pathways produce oxidants that affect both neurons and ambient glia through redox signaling pathways. The result is a combination of pressure on cytoskeletal networks through direct oxidative lesions and indirectly via phosphorylation-driven modifications. The aberrant function of these cells results in release of various paracrine factors, including but not restricted to Aβ peptides, lipid oxidation products and "damage associated molecular patterns" (DAMPs) from dead or dying cells [32]. These paracrine factors then exacerbate the neuroinflammatory forces and accelerate disease progression.

## 3.6 Conclusions

Current knowledge and theory about Alzheimer's disease pathogenesis assumes a significant component of both oxidative stress and neuroinflammatory activation, though there is still much uncertainty about the causal (rather than mere associative) relationships amongst inflammation, oxidative stress, and AD neuropathology. It seems clear that relatively simplistic models of AD pathogenesis that rely solely on consideration of amyloid cascades are grossly inadequate at either explaining the clinical presentation of AD or informing therapeutic development efforts. Taking a broader perspective of the disease by integrating amyloid catabolism in a larger rubric, including components of neuroinflammation and oxidative stress, might yield better predictive models.

Regarding oxidative stress, 20 years of study have revealed that the AD brain is hyperoxidized but perhaps by only 10–25 % beyond typical age-matched bulk brain tissue. On the other hand, individual proteins in the aging brain may be oxidatively damaged far in excess of bulk protein oxidative load, and this damage may be notably exacerbated by AD. Moreover, specific types of oxidative damage (e.g., nitration or dityrosyl cross-linking) may be more severe than general carbonylation. If oxidative stress plays a role in triggering or accelerating AD, the likely proximal targets would include cytoskeletal proteins or proteins that regulate cytoskeletal dynamics.

While there is still much debate over the proper temporal placement and etiological prioritization of inflammation, oxidative stress and classic AD pathological features (plaques and tangles) it appears clear that oxidative stress and neuroinflammation are not separate phenomena but rather represent antipodes of a single "vicious cycle." Neuroinflammatory activation, particularly of microglia and to a lesser extent of astrocytes, results in profligate local production of ROS/RNS and possibly severe focus of these oxidants on ambient neurons. ROS/RNS in turn trigger redox signaling processes, especially through oxidation/glutathionylation cycles in sensitive protein phosphatases that ultimately up-regulate further neuroinflammatory gene expression. This neuroinflammatory gene expression in turn results in further ROS/RNS and aggravating cytokine elaboration, completing a feed-forward cycle.

The degree to which neuroinflammation (and, by extent, redox stress) is damaging vs. protective in AD is still debatable. There are obvious reasons why these phenomena should be deleterious but less obvious reasons why they may be beneficial. Activated microglia phagocytose nascent amyloid plaques and may play a role in structural remodeling of damaged neural circuitry, while low levels of oxidative stress may act as a priming mechanism to induce antioxidant and other cellular defense programs.

Thus far, no conceptualization of AD has resulted in a clinical approach that slows AD degeneration (though there have been several drugs developed that moderate cholinergic symptoms). There is evidence that early and prolonged dosage with nonsteroidal anti-inflammatory drugs, perhaps combined with habitual consumption of natural antioxidants, may reduce AD risk; however, intervention trials have been quite disappointing. It is likely that neuroinflammation theory could be used as a guiding principle to identify specific, novel molecular pathways that are specifically activated or oxidatively damaged and which might be rationally modulated for pharmacological intervention against Alzheimer's disease. Such research efforts would be advised to avoid generalizing oxidative stress concepts but instead focusing on key mediators of oxidative damage, such as redox-sensitive signaling kinases and perhaps cytoskeletal structure-regulating proteins like CRMP2/DPYSL2 and its homologs. Future work in this area likely will benefit from interdisciplinary approaches combining proteomic techniques with protein biology and classical pharmacology approaches once promising lead chemical structures have been identified.

**Acknowledgements** This work was supported in part by grants from the National Institutes of Health (NIH-R01 AG031553 and R21-NS066279).

## References

1. Abbott A. Dementia: a problem for our age. Nature. 2011;475:S2–4.
2. Mancuso C, Siciliano R, Barone E, Butterfield DA, Prziosi P. Pharmacologists and Alzheimer disease therapy: to boldly go where no scientist has gone before. Expert Opin Investig Drugs. 2011;20:1243–61.
3. Duyckaerts C, Delatour B, Potier M-C. Classification and basic pathology of Alzheimer's disease. Acta Neuropathol. 2009;118:5–36.
4. D'Alton S, George DR. Changing perspectives on Alzheimer's disease: thinking outside the amyloid box. J Alzheimers Dis. 2011;25:571–81.
5. Karran E, Mercken M, De Strooper B. The amyloid cascade hypothesis for Alzheimer's disease: an appraisal for the development of therapeutics. Nat Rev Drug Discov. 2011;10:698–711.
6. Wang Q, Woltier RL, Cimino PJ, Pan C, Montine KS, Zhang J, Montine TF. Proteomic analysis of neurofibrillary tangles in Alzheimer disease identifies GAPDH as a detergent-insoluble paired helical filament tau binding protein. FASEB J. 2005;19:869–71.

7. Obulesu M, Venu R, Somashekhar R. Tau mediated neurodegeneration: an insight into Alzheimer's disease pathology. Neurochem Res. 2011;36:1329–35.
8. Takata K, Kitamura Y, Nakata Y, Matsuoka Y, Taomimoto H, Taniguchi T, Shimohama S. Involvement of WAVE1 accumulation in Aβ/APP pathology-dependent tangle modification in Alzheimer's disease. Am J Pathol. 2009;175:17–24.
9. Gu Y, Hamajima N, Ihara Y. Tangle-associated collapsin response mediator protein-2 (CRMP-2) is highly phosphorylated on Thr-509, Ser-518, Ser-522. Biochemistry. 2000;39:4267–75.
10. Hensley K, Christov A, Venkova K, Gunning W, Park J. Collapsin response mediator protein-2: an emerging pathologic feature and therapeutic target for neurodisease indications. Mol Neurobiol. 2011;43:180–91.
11. Lin PC, Chan PM, Hall C, Manser E. CRMPs are a new class of MAP that selectively interact with assembled microtubules via a taxol-sensitive binding interaction. J Biol Chem. 2011; 286:41466–78.
12. Green MS, Kaye JA, Ball MJ. The Oregon brain aging study: neuropathology accompanying healthy aging in the oldest old. Neurology. 2000;54:105–13.
13. Price JL, McKeel Jr DW, Buckles VD, Roe CM, Xiong C, Grundman M, Hansen LA, Petersen RC, Parisi JE, Dickson DW, Smith CD, Davis DG, Schmitt FA, Markesbery WR, Kaye J, Kurlan R, Hullette C, Kurland BF, Higdon R, Kukull W, Morris JC. Neuropathology of non-demented aging: presumptive evidence for preclinical Alzheimer disease. Neurobiol Aging. 2009;30:1026–36.
14. Bierer LM, Hof PR, Purohit DP, Carlin L, Schmeidler J, Davis KL, Perl DP. Neocortical neurofibrillary tangles correlate with dementia severity in Alzheimer's disease. Arch Neurol. 1995;52:81–8.
15. Hanks SD, Flood DG. Region-specific stability of dendritic extent in normal human aging and regression in Alzheimer's disease. I. CA1 of hippocampus. Brain Res. 1991;54:63–82.
16. Scott SA. Dendritic atrophy and remodeling of amygdaloid neurons in Alzheimer's disease. Dementia. 1993;4:264–71.
17. Arendt T. Synaptic degeneration in Alzheimer's disease. Acta Neuropathol. 2009;118:167–79.
18. Scheff SW, Price DA, Schmitt FA, DeKosky ST, Mufson EJ. Hippocampal synaptic loss in early Alzheimer's disease and mild cognitive impairment. Neurology. 2007;68:1501–8.
19. Oddo S, Caccamo A, Shepherd JD, Murphy MP, Golde TE, Kayad R, Metherate R, Mattson MP, Akbari Y, LaFerla FM. Triple-transgenic model of Alzheimer's disease with plaques and tangles: intracellular Aβ and synaptic dysfunction. Nature. 2003;39:409–21.
20. Bertoni-Freddari C, Sensi SL, Giorgetti B, Balletti M, Di Stefano G, Canzonlero LM, Casoll T, Fattoretti P. Decreased presence of perforated synapses in a triple-transgenic mouse model of Alzheimer's disease. Rejuvenation Res. 2008;11:309–13.
21. Gimenez-Llort L, Blazquez G, Canete T, Johansson B, Oddo S, Tobena A, LaFerla FM, Fernandez-Teruel A. Modeling behavioral and neural symptoms of Alzheimer's disease in mice: a role for intraneuronal amyloid. Neurosci Biobehav Rev. 2007;31:125–47.
22. Schliebs R, Arendt T. The cholinergic system in aging and neuronal degeneration. Behav Brain Res. 2011;221:555–63.
23. Smith CD, Carney JM, Starke-Reed PE, Oliver CN, Stadtman ER, Floyd RA, Markesbery WR. Excess brain protein oxidation and enzyme dysfunction in normal aging and in Alzheimer disease. Proc Natl Acad Sci U S A. 1991;88:10540–3.
24. Hensley K, Hall N, Subramaniam R, Cole P, Harris M, Aksenov M, Aksenova M, Gabbita SP, Wu JF, Carney JM, Lovell M, Markesbery WR, Butterfield DA. Brain regional correspondence between Alzheimer's disease histopathology and biomarkers of protein oxidation. J Neurochem. 1995;65:2146–56.
25. Hensley K, Maidt ML, Yu ZQ, Sang H, Markesbery WR, Floyd RA. Electrochemical analysis of protein nitrotyrosine and dityrosine in the Alzheimer brain indicates region-specific accumulation. J Neurosci. 1998;18:8126–32.
26. Williamson KS, Gabbita SP, Mou S, West MS, Pye QN, Markesbery WR, Cooney RV, Grammas P, Reimann-Philipp U, Floyd RA, Hensley K. The nitration product 5-nitro-gamma-tocopherol is increased in the Alzheimer brain. Nitric Oxide. 2002;6:221–7.

27. Sultana R, Butterfield DA. Identification of the oxidative stress proteome in the brain. Free Radic Biol Med. 2011;50:487–94.
28. Sultana R, Perluigi M, Butterfield DA. Oxidatively modified proteins in Alzheimer's disease, mild cognitive impairment and animal models of AD: role of Aβ in pathogenesis. Acta Neuropathol. 2009;118:131–50.
29. Sultana R, Butterfield DA. Role of oxidative stress in the progression of Alzheimer's disease. J Alzheimers Dis. 2009;19:341–53.
30. Reed TT, Pierce WM, Markesbery WR, Butterfield DA. Proteomic identification of HNE-bound proteins in early Alzheimer disease: insights into the role of lipid peroxidation in the progression of AD. Brain Res. 2009;1274:66–76.
31. Perluigi M, Sultana R, Cenini G, Domenico FD, Memo M, Pierce WM, Coccia R, Butterfield DA. Redox proteomics identification of 4-hydroxynonenal modified brain proteins in Alzheimer's disease: role of lipid peroxidation in Alzheimer's disease pathogenesis. Proteomics Clin Appl. 2009;3:682–93.
32. Hensley K. Neuroinflammation in Alzheimer's disease: mechanisms, pathologic consequences and potential for therapeutic manipulation. J Alzheimers Dis. 2010;21:1–14.
33. Gabbita SP, Robinson KA, Stewart CA, Floyd RA, Hensley K. Redox regulatory mechanisms of cellular signal transduction. Arch Biochem Biophys. 2000;376:1–13.
34. Hensley K, Floyd RA, Zheng N-Y, Nael R, Robinson KA, Nguyen X, Pye QN, Stewart CA, Geddes J, Markesbery WR, Patel E, Johnson GV, Bing G. p38 kinase is activated in the Alzheimer's disease brain. J Neurochem. 1999;72:2053–8.
35. Robinson K, Stewart CA, Pye QN, Nguyen X, Kenney L, Salsman S, Floyd RA, Hensley K. Redox-sensitive protein phosphatase activity regulates the phosphorylation state of p38 protein kinase in primary astrocyte culture. J Neurosci Res. 1999;55:724–32.
36. Munoz L, Ammit AJ. Targeting p38 MAPK pathway for the treatment of Alzheimer's disease. Neuropharmacology. 2010;58:561–8.
37. Kamat CD, Hensley K, Gadal S, Mhatre MC, Pye QN, Williamson KS. Antioxidants in central nervous system diseases: preclinical promise and translational challenges. J Alzheimers Dis. 2008;15:473–93.
38. Xiong Y, Uys JD, Tew KD, Townsend KD. S-glutathionylation: from molecular mechanisms to health outcomes. Antioxid Redox Signal. 2011;15:233–70.

# Chapter 4
# GAPDH: β-Amyloid Mediated Iron Accumulation in Alzheimer's Disease: A New Paradigm for Oxidative Stress Induction in Neurodegenerative Disorders

**Michael A. Sirover**

**Abstract** Alzheimer's disease (AD) is an age-related neurodegenerative disorder characterized clinically by dementia, loss of memory, and cognitive dysfunction. Although a number of molecular, biochemical, and cellular defects have been identified, the exact molecular mechanism(s) which underlie this disease are unknown. Of particular interest may be aberrant protein–protein interactions, especially with the amyloid protein which may lead not only to plaque development but also to alterations in cell function due to protein depletion. In that regard, recent evidence suggests a specific interaction of the β-amyloid precursor protein (β-APP) and the β-amyloid protein (β-AP) with glyceraldehyde-3-phosphate dehydrogenase (GAPDH). The latter was thought to be a classical glycolytic protein of seemingly little interest. However, recent studies suggest that it is a multifunctional protein with diverse activities independent of its role in energy production. Further, it exhibits a diverse subcellular localization in the nucleus, membrane, and cytosol which may be not only directly related to its functional diversity but also may represent an intracellular equilibrium in the regulation of GAPDH expression. Accordingly, we shall consider the hypothesis that the formation of β-APP or β-AP–GAPDH protein–protein complexes alters both GAPDH function and its subcellular distribution. In particular, as recent studies indicate a fundamental role for membrane GAPDH in cellular iron uptake, transport, and metabolism, the formation of either β-APP–GAPDH or β-AP–GAPDH complexes may facilitate iron accumulation (a known characteristic of Alzheimer's disease), thereby increasing oxidative stress as a consequence of an intracellular Fenton reaction. This pleiotropic effect of GAPDH binding could serve as a unifying hypothesis providing an initiating event in AD.

---

M.A. Sirover, Ph.D. (✉)
Department of Pharmacology, Temple University School of Medicine,
Philadelphia, PA 19140, USA
e-mail: msirover@temple.edu

D. Praticò and P. Mecocci (eds.), *Studies on Alzheimer's Disease*, Oxidative Stress in Applied Basic Research and Clinical Practice, DOI 10.1007/978-1-62703-598-9_4,

## 4.1 Introduction

*"How does one separate the wheat from the chaff?"*
*An English idiom*

Glyceraldehyde-3-phosphate dehydrogenase (GAPDH, EC 1.2.1.12) was long considered a classical housekeeping gene and protein of little interest. It was relegated to the laboratory shelf, considered useful as an enzymatic or crystallographic reagent or as the obligatory loading control for RNA and protein analyses. That being said, recent evidence indicates that GAPDH is in reality a multifunctional protein with separate activities involving dynamic changes in its subcellular location, is a unique target for oxidative stress and may fulfill a fundamental role in cellular iron uptake, transport, and metabolism [rev. in refs. 1–5].

Recent studies also indicate an association not only between GAPDH and major age-related neurodegenerative disease proteins but also with the structures associated with each disease [rev. in ref. [6]]. These include β-amyloid protein–GAPDH binding in relation to plaque formation in Alzheimer's disease [7–10]; binding to huntingtin [11] facilitating its nuclear translocation in Huntington's disease cells [12] as well as GAPDH binding to other triplet repeat proteins (atrophin, ataxin-1, and the androgen receptor) implicated in DRLPA, spinocerebellar ataxia type-1 and spinobulbar muscular atrophy, respectively [11, 13]. Further, synuclein and GAPDH co-localize in Lewy bodies in Parkinson's disease [14] and drugs which may be used to treat Parkinson's disease bind to and inhibit GAPDH [15, 16].

Plaque formation due in part to GAPDH–β-amyloid protein binding (along with a potential synergy with oxidative stress induced GAPDH modification) represents a now classical mechanism proposed as a foundation for the pathogenesis of Alzheimer's disease [rev. in refs. [17–19]]. However, it is still unclear whether plaque formation is an initiating event in AD development or a characteristic of its clinical progression. In contrast, new data suggests an alternative possibility: a pleiotropic process in which it is possible to interrelate GAPDH–β-AP binding, iron accumulation in Alzheimer's disease [20–22] and the induction of oxidative stress through the Fenton reaction [23, 24] resulting in structural alterations in critical cellular macromolecules. This pleiotropic process has the potential to provide a unifying hypothesis as a key molecular foundation underlying the temporal sequence of cellular perturbations resulting in the development of Alzheimer's disease.

## 4.2 Functional Diversity of GAPDH

As illustrated in Fig. 4.1, GAPDH functions in the control of gene expression [25], membrane trafficking [26–29], receptor mediated cell signaling [30, 31], the maintenance of DNA integrity [32–34], the control of mRNA stability [35–38], and iron uptake, transport, and metabolism [31]. The complexity of GAPDH structure and function is indicated by its role in the posttranscriptional control of gene expression in which it displays three separate activities: stabilization of a mRNA to increase

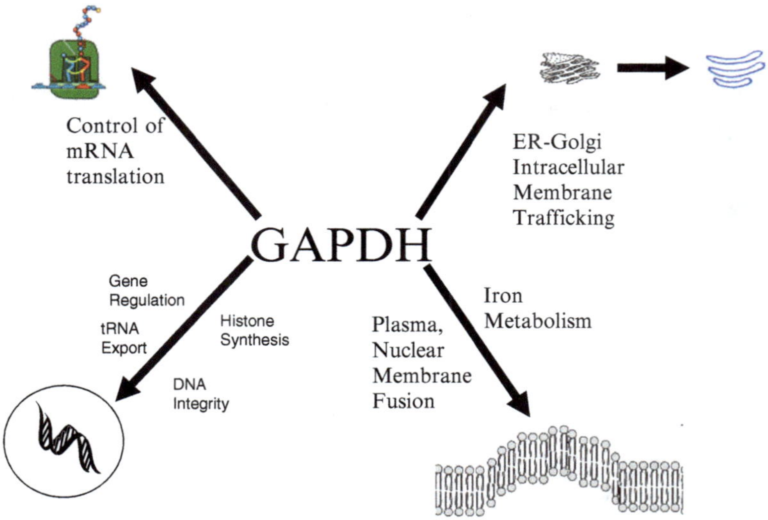


**Fig. 4.1** Functional diversity of GAPDH

protein synthesis [35, 36]; destabilization of an mRNA to promote its decay [37]; and translational inhibition of an mRNA sans degradation solely by its binding [38].

For such purposes, GAPDH displays a diverse subcellular distribution not only as a cytosolic protein but also it is localized in the ER, the Golgi, polysomes, the nucleus and in the cell membrane [rev. in refs. [4, 5]]. However, its intracellular position is not static in time but is ever changing, perhaps depending on changes in the intracellular environment. Three classes of GAPDH subcellular translocations have been identified: termed constitutive, inducible, and semi-constitutive as defined by the temporal sequence of GAPDH intracellular movement in relation to its acquisition of a new activity [rev. in ref. [4]]. It was suggested that posttranslational modification provides the signaling mechanism for GAPDH subcellular transport.

### *4.2.1 GAPDH and Iron Metabolism*

As shown in Fig. 4.2, perhaps of particular relevance to the etiology of Alzheimer's disease, may be a newly discovered property of membrane bound GAPDH, i.e., its role in iron uptake, transport, and metabolism. Iron in vivo represents a now classical intracellular "two edged sword," i.e., it is an essential element required for numerous normal cell functions [39] yet, through the Fenton reaction, may contribute to the pathology of oxidative stress [40]. At low iron concentrations, it is presumed that the former predominate, while at higher iron concentrations the latter may become physiologically significant, providing a source of reactive oxygen species capable of damaging cellular macromolecules.

Previous studies indicated transferrin as the major iron binding protein as well as the the role of transferrin/TfR1 and transferrin/TfR2 receptor binding which

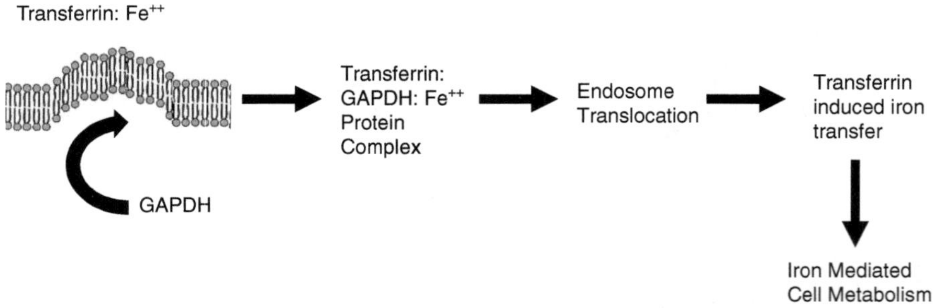


**Fig. 4.2** Iron transport in normal cells

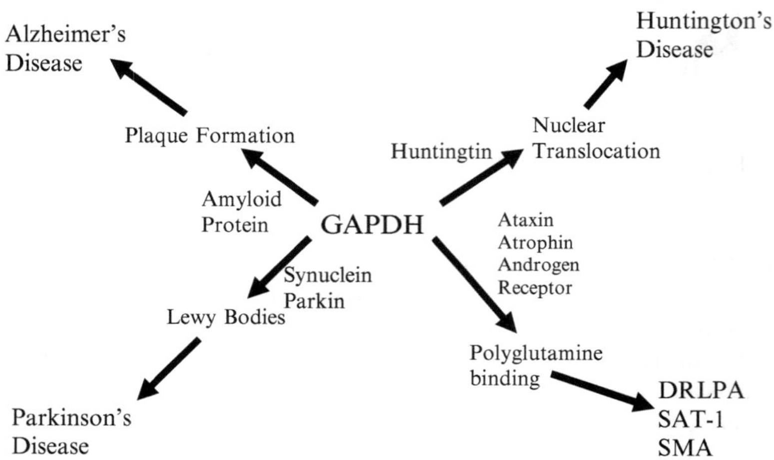


**Fig. 4.3** GAPDH and neurodegenerative disease protein–protein interactions

function to facilitate iron uptake and transfer [41, 42] In contrast, recent studies identified membrane GAPDH as an alternative cell receptor transferrin binding protein [31, 43]; that the GAPDH–transferrin–$Fe^{++}$ protein complex was transported, possibly by endocytosis [44], to the early endosome as was the transferrin–$Fe^{++}$/TfR1 and transferrin–$Fe^{++}$/TfR2 protein complexes; and that GAPDH transport to the membrane was dependent on extracellular iron concentration. The latter suggests the possibility that the cell has an iron-mediated signal transduction pathway which regulates GAPDH transport and inclusion into the cell membrane.

## 4.3 GAPDH: Neurodegenerative Protein Interactions

As the functional diversity of GADPH was being characterized, other studies indicated the potential involvement of GAPDH in the development of age-related neurodegenerative disorders (Fig. 4.3). Both in vitro and in vivo, GAPDH was identified as a binding protein for each of the mutant proteins implicated as causative factors in the development of Alzheimer's disease (β-amyloid precursor protein and the β-amyloid protein$_{1-42}$), Parkinson's disease (α-synuclein and Parkin); Huntington's disease (huntingtin) as well as other triplet repeat disorders: DRLPA, spinocerebellar ataxia type-1, and spinobulbar muscular atrophy,(atrophin, ataxin-1, and the androgen receptor, respectively). The significance of GAPDH binding was indicated by its presence in lesions characteristic of each disease, i.e., senile plaques in Alzheimer's disease or Lewy bodies in Parkinson's disease, or by a disease-related GAPDH activity, i.e., nuclear transport of mutant huntingtin.

### 4.3.1 GAPDH–β-APP and GAPDH–β-AP binding

The β-amyloid precursor protein could be considered as a classical transmembrane protein in view of its extracellular, membrane, and cytosolic domains. Its pathological significance in the development of Alzheimer's disease relates not only to its structure but also to the physiological consequences of its proteolysis to the β-amyloid protein [rev. in refs. [17–19]]. The latter forms fibril structures in vitro and in vivo. Its induction of oxidative stress is considered in section 4.4.3.

Initial studies identified GAPDH as a β-APP binding protein as defined by affinity chromatography utilizing recombinant constructs containing the β-APP C-terminal domain [7]. No diminution in GAPDH catalytic activity was detected as a consequence of β-APP C-terminal domain binding. This lack of effect on GAPDH activity as a function of protein interactions is not unusual, with a similar result observed in studies which defined its multidimensional activities [rev. in refs. 1–5]. Other studies identified the binding of GAPDH to the β-AP. Initial investigations used affinity chromatography with bound β-AP demonstrating the specific retention of GAPDH [8], while subsequent studies defined β-AP–GAPDH interaction by co-immunoprecipitation [9]. Intriguingly, further analysis demonstrated that the β-AP co-precipitated GAPDH from a synaptosomal protein preparation [10]. The significance of the latter is considered below.

### 4.3.2 Functional Consequences of GAPDH–β-APP and GAPDH–β-AP Binding in Alzheimer's Disease

The studies described above indicate the specificity of GAPDH interactions with the β-APP and with the β-AP. The significance of those changes in neuronal cell protein structure could be indicated by subsequent perturbations in neuronal cell function. Two such mechanisms can be identified. The first may be considered as "predictable," while the second may be categorized as "unexpected."

#### 4.3.2.1 Diminution of Glucose Utilization

GAPDH catalyzes one of the most critical reactions in the glycolytic pathway, i.e., the conversion of glyceraldehyde-3- phosphate to 1,3-bisphosphoglycerate. As such, it provides the biochemical starting point not only for the glycolytic synthesis of ATP but also for reducing equivalents available for oxidative phosphorylation. For that reason, catalytic inhibition by β-APP and β-AP binding to GAPDH could result in a reduced level of glycolysis and thus of glucose utilization. Several studies indicate that such is indeed the case using PET analyses of brain metabolism in affected individuals [45]; glucose uptake in transgenic animals [46]; as well as cholinergic deficits due to diminished turnover of glucose [47].

This reduction in glucose utilization would suggest that GAPDH activity could be decreased in Alzheimer's disease which would provide a biochemical foundation for the reduction in glucose utilization. Findings from our laboratory indicated a specific decrease in GAPDH catalysis as a property of AD cells [48–50]. Although normal activity was observed in whole cell extracts, subcellular differences were detected. Cytosolic AD GAPDH activity was decreased, while that observed in the perinuclear and nuclear fractions were equivalent to that observed in normal human cells. Further, in the AD cytosol, a high molecular weight GAPDH species was detected by gradient analysis. No such species was observed in normal cells. It was suggested that this new species resulted from the formation of a β-APP or β-AP–GAPDH protein complex. Alternatively, reductions in GAPDH activity may arise from the formation of disulfide bonds and aggregation induced by β-AP [51].

Subsequent analysis of GAPDH in transgenic models of AD indicated also that GAPDH activity was diminished as part of the phenotype characteristic of animal AD [52]. Perhaps of particular importance was the observation that, although GAPDH activity was decreased; there was no change in GAPDH expression as defined by immunoblot analysis. That finding is consistent with the aforementioned formation of a β-APP or β-AP–GAPDH protein complex. Further analysis demonstrated that different GAPDH isoforms may comprise a portion of the protein complex [53]. Intriguingly, with respect to the latter, two forms of membrane bound GAPDH have been identified [54]. The first has catalytic activity, while the second is enzymatically inactive. Differences in isoelectric point have been also detected.

#### 4.3.2.2 Synapse Dysfunction

Recent evidence suggests that perturbations in synaptic function may be an early characteristic of Alzheimer's disease resulting in the loss of cognition characteristic of this neurodegenerative disorder [rev. in ref. [55]]. This has been ascribed to actions of the β-amyloid protein on synaptic transmission. However, the exact mechanism through which that occurs is unclear. In that regard, other investigations demonstrated the significance of GAPDH in synaptic specific energy production [56]. The latter is considered an a priori functional requirement. ATP synthesis was GAPDH dependent as evidenced by its diminution as a function of GAPDH inhibition. Accordingly, it may be of interest to determine the effect of β-AP on this neuronal source of ATP, i.e., does the formation of β-AP–GAPDH protein complexes diminish this mechanism of localized energy production as a foundation for synaptic dysfunction in Alzheimer's disease?

It has also been suggested that effects on synaptic neurotransmitters may also be of significance as a foundation of AD development [rev. in ref. [55]]. As such, studies indicating the role of GAPDH as a $GABA_A$ receptor phosphorylating agent and the effect of that specific posttranslational modification on neurotransmission [57] suggest an alternative mechanism through which the formation of β-AP–GAPDH protein complexes may underlie AD initiation or progression. In those studies (as with many other investigations identifying the functional diversity of GAPDH), an unknown 38 kDa protein was identified as an endogenous kinase which phosphorylated the $GABA_A$ receptor.

Unexpectedly, that unknown protein was identified as GAPDH which had been previously characterized as a phosphotransferase [rev. in ref. [1]]. The sequential nature of GAPDH-mediated $GABA_A$ receptor phosphorylation, i.e., GAPDH autophosphorylation followed by phosphate group transfer, is consistent with known instances of GAPDH phosphotransferase activity [1] as well as the recently identified transnitrosylase activity of GAPDH, i.e., S-nitrosylation of GAPDH followed by nitroso group transfer to an acceptor protein [58, 59]. With respect to Alzheimer's disease, these findings indicate the possibility of an unexpected effect of β-APP or β-AP binding to GAPDH, i.e., a reduction in neurotransmission arising from inhibition of GAPDH phosphorylating activity. En toto, the findings that indicate β-AP binding to synaptosomal GAPDH may have a dual effect on both ATP production and on neurotransmission. This may be tested experimentally.

## 4.4 GAPDH as a Target of Oxidative Stress: Relationship to AD Progression

The formation of reactive oxygen species (ROS) in vivo represents a classical "double-edge sword" with respect to cell function and to cell pathology. With respect to the former, reactive oxygen species are required for diverse cell activities yet, with respect to the latter, their presence may threaten cell viability. For example, recent evidence demonstrates that nitric oxide is a cellular messenger capable of transmitting critical cell information yet it may affect cell viability by functioning as an apoptotic signal.

This duality of function is illustrated in its specific modification of GAPDH at its active site cysteine, forming SNO-GAPDH. The latter may be required for normal cell function (Fig. 4.4, left) affecting vasodilation/vasoconstriction equilibrium [37] and may be necessary for the regulation of heme metabolism [60]. Yet, as indicated in Fig. 4.4, right, SNO-GAPDH may act to initiate programmed cell death which may be linked to the pathogenesis of Alzheimer's disease [61]. Specifically, induction of oxidative stress has long been considered a unique hazard in AD

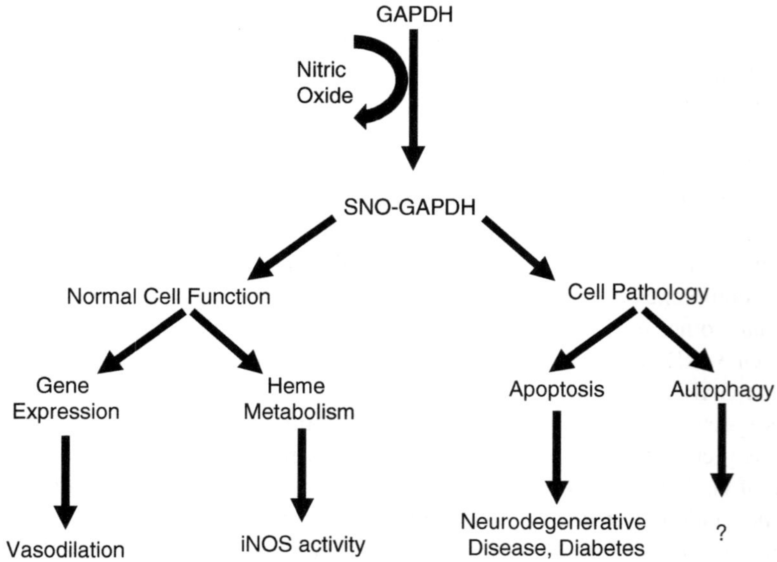


**Fig. 4.4** Pleiotropic effects of oxidative GAPDH modification

development as a foundation for neuronal cell death and the subsequent loss of cognitive function which it underlies [61]. Of particular interest may be those studies which indicate oxidative GAPDH modification by the β-AP [62] as well as its protection by drugs which inhibit β-AP oxidative activity [63].

### *4.4.1 Role of Oxidative Damaged GAPDH in Neuronal Apoptosis and Alzheimer's Disease*

Nitric oxide modification of GAPDH$^{cys149}$ is an apriori requirement for the initiation of programmed cell death [64]. As comprehensively described [65–68], this involves its cytosolic binding to the E3 ubiquitin ligase Siah1, nuclear translocation of the SNO-GAPDH–Siah1 complex, initiation of nuclear protein degradation, downstream gene regulation and, most recently, the observation that SNO-GAPDH can act as a transnitrosylase [58, 59] further demonstrating the pleiotropic effects of the initial oxidative GAPDH modification. En toto, this sequential apoptotic program in neuronal cells would provide a requisite molecular foundation for AD progression. Of further interest are recent observations that the identical pathways appear to exist in endothelial cells, suggesting their physiological relevance to diabetic pathogenesis as well [69, 70]. In contrast, GAPDH may function in a protective capacity during autophagic-related cell survival [71, 72]. However, in this instance, the role of SNO-GAPDH is unknown.

### 4.4.2 Role of SNO-GAPDH in GAPDH: Neuronal Protein Interactions

As indicated, recent studies defined the specific interaction of GAPDH with neurodegenerative disease proteins. As such interactions could be correlated with the pathological features of each disease, it was reasonable to suggest a causative relationship may exist. That being said, although the role of SNO-GAPDH in neuronal apoptosis was in the process of both identification and characterization, the potential interrelationship of that oxidative posttranslational modification with respect to those protein–protein interactions appears to remain uncertain.

Thus, it remains unclear whether such modifications may also be directly relevant to the role of GAPDH as a β-APP/β-AP binding partner (as well as its binding to huntingtin, synuclein and other neurodegenerative disease proteins). Differences may exist in the binding constants of GAPDH vs. oxidatively modified GAPDH to the β-APP or to β-AP. Further, as GAPDH exists as a tetramer, oxidation of a single active site cysteine in one of the 37 kDa monomers may facilitate binding. As high molecular weight β-APP or β-AP/GAPDH species may exist; there may be a preferential deposition of oxidatively modified GAPDH into senile plaques. En toto, such analyses may indicate a previously unknown role of oxidative GAPDH modification as a determinant not only for GAPDH: neuronal disease protein binding but also for the development of the physiological structures characteristic of each disorder.

### 4.4.3 β-AP Induced Oxidative GAPDH Modification in Alzheimer's Disease: Dysregulation of Normal Cell Function

As indicated, recent studies demonstrate the apoptotic significance of SNO-GAPDH as a cell signaling molecule to initiate programmed cell death. Accordingly, β-AP induced oxidative GAPDH modification may be readily considered as a mechanism for Alzheimer's disease pathogenesis. In contrast, it may be more difficult to conceptualize how subtle changes in normal cell activity as a function of β-AP induced oxidative GAPDH modification may also provide a significant contribution to either the initiation or to the progression of this neurodegenerative disease.

#### 4.4.3.1 β-AP Dysregulation of Vasodilation/Vasoconstriction Equilibrium by Oxidative Modification of GAPDH

Cerebral vascular dysfunction has been implicated in the development of Alzheimer's disease [rev. in refs. [73, 74]]. The supposition is that interruption in normal blood supply as a function of β-AP activity represents a severe threat to neuronal viability potentially leading to loss of cognitive function. Accordingly, any such effect on cell molecules which are involved in the regulation of the equilibrium which exists between vasodilation and vasoconstriction would be of significance.

Endothelin-1 (ET-1) is an endothelial vasoconstrictor whose posttranscriptional expression is regulated by GAPDH [37]. GAPDH is a negative regulator of ET-1, i.e., as GAPDH activity is diminished, ET-1 expression persists shifting the vasodilation/vasoconstriction equilibrium to the latter. Accordingly, cellular mediators that diminish GAPDH activity would by definition increase vasoconstriction with the attendant physiological consequences.

In that regard, previous studies demonstrated that both glutathione and *S*- nitrosoglutathione bind to the Rossman fold of GAPDH [37, 75]. The binding of the latter inhibits GAPDH activity thereby increasing ET-1 activity facilitating vasoconstriction. [37]. These new findings suggest the possibility that β-AP induced GAPDH oxidative modification could provide a foundation for cerebral vascular constriction as an element of AD pathogenesis.

#### 4.4.3.2 β-AP Induced Dysregulation of Heme Metabolism by Oxidative Modification of GAPDH

Heme is yet another example of a "two edged sword" cellular molecule. On the one hand, it is perhaps the quintessential functional group required for any number of critical cell activities. In contrast, heme, through its iron moiety may be involved in the induction of oxidative stress. For that reason, "free" cellular heme may be a contradiction in terms with heme distribution under strict regulatory control.

Inducible nitric oxide synthetase (iNOS) is a well established, now classical, protein which functions in the regulation of NO homeostasis. Previous studies indicated that NO itself could function as a feedback inhibitor of iNOS function, i.e., given a sufficient concentration of NO, excess nitric oxide could bind to an iNOS regulatory site thereby inhibiting activity [76]. This role for NO would not be unusual as this type of inhibition is a well established mechanism through which cellular metabolic pathways may be regulated.

As a heme protein, iNOS requires the insertion of that metal moiety for activity. What may be unusual, and which may have relevance to the role of oxidative stress in AD pathogenesis, is not only the recent finding that GAPDH functions as an iNOS heme insertion protein thereby regulating iNOS activity but also that SNO modification of GAPDH prevents heme insertion thereby reducing iNOS function [60, 77]. The basis for each effect is that the GAPDH active site cysteine$^{152}$ is required for heme insertion. The latter's modification by NO would eliminate this new GAPDH function. By definition, this formation of SNO-GAPDH would result not only in the diminution of iNOS activity but also would increase in the concentration of "free" heme in vivo.

It is thus possible to suggest that β-AP may not only perturb iNOS regulation by its formation of SNO-GAPDH but also may be able to utilize the "free" heme produced for the further production of reactive oxygen species. Although this hypothesis is quite speculative at the present time, these new findings suggest a possible new effect of β-AP/GAPDH interactions as a contributory process for AD pathogenesis.

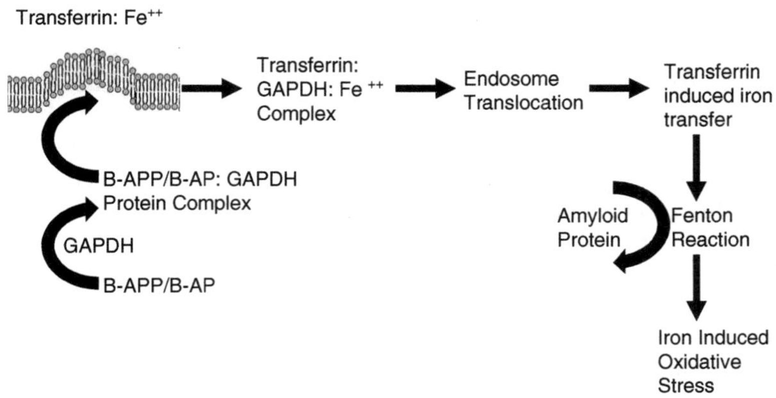


**Fig. 4.5** Iron transport in AD cells

## 4.5 Discussion

The development of unifying theories for multifaceted diseases such as Alzheimer's disease and other age-related neurodegenerative disorders is, at a minimum challenging, and, at its maximum, perhaps impossible. It may be compared to the proverbial field of cut grain in which a mechanism is required to separate the wheat from the chaff. For example, a now classical feature of age-related neurodegenerative disorders is the formation of protein aggregates which characterize each disease. Each aggregate is based on specific protein–protein interactions unique to each disease, i.e., senile plaques in Alzheimer's disease, Lewy bodies in Parkinson's disease, huntingtin aggregates in Huntington's disease as well as high molecular weight structures in other triplet repeat disorders. Yet, their contribution to each respective disorder remains uncertain. Do these protein–protein interactions precede or succeed the fundamental changes in cell structure which ultimately lead to clinical presentation? Are they causal in nature or subsequent effects of other, perhaps more subtle changes which occur far earlier and represent seminal initiating events?

GAPDH is recognized as a contributory protein in Alzheimer's disease based primarily on its localization in senile plaques, its binding to the β-APP and to β-AP as well as its oxidative modification by the β-amyloid protein [rev. in refs. [6, 61]]. Each has the potential to provide a basis for multiple clinical manifestations of Alzheimer's disease including diminution of glucose utilization [45–47], decrease in energy production [56], effects on synaptic transmission [57] and the induction of programmed cell death [65–68]. Each may be explained by the loss of GAPDH function due to β-amyloid protein binding or a gain of function as it pertains to the formation of SNO-GAPDH.

That being said, it is possible to suggest that there is an alternative, or perhaps complementary, pathway through which GAPDH may have the potential to provide a unifying hypothesis for the initiation of AD pathogenesis (Fig. 4.5). In this model, binding of GAPDH to the C-terminal domain of the β-APP [7] facilitates its membrane association in concentrations far greater than normal. In its newly identified capacity as a transferrin receptor binding protein [31, 43] cellular iron uptake would be increased far above that needed for normal metabolism [20–22]. The excess iron would be transported by GAPDH to the endosome, a location in which the β-AP is formed [78]. The reactive oxygen species produced by the latter would now have an excess iron concentration to accelerate the Fenton reaction thereby increasing oxidative modification of critical cellular molecules [23, 40].

Is this model feasible? Current studies indicating the role of GAPDH in iron transfer may present an opportunity to propose this quite simple pathway to provide an initiating event for the pleiotropic cascade which is usually associated with AD pathogenesis. It encompasses the special nature of GAPDH–β-amyloid protein binding, the unique membrane localization of that protein complex, and the postulated increase in iron transfer. Further, each of the four to five steps were not only independently reported but also, when taken in sequence, provide a well-defined cellular pathway.

In its disfavor, it is not unreasonable to question whether, given the added increase in iron concentration in vivo, the resultant increase in oxidative stress would be of sufficient magnitude to initiate the pattern of pathological changes which result in AD development. However, the increase in iron uptake and the resultant increase in oxidative stress need not be dramatic. It may occur at a constant rate which over time may become physiologically significant, in accord with a disease which takes decades to present. In either instance, this may be tested experimentally.

**Acknowledgment** Work in the author's laboratory was funded by a grant from the National Institutes of Health (CA 119285).

**Note Added in Proof** A recent study reported the intriguing finding that GAPDH may be secreted externally to "search and locate" $Fe^{++}$ for intracellular transport and metabolism. (Sheokand et al., Secreted glyceraldehyde-3-phosphate dehydrogenase is a multifunctional autocrine transferrin receptor for cellular iron acquisition, Biochim Biophys Acta, 1830, 3818–3827, 2013). The latter emphasizes further the role of GAPDH in cellular $Fe^{++}$ function.

## References

1. Sirover MA. New insight into an old protein: the functional diversity of mammalian glyceraldehyde-3-phosphate dehydrogenase. Biochim Biophys Acta. 1999;1432:159–84.
2. Sirover MA. New nuclear functions of the glycolytic protein glyceraldehyde-3-phosphate dehydrogenase. J Cell Biochem. 2005;95:45–52.
3. Sirover MA. On the functional diversity of glyceraldehyde-3-phosphate dehydrogenase: Biochemical mechanisms and regulatory control. Biochim Biophys Acta. 2011;1810:741–51.
4. Sirover MA. Subcellular dynamics of multifunctional protein regulation: mechanisms of GAPDH intracellular translocation. J Cell Biochem. 2012;113:2193–200.
5. Tristan C, Shahani N, Sedlak TW, Sawa A. The diverse functions of GAPDH: views from different subcellular compartments. Cell Signal. 2011;23:317–23.

6. Mazzola JL, Sirover MA. Alteration of intracellular structure and function of glyceraldehyde-3-phosphate dehydrogenase: a common phenotype of neurodegenerative disorders? Neurotoxicology. 2002;23:603–9.
7. Shulze H, Schuyler A, Stuber D, Dobeli H, Langen H, Huber G. Rat brain glyceraldehyde-3-phosphate dehydrogenase interacts with the recombinant cytoplasmic domain of Alzheimer's β-amyloid precursor protein. J Neurochem. 1993;60:1915–22.
8. Oyama R, Yamamoto H, Titani K. Glutamine synthetase, hemoglobin K-chain, and macrophage migration inhibitory factor binding to amyloid L-protein: their identification in rat brain by a novel affinity chromatography and in Alzheimer's disease brain by immunoprecipitation. Biochim Biophys Acta. 2000;1479:91–102.
9. Verdier Y, Huszar E, et al. Identification of synaptic plasma membrane proteins co-precipitated with fibrillar β-amyloid peptide. J Neurochem. 2005;94:617–28.
10. Verdier Y, Foldi I, Sergeant N, Fulop L, Penke Z, Janaky T, Szucs M, Penke B. Characterization of the interaction between Aβ 1–42 and glyceraldehyde-3- phosphate dehydrogenase. J Pept Sci. 2008;14:755–62.
11. Burke JR, Enghild JJ, Martin ME, Jou Y-S, Myers RM, Roses AD, Vance JM, Strittmatter WJ. Huntingtin and DRPLA proteins selectively interact with the enzyme GAPDH. Nat Med. 1996;2:347–50.
12. Bae BI, Hara MR, Cascio MB, Wellington CL, Hayden MR, Ross CA, Ha HC, Li X- J, Snyder SH, Sawa A. Mutant huntingtin: nuclear translocation and cytotoxicity mediated by GAPDH. Proc Natl Acad Sci USA. 2006;103:3405–9.
13. Koshy B, Matilla T, Burright EN, Merry DE, Fischbeck KH, Orr HT, Zoghbi HY. Spinocerebellar ataxia type-1 and spinobulbar muscular atrophy gene products interact with glyceraldehyde-3-phosphate dehydrogenase. Hum Mol Genet. 1996;5:1311–8.
14. Tatton NA. Increased caspase 3 and Bax immunoreactivity accompany nuclear GAPDH translocation and neuronal apoptosis in Parkinson's disease. Exp Neurol. 2000;166:29–43.
15. Kragten E, Lalande I, Zimmerman K, Roggo S, Schindler P, Muller D, van Oostrum J, Waldmeier P, Furst P. Glyceraldehyde-3-phosphate dehydrogenase, the putative target of the antiapoptotic compounds CGP 3466 and R-(−)deprenyl. J Biol Chem. 1998;273:5821–8.
16. Maruyama W, Akao Y, Youdim MBH, Davis BA, Naoi M. Transfection- enforced Bcl-2 overexpression and an anti-Parkinson drug, rasagiline, prevent nuclear accumulation of glyceraldehyde-3-phosphate dehydrogenase induced by an endogeneous dopaminergic neurotoxin, N-methyl(R)salsolinol. J Neurochem. 2001;78:727–35.
17. Ling Y, Morgan K, Kalsheker N. Amyloid precursor protein (APP) and the biology of proteolytic processing: relevance to Alzheimer's disease. Int J Biochem Cell Biol. 2003;35:1505–35.
18. Walsh DM, Selkoe DJ. Aβ Oligomers—a decade of discovery. J Neurochem. 2007;101: 1172–84.
19. Yanker BA, Lu T. Amyloid β-protein toxicity and the pathogenesis of Alzheimer disease. J Biol Chem. 2009;284:4755–9.
20. Connor JR, Menzies SL, St. Martin SM, Mufson EJ. A histochemical study of iron, transferrin and ferritin in Alzheimer's diseased brain. J Neurosci Res. 1992;31:75–83.
21. Connor JR, Snyder BS, Beard JL, Fine RE, Mufson EJ. Regional distribution of iron and iron-regulatory proteins in the brain in aging and Alzheimer's disease. J Neuorsci Res. 1992; 31:327–35.
22. Smith MA, Harris PLR, Sayre L, Perry G. Iron accumulation in Alzheimer disease is a source of redox-generated free radicals. Proc Natl Acad Sci USA. 1997;94:9866–8.
23. Rival T, Page RM, et al. Fenton chemistry and oxidative stress mediate the toxicity of the β-amyloid peptide in a *Drosophila* model of Alzheimer's disease. Eur J Neurosci. 2009; 29:1335–47.
24. Jomova K, Vondrakova D, Lawson M, Valko M. Metals, oxidative stress and neurodegenerative disorders. Mol Cell Biochem. 2010;345:91–104.
25. Zheng I, Roeder RG, Luo Y. S phase activation of the histone H2B promoter by OCA-S, a coactivator complex that contains GAPDH as a key component. Cell. 2003;114:255–66.
26. Tisdale EJ. Glyceraldehyde-3-phosphate dehydrogenase is required for vesicular transport in the early secretory pathway. J Biol Chem. 2001;276:2480–6.

27. Tisdale EJ. Glyceraldehyde-3-phosphate dehydrogenase is phosphorylated by protein kinase Cί/λ and plays a role in microtubule dynamics in the early secretory pathway. J Biol Chem. 2002;277:3334–41.
28. Tisdale EJ, Artalejo CR. Src-dependent aprotein kinase C ί/λ (aPKCί/λ) tyrosine phosphorylation is required for aPKCί/λ association with Rab2 and glyceraldehyde-3-phosphate dehydrogenase on pre-golgi intermediates. J Biol Chem. 2006;281:8436–42.
29. Tisdale EJ, Azizi F, Artalejo CR. Rab2 utilizes glyceraldehyde-3-phosphate dehydrogenase and protein kinase Cί to associate with microtubules and to recruit dynein. J Biol Chem. 2009;284:5876–84.
30. Harada N, Yasunaga R, Higashimura Y, Yamaji R, Fujimoto K, Moss J, Inui H, Nakano Y. Glyceraldehyde-3-phosphate dehydrogenase enhances transcriptional activity of androgen receptor in prostate cancer cells. J Biol Chem. 2007;282:22651–61.
31. Raje CI, Kumar S, Harle A, Nanda JS, Raje M. The macrophage cell surface glyceraldehyde-3-phosphate dehydrogenase is a novel transferrin receptor. J Biol Chem. 2007;282:3252–61.
32. Azam S, Jouvet N, Jilani A, Vongsamphanh R, Yang X, Yang S, Ramotar D. Human glyceraldehyde-3-phosphate dehydrogenase plays a direct role in reactivating oxidized forms of the DNA repair enzyme APE1. J Biol Chem. 2008;283:30632–41.
33. Sundararaj KP, Wood RE, Ponnusamy S, Salas AM, Szulc Z, Bielawska A, Obeid LM, Hannun YA, Ogretmen B. Rapid shortening of telomere length in response to ceramide involves the inhibition of telomere binding activity of nuclear glyceraldehyde-3-phosphate dehydrogenase. J Biol Chem. 2004;279:6152–62.
34. Demarse NA, Ponnusamy S, Spicer EK, Apohan E, Baatz JE, Ogretman B, Davies C. Direct binding of glyceraldehyde-3-phosphate dehydrogenase to telomeric DNA protects telomeres against chemotherapy-induced rapid degradation. J Mol Biol. 2009;394:789–803.
35. Bonafe N, Gilmore-Hebert M, Folk NL, Azodi M, Zhou Y, Chambers SK. Glyceraldehyde-3-phosphate dehydrogenase binds to the AU-rich 3′untranslated region of colony-stimulating factor-1 (CSF-1) messenger RNA in human ovarian cancer cells: possible role of CSF-1 post-transcriptional regulation and tumor phenotype. Cancer Res. 2005;65:3762–71.
36. Zhou Y, Yi X, Stofffer JB, Bonafe N, Gilmore-Hebert M, McAlpine J, Chambers SK. The multifunctional protein glyceraldehyde-3-phosphate dehydrogenase is both regulated and controls colony-stimulating factor-1 messenger RNA stability in ovarian cancer. Mol Cancer Res. 2008;6:1375–80.
37. Rodriguex-Pascual F, Redondo-Horcajo M, Magan-Marchal N, Lagares D, Martinez- Ruiz A, Kleinert H, Lamas S. Glyceraldehyde-3-phosphate dehydrogenase regulates endothelin-1 expression by a novel, redox-sensitive mechanism involving mRNA stability. Mol Cell Biol. 2008;28:7139–55.
38. Backlund M, Paukku K, Daviet L, De Boer RA, Valo E, Hautaniemi S, Kalkkinen N, Ehsan A, Kontula KK, Lehtonen JA. Posttranscriptional regulation of angiotensin II type 1 receptor expression by glyceraldehyde-3-phosphate dehydrogenase. NucleicAcids Res. 2009;37:2346–235.
39. Dunn LL, Rahmanto YS, Richardson DR. Iron uptake and metabolism in the new millennium. Trends Cell Biol. 2007;17:93–100.
40. Winterbourn CC. Toxicity of iron and hydrogen peroxide: the Fenton reaction. Toxicol Lett. 1995;82:969–74.
41. Fuchs H, Lucken U, Tauber R, Engel A, Geβner R. Structural model of phospholipid-reconstituted human transferrin receptor derived by electron microscopy. Structure. 1998;6: 1235–43.
42. Aisen P. Transferrin receptor 1. Int J Biochem Cell Biol. 2004;36:2137–43. Raje.
43. Kumar S, Sheokand N, Mhadeshwar MA, Raje CI, Rage M. Characterization of glyceraldehyde-3-phosphate dehydrogenase as a novel transferrin receptor. Int J Biochem Cell Biol. 2012; 44:189–99.
44. Robbins AR, Ward RD, Oliver C. A mutation in glyceraldehyde-3-phosphate dehydrogenase alters endocytosis in CHO cells. J Cell Biol. 1995;130:1093–104.
45. Pietrini P, Alexander GE, Furey ML, Hampel H, Guazzelli M. The neurometabolic landscape of cognitive decline: in vivo studies with positron emission tomography in Alzheimer's disease. Int J Psychophysiol. 2000;37:87–98.

46. Dodart JC, Mathisa C, Bales KR, Paul SM, Ungerera A. Early regional cerebral glucose hypometabolism in transgenic mice overexpressing the V717F b- amyloid precursor protein. Neurosci Lett. 1999;277:49–52.
47. Meier-Ruge W, Bertoni-Freddari C. The significance of glucose turnover in the brain in the pathogenic mechanisms of Alzheimer's disease. Rev Neurosci. 1996;7:1–19.
48. Mazzola JL, Sirover MA. Reduction of glyceraldehyde-3-phosphate dehydrogenase activity in Alzheimer's disease and Huntington's disease fibroblasts. J Neurochem. 2001;76:442–9.
49. Mazzola JL, Sirover MA. Subcellular alteration of glyceraldehyde-3-phosphate dehydrogenase in Alzheimer's disease fibroblasts. J Neurosci Res. 2003;71:279–85.
50. Mazzola JL, Sirover MA. Subcellular analysis of aberrant protein structure in age-related neurodegenerative disorders. J Neurosci Methods. 2004;137:241–6.
51. Cumming RC, Schubert D. Amyloid-β induces disulfide bonding and aggregation of GAPDH in Alzheimer's disease. FASEB J. 2005;19:2060–2.
52. Shalova IN, Cechalova K, et al. Decrease of dehydrogenase activity of dehydrogenase activity of cerebral glyceraldehydes-3-phosphate dehydrogenase in different animal models of Alzheimer's disease. Biochim Biophys Acta. 2007;1770:826–32.
53. Naletova I, Schmalhausen E, Kharitonov A, Katrukha A, Saso L, Caprioli A, Muronetz V. Non-native glyceraldehyde-3-phosphate dehydrogenase can be an intrinsic component of amyloid structures. Biochim Biophys Acta. 2008;1784:2052–8.
54. Glaser PE, Gross RW. Rapid plasmenylethanolamine-selective fusion of membrane bilayers catalyzed by an isoform of glyceraldehyde-3 phosphate dehydrogenase: discrimination between glycolytic and fusogenic roles of individual isoforms. Biochemistry. 1995;34: 12194–203.
55. Selkoe DJ. Alzheimer's disease is a synaptic failure. Science. 2002;298:789–91.
56. Wu K, Aoki C, Else A, Rogalski-Wilk AA, Siekevtz P. The synthesis of ATP by glycolytic enzymes in the postsynaptic density and the effect of endogenously generated nitric oxide. Proc Natl Acad Sci USA. 1997;94:13273–8.
57. Laschet JJ, Minier F, et al. Glyceraldehyde-3-phosphate dehydrogenase is a $GABA_A$ receptor kinase linking glycolysis to neuronal inhibition. J Neurosci. 2004;24:7614–22.
58. Kornberg MD, Sen N, Hara MR, Juluri KR, Nguyen JVK, Snowman AM, Law L, Hester LD, Snyder SH. GAPDH mediates nitrosylation of nuclear proteins. Nat Cell Biol. 2010;12:1094–100.
59. Stamler JS, Hess DT. Nascent nitrosylases. Nat Cell Biol. 2010;12:1024–6.
60. Chakravarti R, Aulak KS, Fox PL, Stuehr DJ. GAPDH regulates cellular heme insertion into inducible nitric oxide synthase. Proc Natl Acad Sci USA. 2010;107:18004–9.
61. Butterfield DA, Hardas SS, Bader Lange ML. Oxidatively modified glyceraldehyde-3-phosphate dehydrogenase (GAPDH) and Alzheimer's disease: many pathways to neurodegeneration. J Alzheimers Dis. 2010;20:369–93.
62. Boyd-Kimball D, Sultana R, et al. Proteomic identification of proteins specifically oxidized by intracerebral injection of amyloid β-peptide (1–42) into rat brain: Implications for Alzheimer's disease. Neuroscience. 2005;132:313–24.
63. Sultana R, Newman SF, et al. Protective effect of D609 against amyloid- $beta_{1-42}$-inducded oxidative modification of neuronal proteins: redox proteomics study. J Neurosci Res. 2006; 84:409–17.
64. Nakajima H, Amano W, et al. The active site cysteine of the proapoptotic protein glyceraldehyde-3-phosphate dehydrogenase is essential in oxidative stress-induced aggregation and cell death. J Biol Chem. 2007;282:26562–74.
65. Hara MR, Agrawal N, et al. S-nitrosylated GAPDH initiates apoptotic cell death by nuclear translocation following Siah1 binding. Nat Cell Biol. 2005;7:665–74.
66. Hara MR, Snyder SH. Nitric oxide-GAPDH-siah: a novel cell death cascade. Cell Mol Neurobiol. 2006;26:525–36.
67. Hara MR, Cascio MB, Sawa A. GAPDH as a sensor of NO stress. Biochim Biophys Acta. 2006;1762:502–9.
68. Sen N, Hara MR, et al. Nitric oxide-induced nuclear GAPDH activates p300/CBP and mediates apoptosis. Nat Cell Biol. 2008;10:866–73.

69. Yego ECK, Vincent JA, Sarthy VP, Busik I, Mohr S. Differential regulation of high glucose-induced glyceraldehyde-3-phosphate dehydrogenase nuclear accumulation in muller cells by IL-1beta and IL-6. Invest Opthalmol Vis Sci. 2009;50:1920–8.
70. Yego ECK, Mohr S. Siah-1 protein is necessary for high glucose-induced glyceraldehyde-3-phosphate dehydrogenase nuclear accumulation and cell death in muller cells. J Biol Chem. 2010;285:3181–90.
71. Colell A, Ricci JE, et al. GAPDH and autophagy preserve survival after cytochrome c release in the absence of caspase activation. Cell. 2007;129:983–97.
72. Colell S, Green DR, Ricci JE. Novel roles for GAPDH in cell death and carcinogenesis. Cell Death Differ. 2009;16:1573–81.
73. Townsend KP, Obregon D, Quadros A, Patel N, Volmer CH, Paris D, Mullan M. Proinflammatory and vasoactive effects of Aβ in the cerebrovasculature. Ann N Y Acad Sci. 2002;977:65–76.
74. Iadecola C. Cerebrovascular effects of amyloid-β peptides: mechanisms and implications for Alzheimer's dementia. Cell Mol Neurobiol. 2003;23:681–9.
75. Puder M, Soberman RJ. Glutathione conjugates recognize the Rossmann fold of glyceraldehyde-3-phosphate dehydrogenase. J Biol Chem. 1997;272:10936–40.
76. Albakri QA, Stuehr DJ. Intracellular assembly of inducible NO synthase is limited by nitric oxide-mediated changes in heme insertion and availability. J Biol Chem. 1996;271:5414–21.
77. Chakravarti R, Stuehr DJ. Thioredoxin-1 regulates cellular heme insertion by controlling S-nitrosation of glyceraldehyde-3-phosphate dehydrogenase. J Biol Chem. 2012;287:16179–86.
78. Koo EH, Squazzo SL. Evidence that production and release of amyloid β- protein involves the endocytic pathway. J Biol Chem. 1994;269:17386–9.

# Chapter 5
# Aβ in Mitochondria—One Piece in the Alzheimer's Disease Puzzle

**Maria Ankarcrona**

**Abstract** Although great advances have been made in our understanding of the neurodegenerative process in Alzheimer's disease (AD), the complete picture has not emerged and there are still pieces missing. One attractive hypothesis is that mitochondrial failure is a cause of synapse loss and cognitive impairment in AD. ATP generation by mitochondria is crucial for proper synaptic function and therefore neurons are highly sensitive to mitochondrial damage potentially leading to synapse loss and cognitive dysfunction. Several evidences indicate that mitochondria are indeed damaged and dysfunctional in the AD brain; these include mitochondrial accumulation of amyloid β-peptide (Aβ), impaired brain glucose metabolism, impaired mitochondrial fusion/fission, and increased generation of reactive oxygen species (ROS). In this chapter we will focus on the role of Aβ in mitochondria and discuss mitochondrial uptake mechanisms and interactions with mitochondrial proteins. Several evidences point towards a central role of Aβ initiating mitochondrial damage and generation of ROS in turn leading to synaptic and neuronal degeneration. Therefore, it would be of high importance to develop drugs that maintain mitochondrial integrity and prevent mitochondrial failure otherwise leading neuronal dysfunction.

## 5.1 Introduction

Alzheimer's disease (AD) is a multifactor disorder resulting in neuronal degeneration and memory loss. The current lack of disease modifying drugs for this detrimental disorder is an increasing problem which leaves patients, relatives, caregivers, and society with an enormous burden. In order to develop such drugs it is mandatory

M. Ankarcrona, Ph.D. (✉)
Department of Neurobiology, Care Sciences and Society, KI-Alzheimer's Disease Center (KI-ADRC), Karolinska Institutet, Novum 5th floor, Stockholm SE-141 86, Sweden
e-mail: maria.ankarcrona@ki.se

D. Praticò and P. Mecocci (eds.), *Studies on Alzheimer's Disease*, Oxidative Stress in Applied Basic Research and Clinical Practice, DOI 10.1007/978-1-62703-598-9_5,

to elucidate the underlying disease mechanisms and one hypothesis that has been put forward is that mitochondrial failure is the cause of synapse loss and cognitive impairment in AD [1]. The amyloid β-peptide (Aβ) is one of the pathological hallmarks in AD and has been suggested to exert its toxicity both extra- and intra-cellulary [2]. Oligomeric forms of Aβ secreted from cells has for example been shown to bind to synapses and inhibit long-term potentiation [3], while intracellular Aβ accumulate in mitochondria and negatively affect mitochondrial function [4]. Aβ has been detected in mitochondria both in humans and animals [4–7]. In vitro studies show that Aβ is transported into mitochondria via the translocase of the outer membrane (TOM) machinery and localize to the mitochondrial cristae [7]. Interestingly, it has also been shown that Aβ is accumulating specifically in synaptic mitochondria in young tg AD mice [8]. In addition, a thorough study on synaptic mitochondria isolated from different brain regions from wt or AD tg mice show that hippocampal and cortical mitochondria show the highest levels of mitochondrial dysfunction (including increased ROS production and complex IV activity and decreased mitochondrial membrane potential) [9]. Together these data further support the mitochondrial hypothesis and suggest that synaptic failure detected early in the AD disease process may be caused by mitochondrial Aβ. In this chapter this hypothesis is further reviewed and mitochondrial targeting possibilities discussed. It is becoming evident that we have to treat AD early on in the disease process in order to prevent/decrease synapse loss and neurodegeneration and mitochondria emerge as one important drug target.

## 5.2 The γ-Secretase Complex, APP, and Aβ Are Localized to Mitochondria and Mitochondria-Associated ER Membranes

Aβ is cleaved out from the amyloid β-precursor protein (APP) by the subsequent cleavage by β- and γ-secretases. β-Secretase cleavage of APP results in the formation of a C-terminal membrane bound fragment referred to as C99. C99 is one of many substrates for the γ-secretase complex. The γ-secretase complex is membrane bound and consists of at least four different proteins, i.e., presenilin (PS1 or PS2), Nicastrin, anterior pharynx defective 1 (Aph-1), and presenilin enhancer-2 (Pen-2) [10]. APP is a type I transmembrane protein located at the plasma membrane, endosomes, Golgi network and ER with its C-terminus facing the cytosol. Depending on the exact cleavage site on the APP molecule γ-secretase complex cleavage results in different lengths of the Aβ peptide. In addition to Aβ γ-secretase cleavage of APP also generates APP intracellular domain (AICD). In the non-amyloidogenic pathway α-secretase cleaves APP within the Aβ sequence leading to generation of C83. Subsequent cleavage of C83 by γ-secretase results in production of AICD and a non-amyloidogenic p3-fragment. $A\beta_{40}$ is most abundant while the longer forms, $A\beta_{42\text{-}48}$ are more prone to aggregate and form fibrils and plaques. The longer species are also more neurotoxic as compared to $A\beta_{40}$. Mutations linked to familiar forms of AD have been identified in APP, PS1, and PS2 and are associated with an increased $A\beta_{42}/A\beta_{40}$ ratio resulting in neurotoxicity and extensive plaque formation [11]. Aβ is generated at the plasma membrane and in the ER, Golgi, endosomal/lysosomal system following the pattern of intracellular localization of APP. Interestingly, Aβ is accumulating inside mitochondria both in human AD brain and in animal models indicating that Aβ is either produced inside mitochondria and/or taken up from the outside. In transgenic mice overexpressing mutant APP (V717/F, K670M, N671L from L Mucke) $A\beta_{40}$ and $A\beta_{42}$ start to accumulate in mitochondria from 4 months old animals before formation of plaques [5, 8]. Of particular interest is that Aβ accumulation starts in synaptic mitochondria where it causes mitochondrial dysfunction by interfering with respiratory function, mitochondrial permeability transition (mPT), and mitochondrial trafficking and transport [8]. Neurons heavily rely on oxidative phosphorylation (OXPHOS) for ATP production and a large part of this ATP is used during propagation of signals at synapses and required to drive $Na^+/K^+$- and $Ca^{2+}$-pumps. Therefore, proper mitochondrial function including ATP production is essential for synaptic function and signaling.

Whether Aβ is produced inside mitochondria or taken up from the outside is not yet fully clarified. Since both APP [12] and active γ-secretase complexes [13] have been detected in mitochondria it is theoretically possible that Aβ is produced locally in mitochondria. APP has been shown to accumulate in AD brain mitochondria via arrested import leaving a large C-terminal part outside [12, 14]. Under these circumstances APP is stuck in the mitochondrial protein import pore, consisting of the translocases of the outer (TOM) and inner membrane (TIM), causing impairment of mitochondrial function and eventually cell toxicity. The import of APP is arrested due to an acidic domain at amino acids 220–290 leaving the Aβ-region outside the import pore (Fig. 5.1). Recent data from our laboratory show that the C-terminal part of APP can be inserted into the outer mitochondrial membrane (OMM) and that the mitochondrial γ-secretase cleaves APP to generate AICD which was detected in the inter membrane space [15]. However, we detected only C83 (generated by α-secretase cleavage) and not C99 (generated by β-secretase cleavage) in the mitochondrial membrane. Subsequent γ-secretase cleavage of C83 results in formation of the p3-fragment and not Aβ formation. Thus, it is more likely that Aβ is taken up from the outside of mitochondria rather than produced inside mitochondria. Aβ coming from the outside of mitochondria could either be transported to mitochondria via vesicles [16] or produced at mitochondria-associated ER membranes (MAM) [17].

MAM is a specialized region of the endoplasmatic reticulum (ER) enriched in cholesterol and the membrane composition thus similar to lipid rafts. MAM is in contact with mitochondria and connects ER and mitochondria both physically and biochemically (Fig. 5.1). MAM has a central role in phospholipid, glucose, sphingolipid, ganglioside, cholesterol, and fatty acid metabolism and also regulates calcium homeostasis and apoptosis [18]. Interestingly, all four components in the γ-secretase as well as APP has been detected in MAM. γ-Secretase activity was

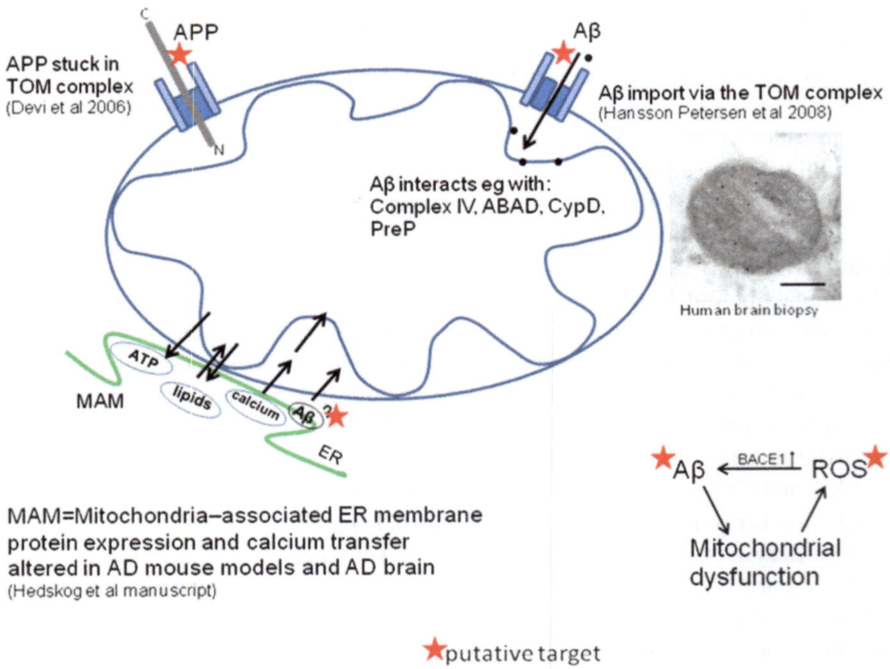


**Fig. 5.1** Schematic illustration of the current knowledge of Aβ in mitochondria. APP is partly imported via the TOM complex and then stuck in the import pore causing impairment of mitochondrial function [14]. Aβ is also imported via the TOM complex and accumulate in the inner membrane where it interacts with proteins of for example the electron transport chain and mitochondrial membrane transition pore [7]. Aβ might be generated in mitochondria-associated ER membranes (MAM), where both APP and the γ-secretase complex have been detected [17], and then transported into mitochondria via TOM

detected using both a fluorescence based energy transfer-based assay and Western blotting to detect AICD [17]. The results showed that the highest γ-secretase activity were detected in MAM as compared to plasma membrane, ER and mitochondrial fractions. We have previously detected active γ-secretase complexes in mitochondria [13] and calculated that a few percent of the total γ-secretase activity in tissue is executed by the mitochondrial γ-secretase (Ankarcrona M *unpublished data*). As described above we do not have data supporting local Aβ production in mitochondria; however, we have detected AICD formation in mitochondria [15]. The function of AICD in mitochondria is presently unknown. Since rather high levels of γ-secretase activity were detected in MAM [17] it is tempting to speculate that Aβ produced at MAM can reach mitochondria via the TOM import machinery in the outer mitochondrial membrane (Fig. 5.1). This uptake mechanism was shown by in vitro import studies in isolated mitochondria performed in our laboratory [7] and further described below.

## 5.3 Mitochondrial Aβ Uptake Mechanisms and Submitochondrial Localization

As discussed above many studies have shown the accumulation of Aβ in mitochondria both from human AD brain and tg mutant APP mice. Several studies have also shown that Aβ cause mitochondrial toxicity and it would be presumably beneficial to block mitochondrial Aβ uptake as a treatment strategy for AD. Therefore, we undertook a study investigating the mechanisms for mitochondrial Aβ uptake (Fig. 5.1). The rationale was to use purified rat liver mitochondria treated with 0.1 μM $A\beta_{1-40}$ or $A\beta_{1-42}$ in the absence or presence of antibodies or inhibitors directed to various mitochondrial translocases, pores and channels [7]. Both $A\beta_{1-40}$ and $A\beta_{1-42}$ were taken up by mitochondria during the 30 min incubation period. The Aβ uptake was not affected by the presence of antibodies directed towards the voltage-dependent anion channel (VDAC) nor in the presence of Cyclosporine A, which is an inhibitor of the mitochondrial permeability transition pore (mPTP). In contrast, import of both $A\beta_{1-40}$ and $A\beta_{1-42}$ was prevented when import competent mitochondria were pre-incubated with antibodies directed towards proteins of the TOM complex, i.e., TOM20, TOM40, TOM70. Aβ import was not affected by the addition of valinomycin, an ionophore which cause depolarization of the mitochondrial inner membrane, indicating that the Aβ import was not dependent on the $\psi_{mit}$. After import Aβ was mostly localized to mitochondrial cristae and associated with the inner membrane fraction. It may be a hydrophobic interaction between Aβ and the TOM receptors leading to import over the outer mitochondrial membrane (OMM). Since Aβ has no classical import signaling sequence Aβ is not further imported into the matrix via the translocase of the inner membrane (TIM). It was previously reported that Aβ co-localizes with the mitochondrial matrix protein Hsp60 in mouse and human samples [5]. One explanation to the discrepancy between this and our study might be that in our in vitro assay we studied Aβ localization after 30 min of import, whereas Caspersen et al. report data from postmortem AD brains and 8-months-old transgenic APP mice. However, our data from human brain biopsies obtained from living subjects, displaying Aβ aggregates in the neuropil, show Aβ immuno-gold labeling in association with mitochondrial inner membranes [7]. Moreover, Singh et al. have in a bioinformatic study predicted that Aβ is localized to the inner membrane and rule out the presence of Aβ in the matrix [19]. Still we cannot exclude that a fraction of Aβ can be released or escapes from the membrane and into the matrix. In summary, these data show that Aβ is imported via the TOM complex where TOM 20 and TOM70 are receptors and TOM40 forms a pore in the OMM.

Recently Roses and colleagues reported that a polymorphic poly-T variant in the *TOMM40* gene (rs10524523) can be used to estimate the age of LOAD onset for *APOEε3* carriers. *APOEε3/4* carriers with very long/long poly-T repeats linked to *APOEε3* had an age of onset 7 years earlier as compared to individuals with shorter repeats [20]. *TOMM40* and *APOE* genes are separated by only ~2 kb on chromosome 19. In a novel study we investigated the effect of different poly-T lengths in

*TOMM40* on the mRNA, protein and mitochondrial levels using fibroblasts from healthy *APOEε3/4* individuals carrying either short/long poly-T or very long/long poly-T (*APOEε4* always brings a long poly-T in *TOMM40*) [21]. A modified protein with potentially impaired function could for example negatively influence protein import into mitochondria which in turn would lead to mitochondrial deficiency and neuronal death explaining the earlier age of onset in *APOEε3/4* carriers with very long/long poly-T repeats. However, in our study we detected no differences in any of the parameters measured (e.g., mRNA splicing/exon skipping, TOM40 expression levels, mitochondrial membrane potential, mitochondrial area and morphology) [21]. Thus, these data, obtained from a rather limited sample set, do not support the hypothesis that the polymorphism rs10524523 directly influence the function of Tom40 and mitochondria. So far we have not investigated if the rate of mitochondrial Aβ import is affected in cells carrying this polymorphism. This may be worthwhile pursuing both for rs10524523 and other polymorphisms in *TOMM40* linked to AD.

To interfere with the TOM import machinery as a treatment strategy for AD is not trivial since this machinery cannot be blocked for import of proteins required for mitochondrial function. One possibility would be to identify the binding site for Aβ on for example TOM20 or TOM70 receptors and screen for compounds that specifically blocks this interaction without affecting import of other proteins.

## 5.4 Aβ Interaction with Mitochondrial Proteins

Within mitochondria Aβ has been shown to interact with several different proteins causing mitochondrial dysfunction and cell toxicity (Fig. 5.1). Here some examples of Aβ-protein interactions will be discussed (for an additional review see ref. [22]).

### *5.4.1 Electron Transport Chain Enzyme Complexes*

Analyses of AD postmortem brain have shown decreased activity of cytochrome c oxidase (COX) also known as complex IV [23–25]. Also platelets from AD patients and AD cybrid cells have a complex IV deficiency [26, 27]. In vitro studies with mitochondria isolated from human leukocytes suggest that $A\beta_{1-42}$ inhibits complex IV activity in a copper-dependent manner [4]. The complex IV was specifically damaged in line with other studies [28–30] reporting that $A\beta_{25-35}$ selective damage complex IV and not complex I, II, or III. Crouch et al. [4] showed that low molecular weight oligomers were the toxic $A\beta_{1-42}$ species responsible for complex IV inhibition. As mentioned above complex IV activity was also shown to be decreased in mitochondria isolated from tg $APP_{swe}$ and tg $APP_{swe}/PS1_{M146V}$ mice [9]. Further evidence for Aβ induced inhibition of complex IV activity also comes from 3× Tg-AD mice expressing mutations in APP, PS1 and Tau. In this model compromised energy production including decreased complex IV activity preceded plaque formation [31].

In another study complex IV activity was decreased in 2× APP/PS2 and 3× APP/PS2/Tau Tg AD cortices but not in mice with a Tau mutation. Instead the tau mice (pR5) had impairments in complex I activity [32]. These data confirm that it is mainly Aβ pathology that affects complex IV activity.

### 5.4.2 *The Mitochondrial Permeability Transition Pore*

Opening of the mitochondrial permeability transition pore (mPTP) results in loss of mitochondrial membrane potential, swelling of mitochondria and the release of pro-apoptotic proteins from the intermembrane space (IMS). The protein composition of mPTP has not been fully elucidated and different models have been proposed. One model is that the voltage-dependent anion channel (VDAC) in the OMM forms the pore together with adenine nucleotide translocase (ANT) and inorganic phosphate carrier (PiC) in the IMM [33]. Aβ has been shown to specifically interact with cyclophilin D (CypD), a mitochondrial matrix protein that associates with the inner membrane during opening of the mitochondrial permeability transition pore (mPTP) [34]. Translocation of matrix Cyclophilin D (CypD) to the inner membrane and CypD binding to PiC has also been proposed to trigger opening of calcium-sensitive nonspecific channels [19]. Cortical mitochondria from CypD deficient mice are resistant to Aβ- and calcium-induced mitochondrial swelling and permeability transition. Moreover, Tg mAβPP/CypD-null mice had improved learning and memory and synaptic function both in 12 and 24 months old animals [34, 35]. Aβ has also been predicted to interact with ANT in the inner membrane [19]. Simulation of protein-protein interactions suggested that the ANT-Aβ interaction is stronger that the CypD-Aβ interaction. At present it is not known what function the ANT-Aβ interaction has; however, it may affect the normal physiological function of ANT which is transport of ATP and ADP.

### 5.4.3 *Mitochondrial Aβ-Binding Alcohol Dehydrogenase and Presequence Protease*

Two different Aβ-binding proteins have been identified in the mitochondrial matrix, i.e., mitochondrial Aβ-binding alcohol dehydrogenase (ABAD) and Presequence Protease (PreP). Our data as described above show that Aβ is located to the inner membrane after import via the TOM40 pore. To what extent this Aβ fraction is available for ABAD and Prep P interactions in the matrix is not clear at present. ABAD has been found to be up-regulated in neurons from AD patients [36] and Aβ has been shown to interact with ABAD resulting in free radical production and neuronal apoptosis. ABAD was identified as an Aβ-binding protein in a yeast two-hybrid screen [36]. ABAD is localized to the mitochondrial matrix and has an essential physiological role in mitochondria. ABAD-Aβ complexes were detected in AD brain and in Tg mutant AβPP/ABAD (Tg mAβPP/ABAD) mice. Cortical neurons cultured from Tg mAβPP/ABAD mice show increased production of ROS and decreased mitochondrial membrane potential, ATP levels, and activity of respiratory chain complex IV. Consistently, these neurons displayed DNA-fragmentation and caspase-3 activity resulting in cell death by day 5–6 in culture [37]. ABAD uses $NAD^+$ and/or NADH as its cofactor and catalyzes the reversible oxidation and/or reduction of alcohol group in its substrates [38]. The crystal structure of ABAD-Aβ complexes has been determined showing that the $NAD^+$ binding pocket is distorted, hindering $NAD^+$ from binding to ABAD in the presence of Aβ [36, 38]. Thus, Aβ blocks ABAD activity causing mitochondrial dysfunction and ultimately cell death. Two stretches of ABAD residues in the $L_D$ loop region (amino acids 95–113) have been shown to be important for Aβ binding. Cell permeable peptides ABAD-DP (ABAD-decoy peptide fused to the Tat protein and a mitochondrial targeting signal) administrated to transgenic APP mice blocked formation of ABAD-Aβ complexes in mitochondria, attenuated oxidative stress, increased mitochondrial respiration, and also importantly improved spatial memory [39]. Thus, the use of inhibitors of ABAD-Aβ interaction emerges as a novel therapeutic strategy for AD.

PreP is also localized to the mitochondrial matrix and putatively responsible for the degradation of the accumulated Aβ in mitochondria [40]. PreP was originally found and characterized in *Arabidopsis thaliana* [41] as a protease degrading targeting peptides that are cleaved off in mitochondria after completed protein import as well as other unstructured peptides up to 65 amino acid residues in length, but not small proteins [42, 43]. Recombinant hPreP completely degrades both Aβ40 and Aβ42 as well as Aβ Arctic protein (42, E22G) at unique cleavage sites including several sites in a very hydrophobic C-terminal Aβ (29–42) segment that is prone to aggregation. Interestingly, PreP is an organellar functional analogue of the human Insulin Degrading Enzyme (IDE), implicated in AD as it cleaves Aβ before insoluble amyloid fibers are formed [44–46]. A recent study using human and transgenic mouse brain show that PreP activity is reduced in human postmortem AD brain (temporal lobe) and in mice overexpressing mutant AβPP (m AβPP, J-20 line) or mutant AβPP together with ABAD (AβPP/ABAD) [47] Enhanced production of ROS may cause the decreased PreP proteolytic activity resulting in mitochondrial Aβ accumulation and in turn leading to toxicity and neuronal degeneration. Interestingly, it has also been shown that the Aβ degrading enzymes neprilysin and IDE are subject to oxidative inactivation [48]. Decreased degradation of Aβ in combination with ROS induced BACE1 activity [49], as discussed below, would result in increased Aβ generation and accumulation.

## 5.5 The Vicious Cycle of Aβ and ROS Generation: A Putative Target?

So far we have discussed data showing that Aβ can enter mitochondria, bind to various proteins and thus induce for example ROS production or disruption of mitochondrial integrity. However, in a recent publication Leuner et al. [49] show that cells treated with toxins (i.e., rotenone and antimycin) inhibiting respiratory complex I and III respectively and subsequently inducing ROS production trigger upregulation of BACE1 activity and increased secretion of Aβ. Also animals with deficiency in complex I (Ndufs3 KO mice) show high production of ROS and increased levels of secreted Aβ. These data suggest that dysfunction in the respiratory chain trigger an increased generation of Aβ. The secreted Aβ may then be toxic by binding to synapses or internalized and for example transported into mitochondria where it further impairs respiratory function initiating a vicious cycle of ROS and Aβ generation (Fig. 5.1). Mitochondria are the main source of ROS in the cell and these free radicals can affect targets (proteins, lipids, RNA, DNA) both inside and outside mitochondria. The study by Leuner et al. [49] reinforces the importance of controlling ROS production in cells as one possible treatment for AD. Indeed it was recently published that MitoQ a mitochondria targeted antioxidant had positive effects on cognition in mice after 4–5 months treatment [50]. MitoQ is accumulating 500–1,000× inside mitochondria and efficiently scavenging ROS at the spot of its production. MitoQ has been through two phase II trials for Parkinson's disease [51]. For an extended review of other antioxidants tested as AD treatment [52]. No matter what is "the hen or the egg" selective modulation of BACE1 and/or γ-secretase activity and antioxidants targeted to mitochondria are two treatment strategies worth pursuing in order to maintain proper mitochondrial function and synapse activity.

At present Alzheimer's disease researchers are questioning what went wrong when the clinical trials of different compounds and antibodies designed to interfere with Aβ production/aggregation/clearance failed. The common sense is that (a) inhibitors of γ-secretase are not useful; instead we need molecules that modulates γ-secretase activity specifically towards APP (b) BACE1 is a difficult target (c) we need to enrol patients who are in early stages of the disease. The last point put high demands on reliable diagnosis with a combination of validated biochemical biomarkers, brain imaging and neuropsychological testing. Most clinical trials conducted so far have enrolled patients with mild to moderate or even severe AD and no consistent improvement of cognition has been reported. Probably these patient's neurons have already started to degenerate in high degree and even though data show a decreased plaque burden the neurons cannot be rescued at this late stage. Our own experience from the Latrepidine (Dimebon) study points in the same direction: to be efficient treatment has to be given before mitochondria and other cell functions are damaged.

Dimebon was originally approved in the former Soviet Union as a nonselective antihistamine for skin allergy and allergic rhinitis [53], but was withdrawn from the market with the advent of more selective treatments. Dimebon attracted renewed interest due to findings suggesting a neuroprotective effect [54–56]. In a Phase II AD trial, dimebon treatment was associated with benefits on cognition, global function, activities of daily living, and behavior [57]. Several Phase III clinical trials were then performed for both AD and Huntington's disease [58]. However, all clinical trials with dimebon have now been terminated since no positive effects of the drug treatment were obtained.

Dimebon exhibits a rich pharmacological profile and binds to histamine-, adrenergic-, dopamine-, and serotonin-receptors [56, 59]. It is known to be a weak inhibitor of: acetylcholinesterase ($IC_{50}$ = 8–42 μM) [55], N-methyl-D-aspartate (NMDA) receptors ($IC_{50}$ = 10 μM) [56, 60], and voltage-gated calcium channels

($IC_{50}$ = 50 μM) [56, 61]. In addition, μM concentrations of dimebon have previously been shown to protect against neuronal cell death induced by $A\beta_{25-35}$ [55] and to modulate the mitochondrial permeability transition pore (10–200 μM) [62]. In a study from our laboratory [63] we show that nM concentrations of dimebon (1–5 days incubation) results in an increase of mitochondrial membrane potential (hyper-polarization) and cellular ATP levels both in mouse cortical neurons and human neuroblastoma cells. Moreover, dimebon pretreatment made cells more resistant to depolarization of mitochondrial membrane potential induced by high intracellular calcium concentrations. Cells were also protected from undergoing cell death induced either by calcium stress or withdrawal of growth factors. Our study suggests that dimebon directly or indirectly affects mitochondria making cells more resistant to cell death stimuli. Based on our in vitro data it is still possible that dimebon might work more efficiently if given to patients early in the disease process.

## 5.6 Conclusions

Accumulating evidence both from human brain as well as AD animal and cell models show that Aβ is imported into mitochondria where it accumulates in the inner membrane and bind to various proteins causing mitochondrial failure and cell toxicity. The consistent inhibition of complex IV (COX) in the respiratory chain by Aβ has been shown by several laboratories. Interestingly, in a recent study [49] it was shown that complex I inhibition or deficiency resulting in increased generation of ROS leads to activation of BACEI and increased secretion of $A\beta_{40}$. Together the data suggests a vicious cycle of ROS and Aβ generation, mitochondrial failure and neuronal degeneration. Importantly, accumulation of Aβ appears to be an early event during the disease process, as Aβ for example has been shown to first accumulate in synaptic mitochondria in young AD tg mice [8]. It is therefore possible that mitochondrial Aβ accumulation is one cause of synaptic failure correlating with cognitive impairment in AD. Future treatment strategies should take this into account and drugs targeting mitochondria developed.

**Note added in proof** For a recent publication about the role of ER-mitochondria interplay in AD see: Modulation of the endoplasmic reticulum-mitochondria interface in Alzheimer's disease and related models. Hedskog L, Pinho CM, Filadi R, Rönnbäck A, Hertwig L, Wiehager B, Larssen P, Gellhaar S, Sandebring A, Westerlund M, Graff C, Winblad B, Galter D, Behbahani H, Pizzo P, Glaser E and Ankarcrona M. Proc Natl Acad Sci USA. 2013;110:7916–21.

## References

1. Swerdlow RH, Khan SM. A "mitochondrial cascade hypothesis" for sporadic Alzheimer's disease. Med Hypotheses. 2004;63:8–20.
2. Selkoe DJ. Soluble oligomers of the amyloid beta-protein impair synaptic plasticity and behavior. Behav Brain Res. 2008;192:106–13.

3. Li S, Jin M, Koeglsperger T, Shepardson NE, Shankar GM, et al. Soluble Abeta oligomers inhibit long-term potentiation through a mechanism involving excessive activation of extrasynaptic NR2B-containing NMDA receptors. J Neurosci. 2011;31:6627–38.
4. Crouch PJ, Blake R, Duce JA, Ciccotosto GD, Li QX, et al. Copper-dependent inhibition of human cytochrome c oxidase by a dimeric conformer of amyloid-beta1-42. J Neurosci. 2005; 25:672–9.
5. Caspersen C, Wang N, Yao J, Sosunov A, Chen X, et al. Mitochondrial Abeta: a potential focal point for neuronal metabolic dysfunction in Alzheimer's disease. FASEB J. 2005;19: 2040–1.
6. Manczak M, Anekonda TS, Henson E, Park BS, Quinn J, et al. Mitochondria are a direct site of A beta accumulation in Alzheimer's disease neurons: implications for free radical generation and oxidative damage in disease progression. Hum Mol Genet. 2006;15:1437–49.
7. Hansson Petersen CA, Alikhani N, Behbahani H, Wiehager B, Pavlov PF, et al. The amyloid beta-peptide is imported into mitochondria via the TOM import machinery and localized to mitochondrial cristae. Proc Natl Acad Sci U S A. 2008;105:13145–50.
8. Du H, Guo L, Yan S, Sosunov AA, McKhann GM, et al. Early deficits in synaptic mitochondria in an Alzheimer's disease mouse model. Proc Natl Acad Sci U S A. 2010;107:18670–5.
9. Dragicevic N, Mamcarz M, Zhu Y, Buzzeo R, Tan J, et al. Mitochondrial amyloid-beta levels are associated with the extent of mitochondrial dysfunction in different brain regions and the degree of cognitive impairment in Alzheimer's transgenic mice. J Alzheimers Dis. 2010;20 Suppl 2:S535–50.
10. Bergmans BA, De Strooper B. Gamma-secretases: from cell biology to therapeutic strategies. Lancet Neurol. 2010;9:215–26.
11. Benilova I, Karran E, De Strooper B. The toxic Abeta oligomer and Alzheimer's disease: an emperor in need of clothes. Nat Neurosci. 2012;15:349–57.
12. Anandatheerthavarada HK, Biswas G, Robin MA, Avadhani NG. Mitochondrial targeting and a novel transmembrane arrest of Alzheimer's amyloid precursor protein impairs mitochondrial function in neuronal cells. J Cell Biol. 2003;161:41–54.
13. Hansson CA, Frykman S, Farmery MR, Tjernberg LO, Nilsberth C, et al. Nicastrin, presenilin, APH-1, and PEN-2 form active gamma-secretase complexes in mitochondria. J Biol Chem. 2004;279:51654–60.
14. Devi L, Prabhu BM, Galati DF, Avadhani NG, Anandatheerthavarada HK. Accumulation of amyloid precursor protein in the mitochondrial import channels of human Alzheimer's disease brain is associated with mitochondrial dysfunction. J Neurosci. 2006;26:9057–68.
15. Pavlov PF, Wiehager B, Sakai J, Frykman S, Behbahani H, et al. Mitochondrial gamma-secretase participates in the metabolism of mitochondria-associated amyloid precursor protein. FASEB J. 2011;25:78–88.
16. Muresan V, Varvel NH, Lamb BT, Muresan Z. The cleavage products of amyloid-beta precursor protein are sorted to distinct carrier vesicles that are independently transported within neurites. J Neurosci. 2009;29:3565–78.
17. Area-Gomez E, de Groof AJ, Boldogh I, Bird TD, Gibson GE, et al. Presenilins are enriched in endoplasmic reticulum membranes associated with mitochondria. Am J Pathol. 2009;175: 1810–6.
18. Schon EA, Area-Gomez E. Is Alzheimer's disease a disorder of mitochondria-associated membranes? J Alzheimers Dis. 2010;20 Suppl 2:S281–92.
19. Singh P, Suman S, Chandna S, Das TK. Possible role of amyloid-beta, adenine nucleotide translocase and cyclophilin-D interaction in mitochondrial dysfunction of Alzheimer's disease. Bioinformation. 2009;3:440–5.
20. Roses AD, Lutz MW, Amrine-Madsen H, Saunders AM, Crenshaw DG, et al. A TOMM40 variable-length polymorphism predicts the age of late-onset Alzheimer's disease. Pharmacogenomics J. 2010;10(5):375–84.
21. Hedskog L, Brohede J, Wiehager B, Pinho CM, Revathikumar P, Lilius L, et al. Biochemical studies of poly-T variants in the Alzheimer's disease associated TOMM40 gene. J Alzheimers Dis. 2012;31:527–36.

22. Pagani L, Eckert A. Amyloid-Beta interaction with mitochondria. Int J Alzheimers Dis. 2011;2011:925050.
23. Maurer I, Zierz S, Moller HJ. A selective defect of cytochrome c oxidase is present in brain of Alzheimer disease patients. Neurobiol Aging. 2000;21:455–62.
24. Cottrell DA, Blakely EL, Johnson MA, Ince PG, Turnbull DM. Mitochondrial enzyme-deficient hippocampal neurons and choroidal cells in AD. Neurology. 2001;57:260–4.
25. Cottrell DA, Borthwick GM, Johnson MA, Ince PG, Turnbull DM. The role of cytochrome c oxidase deficient hippocampal neurones in Alzheimer's disease. Neuropathol Appl Neurobiol. 2002;28:390–6.
26. Cardoso SM, Santana I, Swerdlow RH, Oliveira CR. Mitochondria dysfunction of Alzheimer's disease cybrids enhances Abeta toxicity. J Neurochem. 2004;89:1417–26.
27. Cardoso SM, Proenca MT, Santos S, Santana I, Oliveira CR. Cytochrome c oxidase is decreased in Alzheimer's disease platelets. Neurobiol Aging. 2004;25:105–10.
28. Canevari L, Clark JB, Bates TE. beta-Amyloid fragment 25–35 selectively decreases complex IV activity in isolated mitochondria. FEBS Lett. 1999;457:131–4.
29. Casley CS, Canevari L, Land JM, Clark JB, Sharpe MA. Beta-amyloid inhibits integrated mitochondrial respiration and key enzyme activities. J Neurochem. 2002;80:91–100.
30. Parks JK, Smith TS, Trimmer PA, Bennett Jr JP, Parker Jr WD. Neurotoxic Abeta peptides increase oxidative stress in vivo through NMDA-receptor and nitric-oxide-synthase mechanisms, and inhibit complex IV activity and induce a mitochondrial permeability transition in vitro. J Neurochem. 2001;76:1050–6.
31. Yao J, Irwin RW, Zhao L, Nilsen J, Hamilton RT, et al. Mitochondrial bioenergetic deficit precedes Alzheimer's pathology in female mouse model of Alzheimer's disease. Proc Natl Acad Sci U S A. 2009;106:14670–5.
32. Rhein V, Song X, Wiesner A, Ittner LM, Baysang G, et al. Amyloid-beta and tau synergistically impair the oxidative phosphorylation system in triple transgenic Alzheimer's disease mice. Proc Natl Acad Sci U S A. 2009;106:20057–62.
33. Miura T, Tanno M. The mPTP and its regulatory proteins: final common targets of signalling pathways for protection against necrosis. Cardiovasc Res. 2012;94:181–9.
34. Du H, Guo L, Fang F, Chen D, Sosunov AA, et al. Cyclophilin D deficiency attenuates mitochondrial and neuronal perturbation and ameliorates learning and memory in Alzheimer's disease. Nat Med. 2008;14:1097–105.
35. Du H, Guo L, Zhang W, Rydzewska M, Yan S. Cyclophilin D deficiency improves mitochondrial function and learning/memory in aging Alzheimer disease mouse model. Neurobiol Aging. 2011;32(3):398–406.
36. Lustbader JW, Cirilli M, Lin C, Xu HW, Takuma K, et al. ABAD directly links Abeta to mitochondrial toxicity in Alzheimer's disease. Science. 2004;304:448–52.
37. Takuma K, Yao J, Huang J, Xu H, Chen X, et al. ABAD enhances Abeta-induced cell stress via mitochondrial dysfunction. FASEB J. 2005;19:597–8.
38. Yan Y, Liu Y, Sorci M, Belfort G, Lustbader JW, et al. Surface plasmon resonance and nuclear magnetic resonance studies of ABAD-Abeta interaction. Biochemistry. 2007;46:1724–31.
39. Yao J, Du H, Yan S, Fang F, Wang C, et al. Inhibition of amyloid-beta (Abeta) peptide-binding alcohol dehydrogenase-Abeta interaction reduces Abeta accumulation and improves mitochondrial function in a mouse model of Alzheimer's disease. J Neurosci. 2011;31:2313–20.
40. Falkevall A, Alikhani N, Bhushan S, Pavlov PF, Busch K, Johnson KA, Eneqvist T, Tjernberg L, Ankarcrona M, Glaser E. Degradation of the amyloid beta-protein by the novel mitochondrial peptidasome, PreP. J Biol Chem. 2006;281:29096–104.
41. Stahl A, Moberg P, Ytterberg J, Panfilov O, Brockenhuus Von Lowenhielm H, et al. Isolation and identification of a novel mitochondrial metalloprotease (PreP) that degrades targeting presequences in plants. J Biol Chem. 2002;277:41931–9.
42. Moberg P, Stahl A, Bhushan S, Wright SJ, Eriksson A, et al. Characterization of a novel zinc metalloprotease involved in degrading targeting peptides in mitochondria and chloroplasts. Plant J. 2003;36:616–28.

43. Stahl A, Nilsson S, Lundberg P, Bhushan S, Biverstahl H, et al. Two novel targeting peptide degrading proteases, PrePs, in mitochondria and chloroplasts, so similar and still different. J Mol Biol. 2005;349:847–60.
44. Kurochkin IV. Insulin-degrading enzyme: embarking on amyloid destruction. Trends Biochem Sci. 2001;26:421–5.
45. Selkoe DJ. Clearing the brain's amyloid cobwebs. Neuron. 2001;32:177–80.
46. Tanzi RE, Moir RD, Wagner SL. Clearance of Alzheimer's Abeta peptide: the many roads to perdition. Neuron. 2004;43:605–8.
47. Alikhani N, Guo L, Yan S, Du H, Pinho CM, et al. Decreased proteolytic activity of the mitochondrial amyloid-beta degrading enzyme, PreP peptidasome, in Alzheimer's disease brain mitochondria. J Alzheimers Dis. 2011;27:75–87.
48. Shinall H, Song ES, Hersh LB. Susceptibility of amyloid beta peptide degrading enzymes to oxidative damage: a potential Alzheimer's disease spiral. Biochemistry. 2005;44:15345–50.
49. Leuner K, Schutt T, Kurz C, Eckert SH, Schiller C, et al. Mitochondrion-derived reactive oxygen species lead to enhanced amyloid Beta formation. Antioxid Redox Signal. 2012;16:1421–33.
50. McManus MJ, Murphy MP, Franklin JL. The mitochondria-targeted antioxidant MitoQ prevents loss of spatial memory retention and early neuropathology in a transgenic mouse model of Alzheimer's disease. J Neurosci. 2011;31:15703–15.
51. Snow BJ, Rolfe FL, Lockhart MM, Frampton CM, O'Sullivan JD, et al. A double-blind, placebo-controlled study to assess the mitochondria-targeted antioxidant MitoQ as a disease-modifying therapy in Parkinson's disease. Mov Disord. 2010;25:1670–4.
52. Ankarcrona M, Mangialasche F, Winblad B. Rethinking Alzheimer's disease therapy: are mitochondria the key? J Alzheimers Dis. 2010;20 Suppl 2:S579–90.
53. Matveeva IA. Action of dimebon on histamine receptors. Farmakol Toksikol. 1983;46:27–9.
54. Lermontova NN, Lukoyanov NV, Serkova TP, Lukoyanova EA, Bachurin SO. Dimebon improves learning in animals with experimental Alzheimer's disease. Bull Exp Biol Med. 2000;129:544–6.
55. Bachurin S, Bukatina E, Lermontova N, Tkachenko S, Afanasiev A, et al. Antihistamine agent Dimebon as a novel neuroprotector and a cognition enhancer. Ann N Y Acad Sci. 2001;939: 425–35.
56. Wu J, Li Q, Bezprozvanny I. Evaluation of Dimebon in cellular model of Huntington's disease. Mol Neurodegener. 2008;3:15.
57. Doody RS, Gavrilova SI, Sano M, Thomas RG, Aisen PS, et al. Effect of dimebon on cognition, activities of daily living, behaviour, and global function in patients with mild-to-moderate Alzheimer's disease: a randomised, double-blind, placebo-controlled study. Lancet. 2008; 372:207–15.
58. www.clinicaltrial.gov Accessed 17 May 2012.
59. Schaffhauser H, Mathiasen JR, Dicamillo A, Huffman MJ, Lu LD, et al. Dimebolin is a 5-HT6 antagonist with acute cognition enhancing activities. Biochem Pharmacol. 2009;78:1035–42.
60. Grigorev VV, Dranyi OA, Bachurin SO. Comparative study of action mechanisms of dimebon and memantine on AMPA- and NMDA-subtypes glutamate receptors in rat cerebral neurons. Bull Exp Biol Med. 2003;136:474–7.
61. Lermontova NN, Redkozubov AE, Shevtsova EF, Serkova TP, Kireeva EG, et al. Dimebon and tacrine inhibit neurotoxic action of beta-amyloid in culture and block L-type Ca(2+) channels. Bull Exp Biol Med. 2001;132:1079–83.
62. Bachurin SO, Shevtsova EP, Kireeva EG, Oxenkrug GF, Sablin SO. Mitochondria as a target for neurotoxins and neuroprotective agents. Ann N Y Acad Sci. 2003;993:334–44. discussion 345–339.
63. Zhang S, Hedskog L, Petersen CA, Winblad B, Ankarcrona M. Dimebon (latrepirdine) enhances mitochondrial function and protects neuronal cells from death. J Alzheimers Dis. 2010;21:389–402.

# Chapter 6
# The Role of the Plasma Membrane Redox System in the Pathogenesis of Alzheimer's Disease

**Sara M. Hancock, David I. Finkelstein, Ashley I. Bush, and Paul A. Adlard**

**Abstract** Biological ageing is characterised by prolonged cellular damage over the lifespan, which results from oxidative stress processes that are implicated in both the initiation and progression of age-related disorders such as Alzheimer's disease (AD). Recently, the role of the plasma membrane redox system, which consists of at least three major components: the lipophilic antioxidants (Coenzyme Q (CoQ) and α-tocopherol), the intracellular cytosolic electron donor (NAD(P)H) and membrane-associated quinone reductases (cytochrome b5 reductase and NADH-quinone), in protecting against oxidative stress has come into focus. Research shows that this redox system plays a protective role during mitochondrial dysfunction by aiding in the alternative glycolytic ATP production pathway and reducing oxidative stress. The different aspects of the plasma membrane operate in tandem to protect the membrane from lipid peroxidation, preventing the formation of semiquinone free radicals and reactive oxygen species, and to ultimately limit oxidative stress while still maintaining cellular levels of each component. Past studies have made a link between the plasma membrane redox system and AD and have revealed that each aspect is affected during the disease.

S.M. Hancock, Ph.D. • P.A. Adlard, Ph.D. (✉)
Synaptic Neurobiology Laboratory, The Florey Institute of Neuroscience and Mental Health, Kenneth Myer Building, 30 Royal Pde, Parkville, VIC 3052, Australia
e-mail: sara.hancock@florey.edu.au; padlard@unimelb.edu.au

D.I. Finkelstein, Ph.D.
Parkinson's Disease Laboratory, The Florey Institute of Neuroscience and Mental Health, Kenneth Myer Building, 30 Royal Pde, Parkville, VIC 3052, Australia
e-mail: difink@unimelb.edu.au

A.I. Bush, Ph.D.
Oxidation Biology Laboratory, The Florey Institute of Neuroscience and Mental Health, Kenneth Myer Building, 30 Royal Pde, Parkville, VIC 3052, Australia
e-mail: ashleyib@unimelb.edu.au

D. Praticò and P. Mecocci (eds.), *Studies on Alzheimer's Disease*, Oxidative Stress in Applied Basic Research and Clinical Practice, DOI 10.1007/978-1-62703-598-9_6,

Moreover, research has revealed that manipulating the system components can protect cells from amyloid-β toxicity, suggesting a key role in cell survival. To date, the relationship of the plasma membrane redox system to the brain changes seen in AD has yet to be fully elucidated, and we will review the current literature here.

## 6.1 Introduction

Ageing is an inevitable process that leads to a build-up of cellular damage over the lifespan of all living organisms. This cellular damage is usually the result of oxidative stress, which occurs when the balance between reactive oxygen species (ROS) and antioxidant molecules is no longer in equilibrium [22]. Within the brain, neurons and astrocytes utilise antioxidant pathways such as the glutathione cycle in which both glutathione and ascorbate (vitamin C) are able to detoxify free radicals. However, neurons are more susceptible to damage as astrocytes have a higher glutathione concentration and are capable of coping with a higher oxidative load [40]. While there remains some controversy as to which brain areas and cell populations are the most vulnerable, regions that have been repeatedly shown to be particularly vulnerable include the hippocampus, association areas of the cortex and the temporal lobe [97]. Of these, the most vulnerable neuronal population is the CA1 region of the hippocampus, with other neuronal populations in the subiculum and pyramidal neurons in layer II of the entorhinal cortex also particularly at risk [97]. This differential capacity for enduring oxidative stress, which is heavily cited as a potential catalyst behind the development and progression of Alzheimer's disease (AD) (discussed in other chapters in this book), may help explain the hierarchy of neurodegeneration seen in AD. Recently, another antioxidant defence system that may participate in the pathogenesis of AD has been suggested. The plasma membrane redox system (PMRS) appears to allay oxidative stress, perhaps as a compensatory mechanism, during the ageing process [32]. Hyun and colleagues [34] have also shown that when certain aspects of the PMRS are overexpressed in vitro, cells are better protected from amyloid-β (Aβ) toxicity, suggesting an important role in cell survival and providing a tangible link to AD. Thus far there has been little research into if and how the PMRS is altered in AD. This chapter will discuss the current literature and the potential role of the PMRS in AD and will also highlight the potential implications of this system as a therapeutic target.

### *6.1.1 PMRS*

Recently, the PMRS has been implicated in neuronal survival [32]. Research has shown that multiple enzymes of the PMRS are expressed in neural cells and are up-regulated during mitochondrial dysfunction to preserve energy metabolism and hence, protect cells from oxidative stress [33, 81]. Here we discuss the structure and function of the PMRS.

#### 6.1.1.1 Structure of PMRS

The PMRS consists of at least three major components: (1) the lipophilic antioxidants, Coenzyme Q (CoQ) and α-tocopherol, (2) the intracellular cytosolic electron donor NAD(P)H and (3) the membrane-associated quinone reductases, cytochrome b5 reductase and NADH-quinone oxidoreductase [31, 34]. Coenzyme Q (CoQ) is a redox-active quinone found in the phospholipid bilayer of cell membranes [6]. It participates as a carrier in membrane-based electron transport chains [20]. Alone or in association with α-tocopherol, CoQ scavenges lipid free radicals thereby inhibiting oxidative stress [32]. Alpha-tocopherol is the only lipid-soluble, chain-breaking antioxidant present in biological membranes [14, 35, 96]. Cytochrome b5 reductase (also called ascorbate free radical reductase) is a monomeric flavoprotein which acts by transferring one electron to CoQ. The other membrane-associated quinine reductase, NADH-quinone oxidoreductase (NQO1) is a cytosolic protein that is translated to the inner surface of the cell membrane during oxidative stress [70]. Collectively these components maintain cellular redox status by maintaining the NAD(P)+/NAD(P)H ratio [51]. The involvement of these components in AD has previously been raised in the literature and has been the focus of past research. A schematic representation of the PMRS can be seen in Fig. 6.1.

#### 6.1.1.2 Function of the PMRS

The PMRS has been shown to play a protective role during mitochondrial dysfunction. This is achieved by production of $NAD^+$ for alternative ATP production via the glycolytic pathway. This transfer of electrons from intracellular reducing equivalents to extracellular acceptors reduces oxidative stress [32, 34].

##### CoQ

CoQ is the only lipid-soluble antioxidant and is synthesised by living organisms [8, 29]. CoQ operates physiologically by maintaining the membrane levels of the reduced form of CoQ (ubiquinol) and α-tocopherol to protect the membrane from lipid peroxidation [5]. CoQ also functions to remove the excess reducing power caused by the glycolytic pathway during decreased mitochondrial respiration [29]. Ubiquinol, the fully reduced form of CoQ, inhibits the peroxidation of membrane lipids by reacting with lipid peroxyl radicals and by reducing tocopheroxyl radicals [12, 27, 39, 44]. Cytochrome b5 reductase reduces CoQ through a one-electron reaction mechanism, thereby fostering the antioxidant potential of the membrane by maintaining ascorbate and α-tocopherol in their reduced states [80, 81] (Fig. 6.2).

In addition to being the only lipid soluble chain breaking protein present in the plasma membrane and the most abundant antioxidant agent [96], α-tocopherol is considered essential for normal neurological function. It has also been suggested that α-tocopherol may act as an anti-inflammatory agent that is both neuroprotective

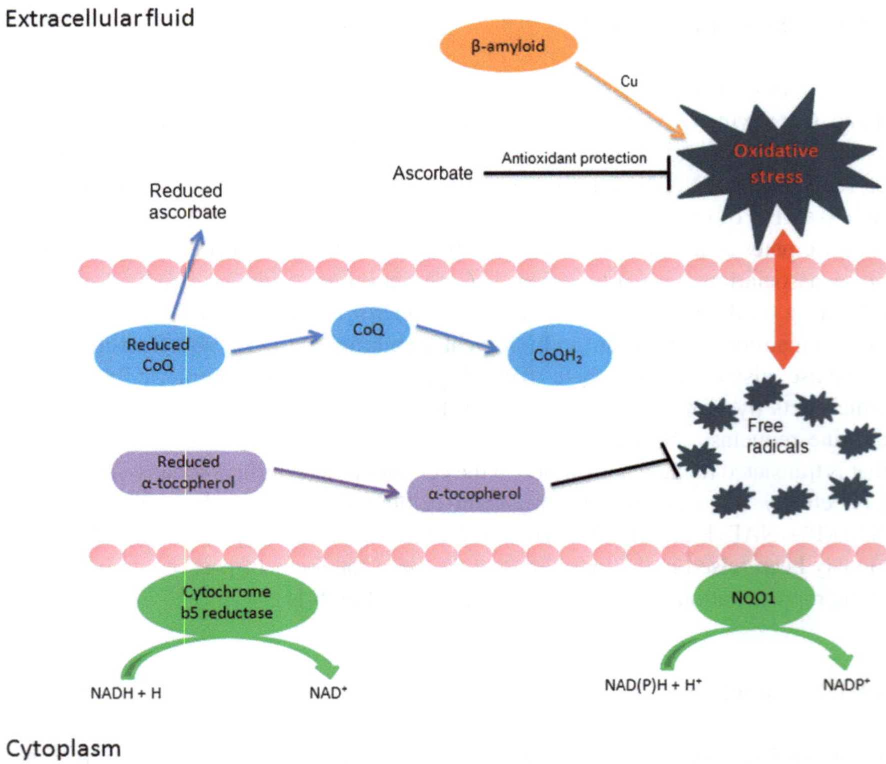


**Fig. 6.1** Simplified schematic showing the different aspects of the plasma membrane redox system and potential protective effects against free radicals and oxidative stress

**Fig. 6.2** Cyclic structure of Coenzyme Q. Adapted from Turunen et al [79] 1.2.2.2 Alpha-tocopherol

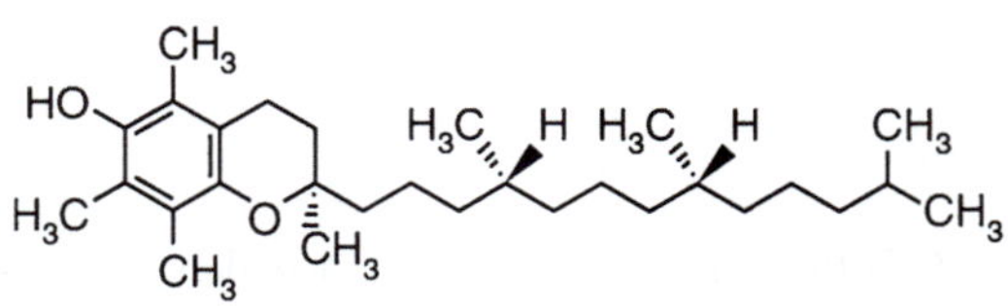

and which can regulate particular enzymes to alter specific membrane properties [48]. The chemical structure of α-tocopherol is shown in Fig. 6.3.

## NADH-Quinone Oxidoreductase

The cytosolic reductase, NQO1 (also known as DT-diaphorase [94]) uses two electrons and NAD(P)H to catalyse the reduction of quinones [32]. The key to this capability is the structure of NQO1 as it consists of two domains; a major catalytic domain and a smaller domain forming part of the binding site for the hydrophilic regions of NAD(P)H [93]. This structure is shown in Fig. 6.4. NQO1 is responsible for reduction of both exogenous and endogenous CoQ [29], thereby preventing the

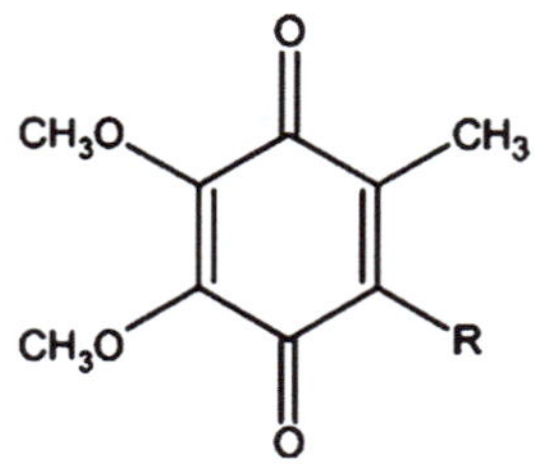


Fig. 6.3 Structure of α-tocopherol. Adapted from [95]

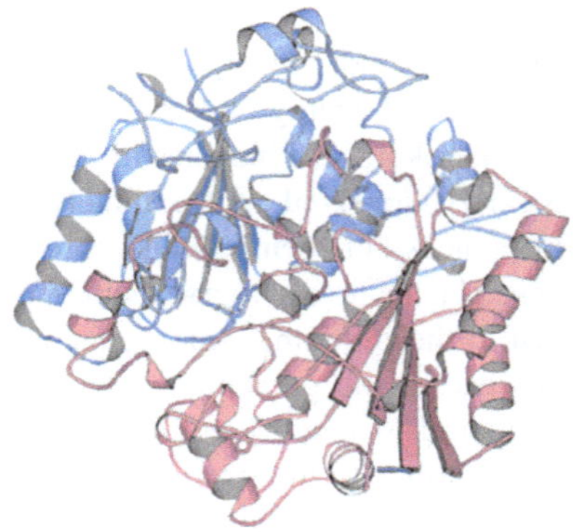

Fig. 6.4 Ribbon diagram of NQO1 structure (imaged with X-ray) showing the two domains. Adapted from [16]

generation of semiquinone free radicals and ROS, thereby limiting oxidative stress [13, 66]. It also acts to both generate and maintain the reduced form of CoQ in the plasma membrane [9].

### *6.1.2 How Is the PMRS Affected in AD*

The effects of oxidative stress on cells in AD are well-known and widely researched with a multitude of studies analysing the various cellular markers of oxidative stress. These markers have been, and are still being tested for their potential diagnostic validity, while also suggesting targets for potential therapies. A marker for lipid peroxidation, 4-hydroxy-nonenal (HNE) has been shown to be elevated in both plasma and ventricular fluid [50]. Similarly, markers of damage to nucleic acids such as DNA and RNA have also been examined in multiple regions of AD brains. Research by Nunomura and colleagues [56, 57] has shown a significant neuronal increase in a marker for RNA damage; 8-hydroxyguanosine (8-OHG), which appears to lessen as the disease progresses and these changes are consistent with the changes in 8-OHG levels in the CSF of AD patients. Interestingly, there is a significant increase in oxidative damage to RNA in the brains of individuals with mild cognitive impairment (MCI), suggesting that oxidative damage may be one of the earliest symptoms of AD [46]. Corresponding DNA changes have also been shown in vulnerable regions of the AD brain [46]. This oxidative stress has been linked to a dyshomeostasis in key transition metals (iron, copper and zinc) in the AD brain, which we have previously reviewed [2]. Moreover, because the precursor for the ß-amyloid (Aβ) protein (Aß is the principal component of the extracellular plaques that characterise the AD brain), is capable of moving into the plasma membrane as part of its role in maintaining cellular iron homeostasis, the PMRS may be crucial in AD.

#### 6.1.2.1 CoQ

The mitochondrial CoQ content decreases across various tissues in the body during both normal ageing and in disease [24]. A study by Mariani and colleagues [49], which took muscle biopsies from both AD patients and controls, reported a decrease in CoQ10 levels of approximately 50 %. However, there does not appear to be any differences between AD and controls in plasma levels of CoQ10 [92], moreover, these plasma levels do not seem to correlate to any clinical parameters [19]. In contrast to these findings, more recent studies have revealed that there is a significant increase in the percentage of oxidised/total CoQ10 in the CSF of AD patients compared to healthy controls, and, that this measure has a significant negative correlation with illness duration [36]. Similarly, there is also evidence for increased brain levels of CoQ of between 30 and 100 % when compared to controls in most regions of the AD brain [21, 74]. These conflicting results highlight the need for further research, particularly as technology and scientific techniques improve, in order to fully elucidate the role of CoQ in AD. Moreover, further studies have shown that high levels of CoQ are maintained within the brain throughout the ageing process

#### 6.1.2.2 Alpha-Tocopherol

The levels of α-tocopherol in different brain regions have also been measured in AD patients and compared to controls. Alpha-tocopherol and its oxidised form "α-tocopherol quinone" were measured in the CSF of AD and vascular dementia patients as well as controls in a study by Toghi and colleagues [78]. They showed a significant increase of α-tocopherol quinone in the vascular dementia cases, as compared to controls. These levels also correlated with the observed changes in cognitive scores, assessed using the mini-mental state examination (MMSE). However, in the CSF of the AD patients they found a significant decrease in the level of α-tocopherol with no changes in its' oxidised form. Serum and CSF levels of α-tocopherol were again measured in a study by Jimenez-Jimenez and colleagues [38], where they showed a significant decrease of α-tocopherol in the AD patients, although these did not correlate with any other parameters that were measured including MMSE score, age, age of onset or duration of disease. Likewise the concentration of α-tocopherol in the plasma has also been shown to be significantly reduced in AD patients [11, 37, 62, 89]. This is consistent with research that shows decreases in α-tocopherol in AD [30]. Cortical samples of AD patients have also been examined as a measure of the brain levels of α-tocopherol; these have shown no differences in AD tissue compared to control subjects [52]. However, it should be noted that research examining the lipid nitration product, 5-nitro-gamma-tocopherol, demonstrated an increase in particular affected regions of the AD brain, when compared to that of controls [84]. This suggests a possible role of gamma-tocopherol in AD.

#### 6.1.2.3 NADH-Quinone Oxidoreductase

There are a number of studies that have examined both the levels and the distribution of NQO1 in AD. In the first study of its kind, Raina and colleagues [65] examined NQO1 levels in the brains of AD individuals and compared the results to both age-matched and young controls. NQO1 was localised to neurofibrillary tangles (NFTs) and to the cytoplasm of hippocampal neurons in AD cases, but there was significantly less NQO1 in the same neuronal populations in both age-matched and young controls, suggesting a specific increase in the AD brain. This finding was supported by a study by Wang and colleagues [83], who demonstrated that NQO1 protein and activity levels were increased in hippocampal pyramidal neurons in tangle-bearing regions of the AD brain, whereas in control cases the hippocampal neurons were devoid of both NQO1 enzyme activity and protein. Finally, SantaCruz and colleagues [73] examined NQO1 enzymatic activity and protein levels in regions of the brain commonly affected in AD, and compared this to regions typically unaffected in AD. They demonstrated a significant increase in the ratio of frontal cortex to cerebellar NQO1 activity in AD. NQO1 protein was localised to astrocytes and neurites surrounding senile plaques, and the extent of staining correlated with the extent of AD pathology within each region examined. Neuronal NQO1 staining in the frontal cortex was absent in the control population; but was found to the same extent in neurons in the substantia nigra of both AD and controls, suggesting a region-specific alteration of NQO1 activity and protein in the AD brain.

### *6.1.3 Research Models*

Thus far, the mechanisms by which the PMRS functions are still not well understood, nor are the ways in which these could be exploited to possibly treat or prevent ageing and AD. It is important to study the PMRS in a variety of different models, but it is vital that future research is well designed, reliable and valid. In this section we will discuss what researchers can learn from the different models presently in use. Increased amounts of membrane associated oxidative stress have been shown in cells in vulnerable brain regions of AD patients, as well as in both animal and cell culture models of AD [34]. Currently, there is strong evidence for the presence of oxidative stress in transgenic (Tg) models of AD [1, 4, 63, 67, 86, 87], however, few studies examining the discrete aspects of the PMRS have been carried out. At present there is only one report of decreased α-tocopherol in a triple transgenic mouse model of AD [67]. Despite this lack of research, other studies have assessed the efficacy of augmenting the different aspects of the PMRS on slowing the progression of AD-like symptoms in Tg models of the disease, particularly with CoQ and α-tocopherol. Similarly, in vitro studies examining the effect of CoQ and α-tocopherol have also been carried out.

#### 6.1.3.1 CoQ

Recently, two studies examined the effect of dietary supplementation with coenzyme Q10 [45, 85]; a form of CoQ. Li and colleagues [45] examined the neuroprotective effect of CoQ10 supplements across both wildtype and Tg mice (APP/PS1 double Tg mice as well as single Tg APP and PS1 mice). Using magnetic resonance imaging (MRI) the researchers determined that animals given CoQ10 displayed less brain atrophy; this effect of CoQ10 on brain volume was most pronounced in APP/PS1 animals, and less so in APP and PS1 transgenics and wild type. In support of this finding, Yang and colleagues [85] used immunohistochemistry in addition to MRI to investigate the effect of supplementing APP/PS1 mice with CoQ10 on amyloid load. They found that the supplemented mice had a reduced amyloid plaque burden and also a decreased intracellular deposition of Aβ. Similarly, a study by Ono and colleagues [58] used an in vitro model to show that CoQ is capable of inhibiting both the formation and extension of Aβ fibrils, as well as destabilising preformed Aβ fibrils in a dose-dependent manner. Moreover, CoQ has been shown to inhibit Aβ-induced mitochondrial deficits in aged diabetic rats [53]. More recently, researchers demonstrated a protective effect of CoQ against amyloid precursor protein C-terminal fragment-induced neurotoxicity in MC65 neuroblastoma cells, also in a dose-dependent manner [82]. It has been suggested that CoQ impacts tau pathology as CoQ has been shown to facilitate tau aggregation and induce the formation of fibrillar tau polymers [71, 72]. Furthermore, CoQ has also been shown to facilitate the interaction of tau with actin to form structures similar to the Hirano bodies found in the AD brain, which also cross-react with antibodies specific to CoQ [71].

#### 6.1.3.2 Alpha-Tocopherol

Transgenic studies have highlighted the relevance of α-tocopherol in the pathogenesis of AD by crossing Tg2576 mice with mice that lack α-tocopherol (α-tocopherol transfer protein knockout mice). These crossed transgenic mice display relatively advanced cognitive dysfunction as well as increased Aβ deposits in the brain [55]. A study by Sung and colleagues [77] investigated the effects of supplementing Tg2576 mice with α-tocopherol and demonstrated that it significantly reduced Aβ load and was associated with a decrease in lipid peroxidation markers, but this decrease only occurred when α-tocopherol was administered prior to Aβ plaque deposition. In another model of AD, the administration of α-tocopherol to APP/PS1 mice was shown to reduce the levels of oxidative stress and to partially reverse the Aβ plaque-associated toxicity [28]. These findings are supported by in vitro studies with a study by Qi and colleagues [64] demonstrating that the exogenous administration of α-tocopherol to SH-SY5Y cells treated with Aβ can partially prevent the neurotoxicity, oxidative stress and alterations in membrane lipid composition caused by Aβ. Likewise, when PC12 cells are treated with Aβ they accumulate ROS and have a subsequent decrease in both redox activity and ATP levels that can be blocked with α-tocopherol [60]. Similarly in cultured neurons, α-tocopherol can both decrease and delay the impairments resulting from exposure to Aβ [7, 10, 15, 41, 88, 90]. Indeed, research has suggested that α-tocopherol may be capable of increasing lifespan in model organisms via this increase in antioxidant protection against ROS [3].

These findings support a role for multiple aspects of the PMRS in the pathogenesis of AD. From these studies, it seems that the PMRS provides particular protection against lipid peroxidation and membrane associated oxidative damage, while other oxidative stress events (damage to nucleic acids and protein oxidation) are targeted by other antioxidant pathways. Further research is essential to determining both the molecular mechanisms of this role and whether it is then possible to manipulate these mechanisms and up-regulate the PMRS and possibly delay the ageing process.

### *6.1.4 How Is the PMRS Clinically Relevant?*

The aspects of the PMRS previously discussed have been shown to be of high clinical relevance to a number of disorders other than AD. For example, CoQ levels can increase in a variety of different tissues, such as heart, muscle and brain, after as little as 14 weeks of treatment [42, 43, 47] and can ameliorate conditions such as hypertension [69] and heart failure [68], as well as be protective in neurodegenerative diseases such as Parkinson's disease (PD) and Huntington's disease (HD) [26, 75, 76].

Given the nature of AD and the potentially key role that the PMRS may have in facilitating multiple aspects of AD symptomatology and pathology, it is clear that the PMRS represents a key therapeutic target. Two main aspects of the PMRS, NQO1 and CoQ currently have not been studied to any great extent. There is very little in the literature regarding the possible antioxidant role of NQO1 in AD and though much has been written about the potential of CoQ as a therapeutic for multiple CNS disorders, there have been no published studies that have tested CoQ in AD as yet. In comparison, α-tocopherol has been quite rigorously studied as an antioxidant treatment in both AD and MCI. Indeed, association studies have shown that low serum levels of α-tocopherol are associated with poorer cognitive performance in the elderly [61], and that supplementing α-tocopherol is associated with less cognitive decline with age [54]. Similarly, data from a prospective cohort study among the elderly (the Rotterdam study), demonstrated that a high dietary intake of α-tocopherol was associated with a lower risk of AD [23]. Despite these findings, intervention trials of α-tocopherol in the clinic have been less than convincing. It is clear that there is a great potential for future research in the possible antioxidant treatment of AD.

### 6.1.5 Conclusion

It is clear that oxidative stress plays a role in AD, and the literature supports the notion that the plasma membrane may represent a key site for both the development of AD pathology and the occurrence of oxidative stress through the dysfunction of a number of systems. To date, only a limited number of studies have specifically examined the plasma membrane redox system in AD, or experimentally manipulated it in transgenic models of the disease. As these studies have demonstrated that aspects of the system can be manipulated with positive, neuroprotective results, there is the opportunity for future research to potentially find an exciting new treatment for AD. Therefore, there is a real need for further research in order to determine the role of the PMRS in AD and indeed, to ascertain whether it represents a logical and viable therapeutic target for preventing or limiting the progression of AD.

**Acknowledgements** P.A.A. and D.I.F. are paid consultants, and shareholders, of Prana Biotechnology Ltd, and A.I.B. is a shareholder of Prana Biotechnology Ltd. P.A.A., D.I.F. and A.I.B. are primarily supported by the National Health and Medical Research Council and the Australian Research Council, and P.A.A. and D.I.F. are also supported by the Joan and Peter Clemenger Trust.

## References

1. Abdul HM, Sultana S, St Clair DK, Markesbery WR, Butterfield DA. Oxidative damage in brain from human mutant APP/PS-1 double knock-in mice as a function of age. Free Radic Biol Med. 2008;45:1420–5.
2. Adlard PA, Bush AI. Metals and Alzheimer's disease. J Alzheimers Dis. 2006;10:145–63.
3. Ames BN, Shigenaga MK, Hagen TM. Oxidants, antioxidants and the degenerative diseases of ageing. Proc Natl Acad Sci U S A. 1993;90:7915–22.
4. Apelt J, Bigl M, Wunderlich P, Schliebs R. Ageing-related increase in oxidative stress correlates with developmental pattern of beta-secretase activity and beta-amyloid plaque formation in transgenic Tg2576 mice with Alzheimer-like pathology. Int J Dev Neurosci. 2004;22:475–84.
5. Arroyo A, Rodriguez-Aguilera JC, Santos-Ocana C, Villalba JM, Navas P. Stabilisation of extracellular ascorbate mediated by coenzyme Q transmembrane electron transport. Methods Enzymol. 2004;378:207–17.
6. Battino M, Ferri E, Gorini A, Frederico Villa A, Rodrigues Huertas JF, Fiorella P, Genova ML, Lenaz G, Marchetti M. Natural distribution and occurrence of coenzyme Q homologues. Membr Biochem. 1990;9(3):179–90.
7. Behl C, Davis J, Cole GM, Schubert D. Vitamin E protects nerve cells from amyloid beta protein toxicity. Biochem Biophys Res Commun. 1992;186:944–50.
8. Bentinger M, Tekle M, Dallner G. Coenzyme Q—biosynthesis and functions. Biochem Biophys Res Commun. 2010;396:74–9.

9. Beyer RE, Segura-Aguilar J, di Bernado S, Cavazzoni M, Fato R, Fiorentini D, Galli MC, Setti M, Landi L, Lenaz G. The two-electron quinone reductase DT-diaphorase generates and maintains the antioxidant (reduced) form of coenzyme Q in membranes. Mol Aspects Med. 1997;18:S15–23.
10. Boothby LA, Doering PL. Vitamin C and vitamin E for Alzheimer's disease. Ann Pharmacother. 2005;39:2073–80.
11. Bourdel-Marchasson I, Delmas-Beauvieux MC, Peuchant E, Richard-Harston S, Decamps A, Reignier B, Emeriau JP, Rainfray M. Antioxidant defences and oxidative stress markers in erythrocytes and plasma from normally nourished elderly Alzheimer patients. Age Ageing. 2001;30:235–41.
12. Boveris A. Determination of the production of superoxide radicals and hydrogen peroxide in mitochondria. Methods Enzymol. 1984;105:429–35.
13. Brunmark A, Cadenas E, Lind C, Segura-Aguilar J, Ernster L. DT-diaphorase-catalysed two-electron reduction of quinone epoxides. Free Radic Biol Med. 1987;3:181–8.
14. Burton GW, Joyce A, Ingold KU. Is vitamin E the only lipid-soluble, chain-breaking antioxidant in human blood plasma and erythrocyte membranes? Arch Biochem Biophys. 1983; 221:281–90.
15. Butterfield DA, Castegna A, Drake J, Scapagnini G, Calabrese V. Vitamin E and neurodegenerative disorders associated with oxidative stress. Nutr Neurosci. 2002;5:229–39.
16. Cavelier G, Amzel LM. Mechanism of NAD(P)H:quinone reductase: ab initio studies of reduced flavin. Proteins. 2001;43:420–32.
17. Clement AB, Gimpl G, Behl C. Oxidative stress resistance in hippocampal cells is associated with altered membrane fluidity and enhanced nonamyloidogenic cleavage of endogenous amyloid precursor protein. Free Radic Biol Med. 2010;48:1236–41.
18. Curtain CC, Ali F, Volitakis I, Cherny RA, Norton RS, Beyreuther K, Barrow CJ, Masters CL, Bush AI, Barnham KJ. Alzheimer's disease amyloid-beta binds copper and zinc to generate an allosterically ordered membrane-penetrating structure containing superoxide dismutase-like subunits. J Biol Chem. 2001;276:20466–73.
19. De Bustos F, Molina JA, Jimenez-Jimenez FJ, Garcia-Redondo A, Gomez-Escalonilla C, Porta-Etessam J, Berbel A, Zurdo M, Barcenilla B, Parrilla G, Enriquez-de-Salamanca R, Arenas J. Serum levels of coenzyme Q10 in patients with Alzheimer's disease. J Neural Transm. 2000;107:233–9.
20. De Cabo R, Cabello R, Rios M, Lopez-Lluch G, Ingram DK, Lane MA, Navas P. Calorie restriction attenuates age-related alterations in the plasma membrane antioxidant system in rat liver. Exp Gerontol. 2004;39:297–304.
21. Edlund C, Soderberg M, Kristensson K, Dallner G. Ubiqionone, dolichol and cholesterol metabolism in ageing and Alzheimer's disease. Biochem Cell Biol. 1992;70:422–8.
22. Egaña JT, Zambrano C, Nuñez MT, Gonzalez-Billault C, Maccioni RB. Iron-induced oxidative stress modify tau phosphorylation patterns in hippocampal cell cultures. Biometals. 2003;16:215–23.
23. Engelhart MJ, Geerlings MI, Ruitenberg A, Van Swieten JC, Hofman A, Witteman JC, Breteler MM. Dietary intake of antioxidants and risk of Alzheimer's disease. J Am Med Assoc. 2002; 287:3223–9.
24. Ernster L, Dallner G. Biochemical, physiological and medical aspects of ubiquionone function. Biochim Biophys Acta. 1995;1271:195–204.
25. Ernster L, Navazio F. Soluble diaphorase in animal tissues. Acta Chem Scand. 1985;12: 595–602.
26. Ferrante RJ, Andreassen OA, Dedeoglu A, Ferrante KL, Jenkins BG, Hersch SM, Beal MF. Therapeutic effects of coenzyme Q10 and remacemide in transgenic mouse models of Huntington's disease. J Neurosci. 2002;22:1529–99.
27. Forsmark-Andree P, Dallner G, Ernster L. Endogenous ubiquinol prevents protein modification accompanying lipid peroxidation in beef heart submitochondrial particles. Free Radic Biol Med. 1995;19:749–57.

28. Garcia-Alloza M, Dodwell SA, Meyer-Luehmann M, Hyman BT, Backsai BJ. Plaque-derived oxidative stress mediates distorted neurite trajectories in the Alzheimer mouse model. J Neuropathol Exp Neurol. 2006;65:1082–9.
29. Genova ML, Pich MM, Biondi A, Bernacchio A, Falasca A, Bovina C, Formiggini G, Castelli GP, Lenaz G. Mitochondrial production of oxygen radical species and the role of coenzyme Q as an antioxidant. Exp Biol Med. 2003;228:506–13.
30. Glaso M, Nordbo G, Diep L, Bohmer T. Reduced concentrations of several vitamins in normal weight patients with late-onset dementia of the Alzheimer type without vascular diseases. J Nutr Health Aging. 2004;8:407–13.
31. Hyun D-H, Emerson SS, Jo D-G, Mattson MP, De Cabo R. Calorie restriction up-regulates the plasma membrane redox system in brain cells and suppresses oxidative stress during ageing. Proc Natl Acad Sci U S A. 2006;103(52):19908–12.
32. Hyun D-H, Hernandez JO, Mattson MP, De Cabo R. The plasma membrane redox system in ageing. Ageing Res Rev. 2006;5:209–20.
33. Hyun D-H, Hunt ND, Emerson SS, Hernandez JO, Mattson MP, De Cabo R. Up-regulation of plasma membrane-associated redox activies in neuronal cells lacking functional mitochondria. J Neurochem. 2007;100:1364–74.
34. Hyun D-H, Mughal MR, Yang H, Lee JH, Ko EJ, Hunt ND, De Cabo R, Mattson MP. The plasma membrane redox system is impaired by amyloid-β peptide and in the hippocampus and cerebral cortex of 3× TgAD mice. Exp Neurol. 2010;225:423–9.
35. Ingold KU, Webb AC, Witter D, Burton GW, Metcalfe TA, Muller DP. Vitamin E remains the major lipid-soluble, chain-breaking antioxidant in human plasma even in individuals suffering severe vitamin E deficiency. Arch Biochem Biophys. 1987;259:224–5.
36. Isobe C, Abe T, Terayama Y. Increase in the oxidised/total coenzyme Q-10 ratio in the cerebrospinal fluid of Alzheimer's disease patients. Dement Geriatr Cogn Disord. 2009;28: 449–54.
37. Jeandel C, Nicolas MB, Dubois F, Nabet-Belleville F, Penin F, Cuny G. Lipid peroxidation and free radical scavengers in Alzheimer's disease. Gerontology. 1989;35:275–82.
38. Jimenez-Jimenez FJ, de Bustos F, Molina JA, Benito-Leon J, Tallon-Barranco A, Gasalla T, Orti-Pareja M, Guillamon F, Rubio JC, Arenas J, Enriquez-de-Salamanca R. Cerebrospinal fluid levels of alpha-tocopherol (vitamin E) in Alzheimer's disease. J Neural Transm. 1997;104(6–7):703–10.
39. Kagan VE, Tyurina YY, Witt E. Role of coenzyme Q and superoxide in vitamin E cycling. Subcell Biochem. 1998;30:491–507.
40. Kim EJ, Park YG, Baik EJ, Jung SJ, Won R, Nahm TS, Lee BH. Dehydroascorbic acid prevents oxidative cell death through a glutathione pathway in primary astrocytes. J Neurosci Res. 2005;79:670–9.
41. Koppal T, Subramanina R, Drake J, Prasad MR, Dhillon H, Butterfield DA. Vitamin E protects against Alzheimer's amyloid peptide (25–35) induced changes in neocortical synaptosomal membrane lipid structure and composition. Brain Res. 1998;786:270–3.
42. Lass A, Forster MJ, Sohal RS. Effects of coenzyme Q10 and alpha-tocopherol administration on their tissue levels in the mouse: elevation of mitochondrial alpha-tocopherol by coenzyme Q10. Free Radic Biol Med. 1999;26:1375–82.
43. Lass A, Kwong L, Sohal RS. Mitochondrial coenzyme Q content and ageing. Biofactors. 1999;9:199–205.
44. Lass A, Sohal RS. Electron transport-linked ubiquinone-dependent recycling of alpha-tocopherol inhibits autooxidation of mitochondrial membranes. Arch Biochem Biophys. 1998;352:229–36.
45. Li G, Jack CR, Yang XF, Yang ES. Diet supplement CoQ10 delays brain atrophy in aged transgenic mice with mutations in the amyloid precursor protein: an in vivo volume MRI study. Biofactors. 2008;32:169–78.
46. Lovell MA, Markesbery WR. Oxidatively modified RNA in mild cognitive impairment. Neurobiol Dis. 2008;29:169–75.

47. Matthews RT, Yang L, Browne S, Baik M, Beal MF. Coenzyme Q10 administration increases brain mitochondrial concentrations and exerts neuroprotective effects. Proc Natl Acad Sci U S A. 1998;95:8892–7.
48. Matsura T, Yamada K, Kawasaki T. Antioxidant role of cellular reduced coenzyme Q homologs and α-tocopherol in free radical-induced injury of hepatocytes isolated from rats fed diets with different vitamin E contents. Biochim Biophys Acta. 1992;1127:277–83.
49. Mariani C, Bresolin N, Farina E, Moggio M, Ferrante C, Ciafaloni E, Sertorelli S, Ciccone A, Scarlato G. Muscle biopsy in Alzheimer's disease: morphological and biochemical findings. Clin Neuropathol. 1991;10:171–6.
50. McGrath LT, McGleenon BM, Brennan S, McColl D, Mc IS, Passmore AP. Increased oxidative stress in Alzheimer's disease as assessed with 4-hydroxynonenal but not malondialdehyde. Q J Med. 2001;94:485–90.
51. Merker MP, Bongard RD, Kettenhofen NJ, Okamato Y, Dawson CA. Intracellular redox status affects transplasma membrane electron transport in pulmonary arterial endothelial cells. Am J Physiol Lung Cell Mol Physiol. 2002;282:L36–43.
52. Metcalfe T, Bowen DM, Muller DP. Vitamin E concentrations in human brain of patients with Alzheimer's disease, fetuses with Down's syndrome, centenarians, and controls. Neurochem Res. 1989;14:1209–12.
53. Moreira PI, Santos MS, Sena C, Nunes E, Seica R, Oliveira CR. CoQ10 therapy attenuates amyloid beta-peptide toxicity in brain mitochondria isolated from aged diabetic rats. Exp Neurol. 2005;196:112–9.
54. Morris MC, Evans DA, Bienias JL, Tangney CC, Wilson RS. Vitamin E and cognitive decline in older persons. Arch Neurol. 2002;59:1125–32.
55. Nishida Y, Yokota T, Takahashi T, Uchihara T, Jishage K, Mizusawa H. Deletion of vitamin E enhances phenotype of Alzheimer disease model mouse. Biochem Biophys Res Commun. 2006;350:530–6.
56. Nunomura A, Perry G, Aliev G, Hirai K, Takeda A, Balraj EK, Jones PK, Ghanbari H, Wataya T, Shimohama S, Chiba S, Atwood CS, Petersen RB, Smith MA. Oxidative damage is the earliest event in Alzheimer disease. J Neuropathol Exp Neurol. 2001;60:759–67.
57. Nunomura A, Perry G, Pappolla MA, Wade R, Hirai K, Chiba S, Smith MA. RNA oxidation is a prominent feature of vulnerable neurons in Alzheimer's disease. J Neurosci. 1999;19:1959–64.
58. Ono K, Hasegawa K, Naiki H, Yamada M. Preformed beta-amyloid fibrils are destabilized by coenzyme Q10 in vitro. Biochem Biophys Res Commun. 2005;330:111–6.
59. Parihar MS, Brewer GJ. Amyloid beta as a modulator of synaptic plasticity. J Alzheimers Dis. 2010;22(3):741–63.
60. Pereira C, Santos MS, Oliveira C. Involvement of oxidative stress on the impairment of energy metabolism induced by A beta peptides on PC12 cells: protection by antioxidants. Neurobiol Dis. 1999;6:209–19.
61. Perkins AJ, Hendrie HC, Callahan CM, Gao S, Unverzagt FW, Xu Y, Hall KS, Hui SL. Association of antioxidants with memory in a multiethnic elderly sample using the Third National Health and Nutrition Examination Survey. Am J Epidemiol. 1999;150:37–44.
62. Polidori MC, Mecocci P. Plasma susceptibility to free radical-induced antioxidant consumption and lipid peroxidation is increased in very old subjects with Alzheimer disease. J Alzheimers Dis. 2002;4:517–22.
63. Pratico D, Uryu K, Leight S, Trojanoswki JQ, Lee VM. Increased lipid peroxidation precedes amyloid plaque formation in an animal model of Alzheimer amyloidosis. J Neurosci. 2001;21:4183–7.
64. Qi XL, Xiu J, Shan KR, Xiao Y, Gu R, Liu RY, Guan ZZ. Oxidative stress induced by beta-amyloid peptide(1–42) is involved in the altered composition of cellular membrane lipids and the decreased expression of nicotinic receptors in human SH-SY5Y neuroblastoma cells. Neurochem Int. 2005;46:613–21.
65. Raina AK, Templeton DJ, Deak JC, Perry G, Smith MA. Quinone reductase (NQO1), a sensitive redox indicator, is increased in Alzheimer's disease. Redox Rep. 1999;4:23–7.

66. Rauth AM, Goldberg Z, Misra V. DT-diaphorase: possible roles in cancer chemotherapy and carcinogenesis. Oncol Res. 1997;9:339–49.
67. Resende R, Moreira PI, Proenca T, Deshpande A, Busciglio J, Pereira C, Oliveira CR. Brain oxidative stress in a triple-transgenic mouse model of Alzheimer disease. Free Radic Biol Med. 2008;44:2051–7.
68. Rosenfeldt F, Hilton D, Pepe S, Krum H. Systematic review of effect of coenzyme Q10 in physical exercise, hypertension and heart failure. Biofactors. 2003;18:91–100.
69. Rosenfeldt FL, Haas SJ, Krum H, Hadj A, Ng K, Leong JY, Watts GF. Coenzyme Q10 in the treatment of hypertension: a meta-analysis of the clinical trials. J Hum Hypertens. 2007; 21:297–306.
70. Rushmore TH, Morton MR, Pickett CB. The antioxidant responsive element. Activation by oxidative stress and identification of the DNA consensus sequence required for functional activity. J Biol Chem. 1991;266:11632–9.
71. Santa-Mara I, Santpere G, MacDonald MJ, Gomez de Barreda E, Hernandez F, Moreno FJ, Ferrer I, Avila J. Coenzyme q induces tau aggregation, tau filaments, and Hirano bodies. J Neuropathol Exp Neurol. 2008;67:428–34.
72. Santa-Maria I, Hernandez F, Martin CP, Avila J, Moreno FJ. Quinones facilitate the self-assembly of the phosphorylated tubulin binding region of tau into fibrillar polymers. Biochemistry. 2004;43:2888–97.
73. SantaCruz KS, Yazlovitskaya E, Collins J, Johnson J, DeCarli C. Regional NAD(P)H:quinone oxidoreductase activity in Alzheimer's disease. Neurobiol Aging. 2004;25:63–9.
74. Soderberg M, Edlund C, Alafuzoff I, Kristensson K, Dallner G. Lipid composition in different regions of the brain in Alzheimer's disease/senile dementia of Alzheimer's type. J Neurochem. 1992;59:1646–53.
75. Shults CW, Haas R. Clinical trials of coenzyme Q10 in neurological disorders. Biofactors. 2005;25:117–26.
76. Shults CW, Oakes D, Kieburtz K, Beal MF, Haas R, Plumb S, Juncos JL, Nutt J, Shoulson I, Carter J, Kompoliti K, Perlmutter JS, Reich S, Stern M, Watts RL, Kurlan R, Molho E, Harrison M, Lew M. Effects of coenzyme Q10 in early Parkinson disease: evidence of slowing of the functional decline. Arch Neurol. 2002;59:1541–50.
77. Sung S, Yao Y, Uryu K, Yang H, Lee VM, Trojanowski JQ, Pratico D. Early vitamin E supplementation in young but not aged mice reduces Abeta levels and amyloid deposition in a transgenic model of Alzheimer's disease. FASEB J. 2004;18:323–5.
78. Tohgi H, Abe T, Nakanishi M, Hamato F, Sasaki K, Takahashi S. Concentrations of alpha-tocopherol and its quinone derivative in cerebrospinal fluid from patients with vascular dementia of the Binswanger type and Alzheimer type dementia. Neurosci Lett. 1994;174:73–6.
79. Turunen M, Olsson J, Dallner G. Metabolism and function of coenzyme Q. Biochim Biophys Acta. 2004;1660:171–99.
80. Villalba JM, Navarro F, Gomez-Diaz C, Arroyo A, Bello RI, Navas P. Role of cytochrome b5 reductase on the antioxidant function of coenzyme Q in the plasma membrane. Mol Aspects Med. 1997;18(Suppl):S7–S13.
81. Villalba JM, Navas P. Plasma membrane redox system in the control of stress-induced apoptosis. Antiox Redox Sign. 2000;2(2):213–30.
82. Wadsworth TL, Bishop JA, Pappu AS, Woltjer RL, Quinn JF. Evaluation of coenzyme Q as an antioxidant strategy for Alzheimer's disease. J Alzheimers Dis. 2008;14:225–34.
83. Wang Y, Santa-Cruz K, DeCarli C, Johnson JA. NAD(P)H:quinone oxidoreductase activity is increased in hippocampal pyramidal neurons of patients with Alzheimer's disease. Neurobiol Aging. 2000;21:525–31.
84. Williamson KS, Gabbita SP, Mou S, West M, Pye QN, Markesbery WR, Cooney RV, Grammas P, Reimann-Philipp U, Floyd RA, Hensley K. The nitration product 5-nitro-gamma-tocopherol is increased in the Alzheimer brain. Nitric Oxide. 2002;6:221–7.
85. Yang X, Dai G, Li G, Yang ES. Coenzyme Q10 reduces beta-amyloid plaque in an APP/PS1 transgenic mouse model of Alzheimer's disease. J Mol Neurosci. 2010;41:110–3.

86. Yao J, Irwin RW, Zhao L, Nilsen J, Hamilton RT, Brinton RD. Mitochondrial bioenergetic deficit precedes Alzheimer's pathology in female mouse model of Alzheimer's disease. Proc Natl Acad Sci U S A. 2009;106:14670–5.
87. Yao Y, Chinnici C, Tang H, Trojanowski JQ, Lee VM, Pratico D. Brain inflammation and oxidative stress in a transgenic mouse model of Alzheimer-like brain amyloidosis. J Neuroinflammation. 2004;1:21.
88. Yatin SM, Aksenov M, Butterfield DA. The antioxidant vitamin E modulates amyloid beta-peptide-induced creatine kinase activity inhibition and increased protein oxidation: implications for the free radical hypothesis of Alzheimer's disease. Neurochem Res. 1999;24:427–35.
89. Zaman Z, Roche S, Fielden P, Frost PG, Niriella DC, Cayley AC. Plasma concentrations of vitamins A and E and carotenoids in Alzheimer's disease. Age Ageing. 1992;21:91–4.
90. Zhou Y, Gopalakrishnan V, Richardson JS. Actions of neurotoxic beta-amyloid on calcium homeostasis and viability of PC12 cells are blocked by antioxidants but not by calcium channel antagonists. J Neurochem. 1996;67:1419–25.
91. Zou K, Gong JS, Yanagisawa K, Michikawa M. A novel function of monomeric amyloid beta protein serving as an antioxidant molecule against metal-induced oxidative damage. J Neurosci. 2002;22:4833–41.
92. Battino M, Bompadre S, Leone L, Devecchi E, Degiuli A, D'Agostino F, Cambie G, D'Agostino M, Faggi L, Colturani G, Gorini A, Villa RF. Coenzyme Q, vitamin E and Apo-E alleles in Alzheimer disease. Biofactor. 2003;18(1–4):277–81.
93. Colucci MA, Couch GD, Moody CJ. Natural and synthetic quinines and their reduction by the quinine reductase enzyme NQO1: from synthetic organic chemistry to compounds with anticancer potential. Org Biomol Chem. 2007;6:637–56.
94. Ernster L, Navazio F. Soluble diaphorase in animal tissues. Acta Chem Scand. 1958;12:595.
95. Jensen SK, Lauridsen C. α-Tocopherol stereoisomers. Vitam Horm. 2007;76:281–308.
96. Rigotti A. Absorption, transport and tissue delivery of vitamin E. Mol Aspects Med. 2007; 28:423–36.
97. Wang X, Michaelis ML, Michaelis EK. Functional genomics of brain aging and Alzheimer's disease: focus on selective neuronal vulnerability. Curr Genomics. 2010;11(8):618–33.

# Part II
# Mechanisms

# Chapter 7
# Metal Dysfunction in Alzheimer's Disease

**Rosanna Squitti, Mariacristina Siotto, Carlo Salustri, and Renato Polimanti**

**Abstract** It has been recently established that oxidative stress plays a key role in neurodegeneration. Consequently, researchers have focused their attention on transition metals, as they are known to participate in biochemical reactions that produce free radicals. In Alzheimer's disease (AD), in particular, in vitro and animal studies have uncovered the role of iron and copper in the disease's pathogenesis, recently confirmed in clinical studies. However, the link between AD and metals has been mostly investigated with a focus on local accumulations in brain areas critical for AD. More recently, a wider view has emerged proposing a relationship between AD and systemic changes of metal metabolism, upon genetic variability. In this chapter, we describe the major functions of iron and copper in the body and summarize the reasons why we should closely monitor their dyshomeostases in AD.

R. Squitti, Ph.D. (✉)
Department of Neuroscience, AFaR—Fatebenefratelli Hospital, Rome, Italy

Department of Neurology, "Campus Bio-medico" University,
Via Alvaro de Portillo 21, 00128 Rome, Italy
e-mail: rosanna.squitti@afar.it

M. Siotto, Ph.D.
Don Carlo Gnocchi Foundation, ONLUS, Milan, Italy
e-mail: mary.siotto@gmail.com

C. Salustri, Ph.D.
Institute of Cognitive Sciences and Technologies (CNR), MEG Unit,
Fatebenefratelli Hospital, Isola Tiberina, 00186 Rome, Italy
e-mail: carlo.salustri@istc.cnr.it

R. Polimanti, Ph.D.
Department of Biology, "Tor Vergata" University, 00133 Rome, Italy
e-mail: renatopolimanti@alice.it

D. Praticò and P. Mecocci (eds.), *Studies on Alzheimer's Disease*, Oxidative Stress in Applied Basic Research and Clinical Practice, DOI 10.1007/978-1-62703-598-9_7,

## 7.1 Introduction

About 90–95 % of Alzheimer's disease (AD) patients suffer from the sporadic form, which results from a complex interaction of genetic and nongenetic factors and normally has a late onset (after age 65). In this form, inheritance of the ε4 allele of the apolipoprotein E (*APOE*) has been established to be a major risk factor.

Recently, oxidative stress has been found to be responsible for major cellular damage, and AD researchers have uncovered the neurodegenerative role of transition metals. Iron and copper, for example, are essential for life but also participate in Fenton-type reactions that generate uncontrollable ROS capable of damaging cells. Iron is a fundamental constituent of hemoglobin, the protein that delivers oxygen to cells, and of myoglobin, which delivers oxygen to muscles. It participates in making myelin and neurotransmitters and plays a role in the immune system and in the production of energy. Copper is a cofactor of many enzymes that contribute to energy generation, oxygen transport, cellular metabolism, signal transduction, and many other processes.

In this chapter, we describe the major functions of the two metals in the body and summarize the reasons why we should closely monitor their dyshomeostases.

## 7.2 Metal Absorption and Distribution: From the Intestine to Circulation

### *7.2.1 Iron*

Iron absorption is highly regulated, since our body does not have a physiological mechanism to excrete it. We lose some iron only via indirect events (e.g., death of enterocytes, sloughing of skin) or occasional situations (e.g., minor hemorrhages, menstruation in women), but in very modest amounts (about 1–2 mg/day). Thus, a healthy body absorbs only as much iron as it needs, and the intake is recycled multiple times. On average, we absorb as little as 1–2 mg/day. The majority enters as heme iron, and the remaining as inorganic, non-heme iron. At any time, 4–5 g of iron is present in our body. Of this amount, about 2.5 g is contained in hemoglobin and 600 mg in reticuloendothelial macrophages, 300 mg is used by proteins for cellular processes or energy production, and 3–4 mg binds to transferrin. The rest is stored in ferritin.

Virtually all food iron is in the ferric oxidation state (ferric iron) $Fe^{3+}$ and must be first reduced to $Fe^{2+}$ (ferrous iron) to be transported through the apical membrane of the duodenum's enterocyte. The reduction is completed by reductases such as the cytochrome b558, Steap 2, and CYBRD1 [1]. This transport is provided by the divalent metal transporter 1 (DMT-1), which accepts only ferrous iron [2].

Absorption levels are regulated by hepcidin, a hormone synthesized in the liver when it detects an excess of iron [3]. Hepcidin is responsible for the internalization and degradation of ferroportin, which exports iron across the enterocyte membrane and thus controls the rate of iron transfer to the blood [4]. After reaching the blood, iron is carried by transferrin into circulation. Normal mean levels of serum iron are 37–164 μg/dL.

Table 7.1 and Fig. 7.1 describe the details of iron absorption and distribution.

### 7.2.2 Copper

We ingest between 0.6 and 1.6 mg of copper per day, and altogether a healthy 70 kg human body contains no more than 110 mg [5]. About 10 mg are found in the liver, 8.8 mg in the brain, 6 mg in the blood, 46 mg in the skeleton (including bone marrow), and 26 mg in the skeletal muscles. The status of our body copper is regulated by duodenal absorption (intestine) and biliary excretion (liver). Food is the main source, but it has been recently proposed that a non-negligible contribution comes from drinking water piped through copper plumbing [6]. Food copper is in the $Cu^{+2}$ oxidation state. In order to be transported across the enterocyte's apical membrane, $Cu^{+2}$ must be first reduced to $Cu^{1+}$ by reductases such as the cytochrome b558, Steap 2, and CYBRD1 [1]. $Cu^{1+}$ is then imported into the enterocyte by hCTR-1 [7] (Table 7.1 and Fig. 7.1). Another protein that is believed to be implicated in copper absorption is the same DMT1 that supervises iron uptake [8].

Once inside the enterocyte, copper is loaded onto copper-dependent enzymes via chaperone proteins, such as the cytochrome C oxidase assembly homolog (COX17), which delivers copper to mitochondria, antioxidant protein 1 homolog (ATOX1) and ATP7A protein, which transport copper to the *trans*-Golgi network, and the copper chaperone (CCS) which delivers copper to superoxide dismutase (SOD1) in the cytosol [7]. The system also includes MT proteins, which trap excess metals. Copper is then pumped out of the enterocyte's basolateral membrane by ATP7A and then transported to the liver by albumin or α2-macroglobulin via the portal vein (Table 7.1 and Fig. 7.1). In the liver, copper is partly stored and partly redistributed to other organs. Absorption and excretion interplay in such a way that an occasional over-ingestion in healthy adults normally results in a downregulation of copper uptake in the duodenum and an upregulation of biliary excretion. Normal mean levels of serum copper are 11–24.4 μmol/L [9].

## 7.3 Liver: The Metabolic Control Unit

### 7.3.1 Iron

The portion of iron that does not participate in erythropoiesis is sent to the liver, where it joins hepatocytes and reticuloendothelial cells. The latter are essential for iron recycling as they ingest old red cells, break down their hemoglobin, and liberate

**Table 7.1** Iron and copper absorption and distribution

| | Iron | | Copper | |
|---|---|---|---|---|
| Enterocyte | Proteins involved | Function | Protein involved | Function |
| Apical surface | *Non-heme* | Reduction of $Fe^{+3}$ to $Fe^{2+}$ | Cytochrome b558 ferric/cupric reductase | Reduction of $Cu^{2+}$ to $Cu^{+}$ |
| | Cytochrome b558 ferric/cupric reductase | | Steap2 | |
| | Steap2 | | CYBRD1 | |
| | CYBRD1 | | | |
| Membrane | *Non-heme* | Channeling of $Fe^{2+}$ via electrochemical gradient | hCTR-1 | Import of $Cu^{+}$ |
| | DMT-1 | | DMT-1 | |
| | *Heme* | Import of heme iron | | |
| | hCP-1 | | | |
| Cytosol | *Non-heme* | Storage | *Copper chaperones* | Transport |
| | Ferritin | | COX17 | |
| | *Heme* | Cleavage of part of heme | ATOX1 | |
| | Heme oxygenase | | CCS | |
| | | | ATP7A | |
| | | | MT | |
| Basolateral membrane | Ferroportin | Transport of $Fe^{2+}$ from cytosol to circulation | ATP7A | Transmigration and export of copper into circulation |
| Basolateral outer surface | Hephaestin | Oxidation of $Fe^{2+}$ to $Fe^{3+}$ | | |
| Regulation | Peptide hormone hepcidin | Excess of iron: it regulates ferroportin degradation and iron storage | | |
| Circulation | Transferrin (apo/holo-Tf) | Binding and transport of 2 atoms of $Fe^{3+}$ | Albumin | Transport of copper |
| | | | α2-macroglobulin | |
| | | | Amino acids | |
| | Hemoglobin | Erythropoiesis | | |

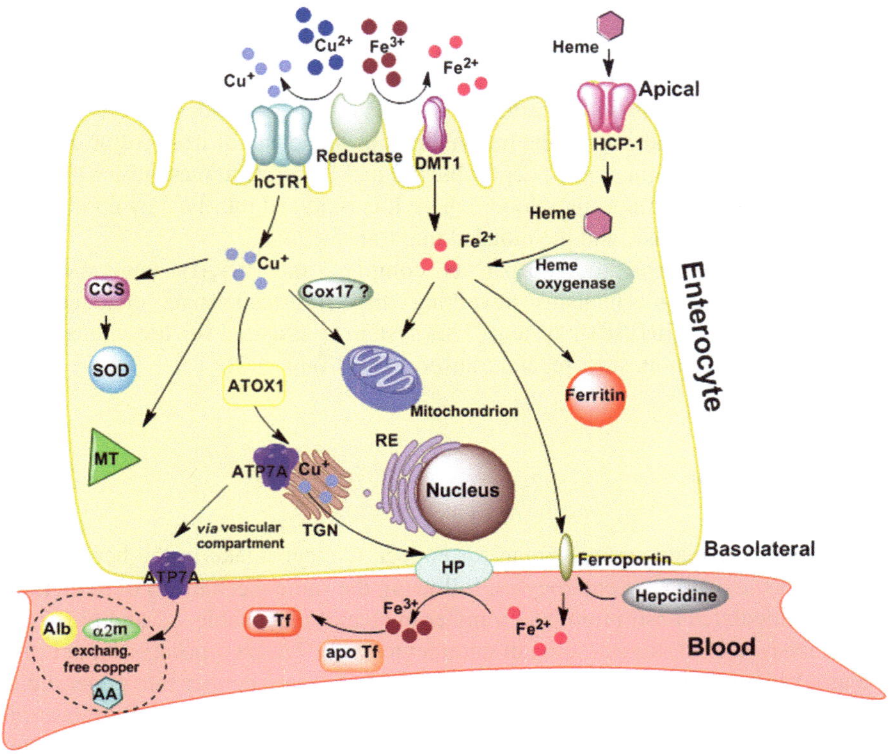


**Fig. 7.1** Absorption of iron and copper through the enterocyte. $Cu^{2+}$ and $Fe^{3+}$ are reduced from reductases before entering the enterocyte. *Iron*: The divalent iron transporter DMT-1 channels $Fe^{2+}$ thanks to an electrochemical gradient (not shown); in a secondary mechanism, iron enters as part of heme (thus as $Fe^{2+}$) via the heme carrier protein 1 (HCP-1), also present on the enterocyte's apical surface. The heme is then cleaved by a heme oxygenase and releases its iron. Iron then is stored into ferritin. On the basolateral outer surface, ferroportin transfers a controlled amount of iron to the portal plasma, and the ferroxidase hephaestin (HP) reoxidizes $Fe^{2+}$ into $Fe^{3+}$, which binds apo-transferrin (apo-Tf) to form holo-transferrin (Tf), which transports iron in the circulation. The hormone hepcidin regulates the iron export from the intestine, via a mechanism of internalization and degradation of ferroportin in case of iron excess. *Copper*: The integral membrane protein hCtr-1 imports $Cu^{+}$ which is loaded onto copper-dependent enzymes, via chaperone proteins: cytochrome C oxidase assembly homolog (COX17), which delivers copper to mitochondrion; antioxidant protein 1 homolog (ATOX1) and ATP7A that transport copper to the *trans*-Golgi network; and the copper chaperone (CCS) which delivers copper to superoxide dismutase (SOD1) in the cytosol. The chaperone system also includes the family of metallothioneins (MT). Copper is then pumped out of the enterocyte's basolateral membrane by ATP7A (via vesicular compartment) and then transported to the liver bound to albumin (Alb) or **α**2-macroglobulin (**α**2m) or amino acids (AA) via the portal vein

their iron for reuse [10]. The hepatocytes absorb iron through an endocytosis mediated by the transferrin receptor 1 (TfR1) located on the cell surface [11] (Table 7.2 and Fig. 7.2).

After endocytosis, an ATP proton pump acidifies the endosome, causing the release of $Fe^{+3}$ from the holo-transferrin. The ferric ions are then reduced to ferrous ones by a reductase, which are hence able to pass through DMT1. In the hepatocyte, iron forms labile bonds with citrate, ATP, AMP, or other peptides. In this form, it reaches the mitochondrion for the biosynthesis of heme and of iron–sulfur clusters or is enclosed in ferritin stores [12] (Table 7.2 and Fig. 7.2). Excess iron travels as $Fe^{+2}$ through ferroportin to the blood, where it is oxidized into $Fe^{+3}$ by ceruloplasmin, a 132 kDa oxidase, and then loaded into transferrin.

Regulation of iron metabolism is complex and strictly controlled by post-trascriptional mechanisms involving the iron-responsive element/iron-regulatory protein (IRE/IRP) system. This system is essential for life as regulates iron uptake, utilization, storage, and transport [13] (Table 7.2).

### *7.3.2 Copper*

The liver is the main storage organ for copper. Copper intake in the hepatocytes resembles the one in the enterocytes, following the same mechanism of reduction, hCTR-1-mediated absorption, and transfer to chaperones (Table 7.2 and Fig. 7.2). In the liver, ATP7B, the homolog of enterocytes' ATP7A, incorporates copper into ceruloplasmin (Table 7.2 and Fig. 7.2). The liver releases copper into general circulation. About 85–95 % of circulating copper is tightly bound to ceruloplasmin, whereas the remainder loosely binds to and is exchanged among albumin, α2-macroglobulin, amino acids, peptides, and several micronutrients. We will refer to the portion that is bound to ceruloplasmin as "ceruloplasmin-bound" copper and to the labile portion that binds to the loose compounds as "free copper." Within the hepatocytes, an excess of copper induces the translocation of ATP7B from the *trans*-Golgi network to the canalicular membrane, where the metal is secreted in the bile (Table 7.2 and Fig. 7.2).

## 7.4 From General Circulation to the Brain

### *7.4.1 Iron*

Transported by transferrin in the blood, $Fe^{3+}$ reaches the brain capillaries, whose endothelial cells, named BCEC, constitute the BBB. On the BCEC's luminal side, iron-loaded transferrin is captured by a TfR1 (Fig. 7.3a), and the complex is internalized into an endosome. Iron remains inside the endosome, while the latter travels through the intracellular space. When the endosome reaches the BCEC's abluminal side, it fuses with the membrane, exposing and then releasing $Fe^{3+}$ to the extracellular interstitial space. Apo-transferrin (i.e., transferrin without iron) remains

**Table 7.2** Iron and copper metabolism in the liver

| | Iron | | Copper | |
|---|---|---|---|---|
| Hepatocyte | Proteins involved | Function | Protein involved | Function |
| Apical membrane | TfR1<br>Holo-Tf | Complex formation for internalization: formation of acidified endosome | Cytochrome b558 ferric/cupric reductase<br>Steap2<br>hCTR-1<br>DMT-1 | Reduction of $Cu^{2+}$ to $Cu^{+}$<br>Import of $Cu^{+}$ |
| Cytosol | Reductases<br>DMT1 | Reduction of $Fe^{3+}$ to $Fe^{2+}$ (electrochemical gradient)<br>Channeling of $Fe^{2+}$ outside of endosome | *Copper chaperones*<br>COX17<br>ATOX1<br>CCS (for SOD)<br>APP<br>MT | Transport and delivery of copper into specific proteins |
| | Ferritin<br>Citrate, AMP, ATP | Storage of labile pool | ATP7B (*trans*-Golgi) | Copper loading into ceruloplasmin |
| Interface with bile canaliculus | | | ATP7B | Transmigration from *trans*-Golgi and export of excess copper into bile |
| Basolateral outer membrane | Ferroportin<br>Ceruloplasmin | Transport of $Fe^{2+}$ from cytosol to circulation<br>Oxidation of $Fe^{2+}$ to $Fe^{3+}$ | Ceruloplasmin | "Transport" of copper structurally bound and not exchangeable (85:95 % of serum copper) |
| Regulation | IRE/IRP | Cellular iron uptake, utilization, storage, and transport | | |
| Circulation | Transferrin (apo/holo-Tf) | Binding and transport of 2 atoms of $Fe^{3+}$ | Albumin<br>α2-macroglobulin<br>Amino acids | Transport of exchangeable free copper |

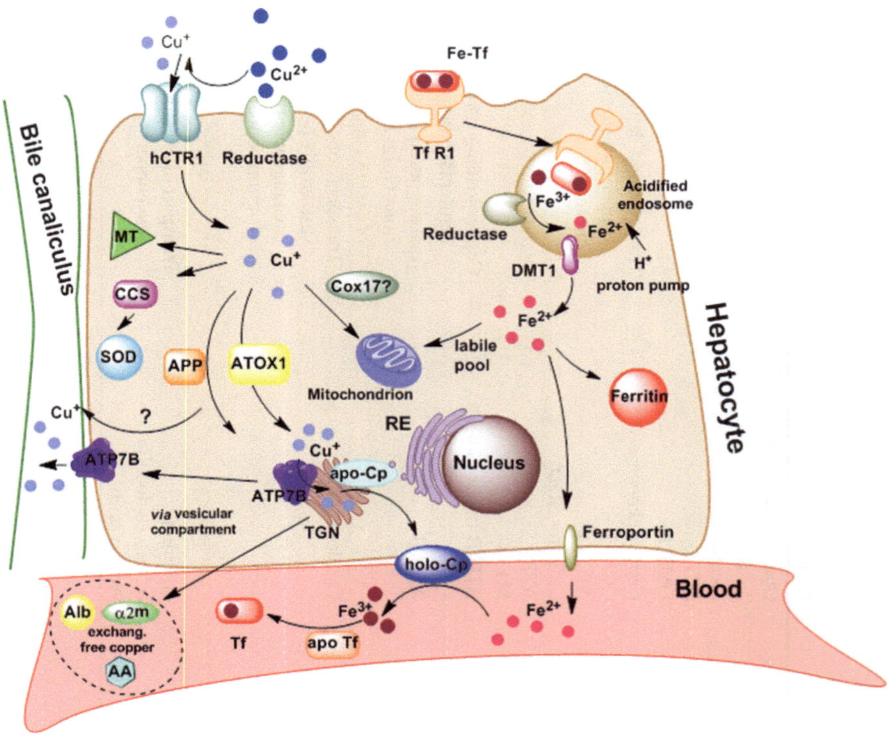


**Fig. 7.2** Iron and copper metabolization in the hepatocyte. *Iron*: The hepatocytes absorb iron through a receptor-mediated endocytosis: the transferrin receptor 1 (TfR1) located on the cell surface binds a holo-transferrin (Fe-Tf). After endocytosis, an ATPase proton pump acidifies the endosome, allowing the release of $Fe^{+3}$ from the holo-Tf. $Fe^{3+}$ is then reduced by a reductase to $Fe^{2+}$ and passes through DMT-1. Once inside, iron is believed to form labile bonds with citrate, ATP, AMP, or other peptides (labile pool) and reaches the mitochondrion for the biosynthesis of heme and of iron–sulfur clusters or is stored in ferritin (the main source of reserve iron). The excess of iron, as $Fe^{+2}$, travels through ferroportin to the blood, where it is oxidized into $Fe^{+3}$ by ceruloplasmin (holo-Cp) and is loaded into transferrin (Tf). *Copper*: The liver is the main storage organ for copper. Copper intake in the hepatocytes resembles the one in the enterocytes, following the same mechanism of reduction by reductase, hCTR-1-mediated absorption and delivery to chaperones. In addition, the amyloid precursor protein (APP) has been shown to be involved in liver as well as neuron copper metabolism. In the liver, ATP7B, the homolog of enterocytes' ATP7A, incorporates copper into ceruloplasmin (Cp). About 85–95 % of copper tightly binds Cp, whereas the remainder (free copper) loosely binds to and is exchanged among albumin (Alb), α2-macroglobulin (α2m), amino acids (AA), peptides, and several micronutrients. An excess of copper induces the translocation of ATP7B from the *trans*-Golgi network to the canalicular membrane (via vesicular compartment), where the metal is secreted in the bile

attached to the receptor, and the two undergo again an endocytosis to travel back to the BCEC luminal side, where apo-transferrin is released into the capillary blood ready to bind again $Fe^{3+}$ ions [14].

This is the mechanism most used by iron to enter the brain. The BCEC luminal side does not normally communicate directly with neurons but rather faces astrocytes. In the BCEC–astrocyte gap, the released $Fe^{3+}$ finds again apo-transferrin and several low-weight elements able to bind it, such as ATP and citrate. They are ubiquitous in the brain's extracellular fluid, but they appear particularly concentrated in the BCEC–astrocyte gap. Traveling through the extracellular space bound to transferrin, iron can easily reach the vicinity of a neuron, which expresses transferrin receptors on its surface. Moreover, both ferroportin and a GPI-anchored form of ceruloplasmin are strongly expressed on the astrocytic foot in the vicinity of neurons: thus, iron can be internalized into the astrocyte, possibly as $Fe^{3+}$ bound to ATP or citrate, transported through the astrocyte, reduced at some point by a yet unidentified ferrireductase and released from the end foot via ferroportin.

On the outside of the astrocyte foot, ceruloplasmin catalyzes the oxidation of $Fe^{2+}$ into $Fe^{3+}$, making iron ready to bind apo-transferrin. The extracellular and intracellular mechanisms are not mutually exclusive and can certainly work in synergy to deliver iron to the neuron (Fig. 7.3b). As for the BCEC, the neuron's membrane displays transferrin receptors, and an endocytosis is again initiated. The key difference between BCEC and the neuronal membrane lies in the fact that the latter displays DMT-1, which becomes engulfed in the endosome and sinks with it [14] (Fig. 7.3b). Interestingly, ferroportin has been also found in neuronal presynaptic vesicles [15], suggesting that ferrous iron may be transported via vesicles to the synaptic cleft, where it is released upon vesicle fusion. This presence of free iron in the synapse cleft could potentially expose pre- and postsynaptic membranes to iron-mediated oxidation and damage.

### *7.4.2 Copper*

Another barrier to consider is the BCB, between blood and CSF. CSF is produced from arterial blood by the choroid plexuses and plays an important function in CNS homeostasis and metabolism. Since CSF composition is similar to the one of extracellular brain fluids to which CSF is connected without a barrier, the amount of elements in CSF closely reflects the situation in neighboring regions [16]. The mechanism by which copper passes in and out of CSF is not fully understood and could be an active transport or a passive diffusion. Normal CSF copper levels range between 0.5 and 2.5 μmol/L as reviewed in Bucossi et al. [17].

Also the role of BBB and BCB in regulating brain copper absorption is still unclear. A recent study in a rat model tested the hypothesis that blood, BBB, and BCB play different roles in transport and homeostasis of free, ceruloplasmin-bound, and albumin-bound copper present in the brain capillaries, parenchyma, and CSF [18]. Perfusion of three species of radioactive $^{64}Cu$ into the rat brain via the internal carotid artery revealed that free copper is the main species transported in the brain. This evidence fits very well with previous results obtained from a mouse model of brain uptake of radio-copper $^{67}Cu^{+2}$, in which a net brain copper uptake occurs and parallels the free copper increase in the injectate starting from a concentration of $Cu^{+2}$ of 3.2 ng/ml (0.05 μmol/L). The highest copper accumulations are in BBB

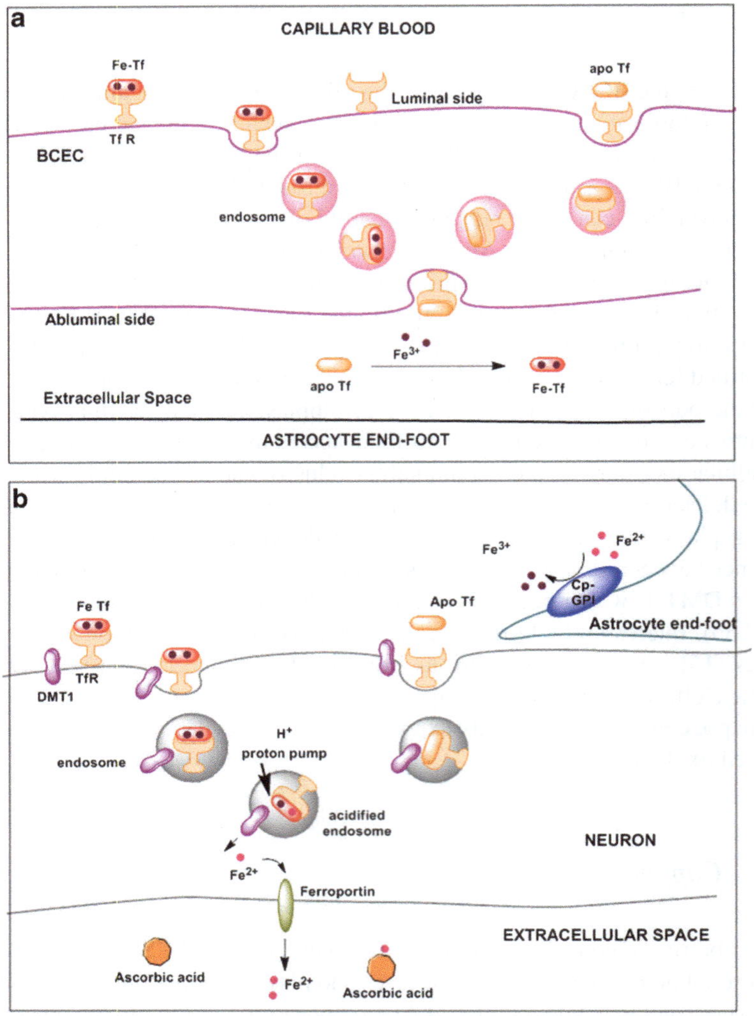


**Fig. 7.3** Iron metabolism in the CNS. (**a**) Transported by transferrin in the blood, $Fe^{3+}$ reaches the brain's capillaries, whose endothelial cells (BCEC) constitute the blood–brain barrier (BBB). On the BCEC's luminal side, iron-loaded transferrin is captured by the transferrin receptor 1(TfR1), and the complex is internalized into an endosome which travels through the intracellular space until it reaches the BCEC's abluminal side; here it fuses with the external membrane, exposing and then releasing $Fe^{3+}$ to the extracellular interstitial space. Apo-transferrin (apo-Tf) remains attached to the receptor undergoing again an endocytosis to travel back to the BCEC luminal side, where apo-Tf is released. (**b**) Traveling through the extracellular space bound to transferrin, iron reaches a neuron, whose membrane displays transferrin receptors, and endocytosis is initiated. The key difference between the BCEC and the neuronal membrane lies in the fact that the latter displays DMT-1 which becomes engulfed in the endosome and sinks with it. Interestingly, ferroportin has been also found in neuronal presynaptic vesicles (not shown), suggesting that ferrous iron may be transported via vesicles to the synaptic cleft, where it is released upon vesicle fusion. This presence of free iron in the synapse cleft could potentially expose pre- and postsynaptic membranes to iron-mediated oxidation and damage. On the outside of the astrocyte end foot, GPI-anchored ceruloplasmin (Cp-GPI) catalyzes the oxidation of $Fe^{2+}$ into $Fe^{3+}$, making iron ready to bind apo-transferrin. The extracellular and intracellular mechanisms are not mutually exclusive and can certainly work in synergy to deliver iron to the neuron

tissues (choroid plexus and cerebral capillaries), brain parenchyma and, to a lesser extent, in CSF. The BBB is the principal gate of copper into the brain parenchyma, whereas the BCB appears more involved in the regulation of copper homeostasis in CSF. Copper uptake into brain cells requires the reduction from $Cu^{2+}$ to $Cu^{1+}$, which is performed by still unidentified reductases. Compared to other cell types, the brain presents an overcontrol on copper uptake. In fact, in addition to the hHCTR1, two further proteins, the APP and the prion protein (PrP), are located on the surface of brain cells (Fig. 7.4).

After entering the brain, $Cu^{1+}$ is delivered to the intracellular compartments by the copper chaperones ATOX1 (or HAH1), CCS, COX17, or metallothioneins (Fig. 7.4). Copper secretion is assigned principally to ATP7A, and its role has been demonstrated in pumping copper into synaptic vesicles at the glutamatergic synapse, where this metal acts as a neuromodulator of neurotransmission [19]. Conversely, ATP7B is expressed in some isolated region of the brain [20].

### 7.4.3 *The Key Role of Ceruloplasmin*

Ceruloplasmin is mainly synthesized in the hepatocytes, where the ATP7B protein loads six copper atoms into apo-ceruloplasmin, generating an active holo-ceruloplasmin that is then transferred to the blood. Ceruloplasmin is also expressed endogenously by the CNS and is strongly present on astrocytes close to neurons. A key process catalyzed by ceruloplasmin is the oxidation of $Fe^{2+}$ to $Fe^{3+}$ via the following chemical reaction:

$$4Fe^{2+} + 4H^{+} + O_2 \rightarrow 4Fe^{3+} + 2H_2O \tag{7.1}$$

This reaction reduces the availability of $Fe^{2+}$, promoting instead iron bussing to various organs by transferrin, which accepts only $Fe^{3+}$. Not insignificantly, the by-product of (7.1) is simply water.

Another important role of ceruloplasmin is to scavenge $H_2O_2$ [21]. This prevents production of dangerous oxygen radicals, since oxidation of $Fe^{2+}$—which takes place at physiological pH—can trigger Fenton or Haber–Weiss reactions. Moreover, it prevents lipid peroxidation. The synthesis of holo-ceruloplasmin strictly depends on the amount of copper available in the liver. Consequently, reduced copper availability results in the production of apo-ceruloplasmin, which has no ferroxidase capability and is rapidly degraded in plasma.

A reduction of ferroxidase activity due to a decreased presence of ceruloplasmin results in more $Fe^{2+}$ available for Fenton reactions. This is why aceruloplasminemia, which is an autosomal recessive trait causing a drop in ceruloplasmin activity, results in severe iron overload and is often misdiagnosed as hemochromatosis. In this case, measurement of transferrin saturation may help reaching a correct diagnosis, as transferrin saturation decreases in aceruloplasminemia and increases hemochromatosis. Patients with aceruloplasminemia show evidence of increased lipid

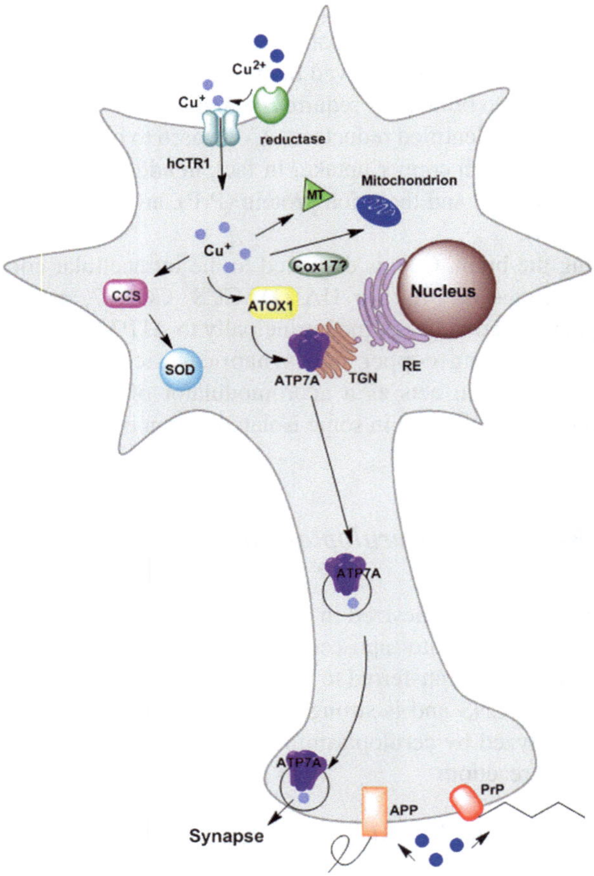


**Fig. 7.4** Copper metabolism in the neuron. Uptake of copper into brain cells requires the reduction from $Cu^{2+}$ to $Cu^{1+}$. This reaction is performed by unidentified reductases probably analogue to those present in other organs. In comparison to other cell types, the brain presents an overcontrol on copper uptake; in fact, in addition to the hCTR1, there are two other proteins located on the surface of brain cells: the APP and the prion protein (PrP). Once $Cu^{1+}$ enters the brain, it is delivered to the intracellular compartments by the abovementioned copper chaperones: ATOX1, which delivers copper to ATP7A; CCS, which deliver copper to SOD; Cox17, to mitochondrion; or metallothioneins (MT). The secretion of copper is assigned principally to ATP7A, via vesicular compartment which secrete copper into the synapse

peroxidation and impaired fatty acid oxidation. Low holo-ceruloplasmin levels can be also caused by genetic mutations of the *ATP7B* gene, as it happens in Wilson's disease which is the paradigmatic disease of free copper toxicosis or accumulation in both liver and brain [22].

Ceruloplasmin affects the rate of serum iron oxidation, which is required for iron mobilization and iron release from tissue storage [23]. Experiments made in

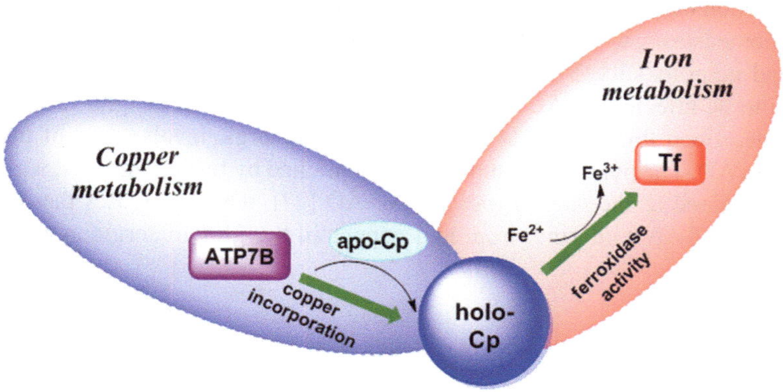


**Fig. 7.5** The cross-talking role of ceruloplasmin between copper and iron metabolisms. Holo-ceruloplasmin (holo-Cp) is schematically depicted in *blue* as cross-talking protein between copper and iron metabolism. ATP7B in the hepatocyte is the protein which incorporates six atoms of copper into apo-ceruloplasmin (apo-Cp), thus allowing the formation of the holo-Cp; only the holo form has a ferroxidase activity, oxidizing $Fe^{2+}$ into $Fe^{3+}$ which can be loaded into transferrin (Tf)

the 1960s showed that addition of ceruloplasmin (and apo-transferrin) markedly stimulates iron efflux in general circulation, suggesting that ceruloplasmin is crucial for iron mobilization.

In summary, ceruloplasmin is a protein essential for iron metabolism and for antioxidant defense and is implicated in the molecular mechanisms of hepatic disease and neurodegenerative disorders [24] (Fig. 7.5).

## 7.5 Iron and Copper Toxicity

The body undergoes sophisticated processes to regulate iron and copper absorption, as any physiological unbalance can easily generate free radicals causing tissue damage.

### *7.5.1 The Role of Iron and Copper in Alzheimer's Disease*

It is well known that the hallmark of the AD brain is the presence of amyloidal plaques and neurofibrillary tangles, respectively outside and inside the neuronal space. The major constituent of the plaques is amyloid-beta (Aβ), a 39–43 amino acid peptide, generated from the cleavage of the transmembrane protein APP (Fig. 7.6). APP has selective copper binding sites, which mediate redox activity causing precipitation of Aβ even at low concentrations.

Enhanced concentrations of iron and copper have been reported within both amyloid plaques and neurofibrillary tangles (although these findings have been recently questioned [25]). The fact that Aβ deposition in plaques is an age-dependent phenomenon but Aβ production does not increase with age suggests that changes in metal homeostasis, which are age-dependent, may play a key role in Aβ transformation and neurotoxicity. Metal deposits have been found also in other brain areas known to be affected by AD, such as for example, the basal ganglia [26]. All the above-described findings led to hypothesizing a synergic involvement of iron and copper in AD. The hypothesis, subsequently enriched by further findings, was eventually proposed as the *metal hypothesis* of AD [27], which proposes that it is the interaction of Aβ with zinc, iron, and copper in the glutamatergic synapse that

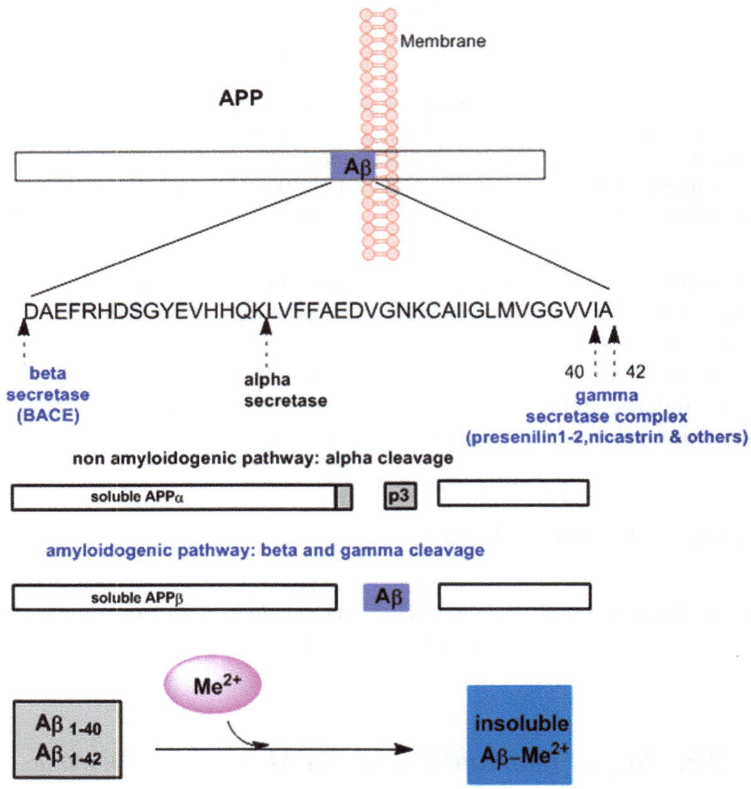


**Fig. 7.6** Amyloidogenic pathway: Aβ formation and its interaction with metals. The major constituent of the amyloidal aggregations is Aβ, a 39–43 amino acid peptide, generated from the cleavage of the transmembrane APP. Specifically, APP can be cleaved in two different pathways. In one, it is first cleaved by β-secretase at a site outside the membrane, leaving a C-terminal fragment which is subsequently cleaved by γ-secretase at a site within the transmembrane section. The product of this process is Aβ. Alternatively, APP is initially cleaved by α-secretase at a different site and then by γ-secretase, resulting in a truncated P3 fragment. It was observed that APP has selective copper and zinc binding sites, which mediate redox activity causing precipitation of Aβ even at low concentrations. Aβ too shows selective copper and zinc sites that in normal conditions bind equimolar amounts of the two metals ($Me^{2+}$). However, in conditions of acidosis, copper completely displaces zinc from Aβ

promotes the aggregation of Aβ in insoluble plaques, short-circuiting the entire neuronal network to which the synapses belong and thus driving AD pathogenesis [27]. In this view, systemic metal abnormalities, even if not severe but over a long period of time, may disturb the processes devoted to regulate protection against oxidative stress, especially at the glutamatergic synapse. Support to the metal hypothesis of AD [27] comes from numerous clinical studies, including some of ours, recently reviewed in Squitti (2012) [28].

### 7.5.2 *Metals and Aβ*

The interaction between transition metals and Aβ can be described by the following reactions [29]:

$$(A\beta)_2 + Me^{n+} \rightarrow A\beta : A\beta^{+} + Me^{(n-1)+} \tag{7.2}$$

where Me stands for metal. This reaction has two negative effects: the first is the production of an insoluble Aβ peptide, and the second the production of a superoxide anion via the following reaction:

$$Fe^{2+} + O_2 \rightarrow Fe^{3+} + O_2^{-} \tag{7.3}$$

The generated superoxide can now participate in a Haber–Weiss reaction with peroxide

$$O_2^{-} + H_2O_2 \rightarrow {}^{\bullet}OH + OH^{-} + O_2 \tag{7.4}$$

which produces ROS. The only way the superoxide anion can be depleted is by a dismutation:

$$O_2^{-} + O_2^{-} + 2H^{+} \rightarrow H_2O_2 + O_2 \tag{7.5}$$

which is generally, but not exclusively, accelerated by superoxide dismutases, as, for example, SOD1. For a review on the role of metallobiology and Aβ in AD, see also [30].

## 7.6 Genetics of Biometals in AD

### 7.6.1 *Iron*

The brain proteins mostly involved in iron homeostasis are HFE, ferritin, transferrin, TfR1, IRP, DMT-1, and ceruloplasmin. However, numerous rarer LoF variants have been discovered in gene encoding for proteins regulating metal metabolism [31], and numerous studies have explored their role in neurodegenerative disorders in general and in AD in particular. Mutations in the gene of HFE (*HFE* gene) have been particularly investigated, since HFE controls iron absorption and mutations in *HFE* gene are known to cause hemochromatosis, i.e., increased absorption of dietary iron and over-deposition in tissues and organs. The first study on the relationship between *HFE* gene mutations and AD risk [32] showed that the mutation frequencies, both in men affected by familial AD and among noncarriers of *APOE* ε4 allele, were significantly higher than in healthy elderly men. Subsequently, other studies [33] have reported significant associations of *HFE* gene mutations with specific AD clinical traits, such as disease onset, cognitive symptoms, severity of neuropsychological deficits, CSF marker, and the rate of progression from MCI to AD. However, these diverse studies did not produce univocal results probably due to interethnic differences in *HFE* gene allele frequencies and to interaction with genetic, environmental, and demographic factors. Associations with AD phenotypes were also found in studies investigating the genes of TfR1, DMT-1, and IREB2.

### *7.6.2 Copper*

Copper homeostasis is regulated instead by ATP7A, ATP7B, SOD1, and the chaperones CCS, ATOX1, COX17, and COMMD1. Manipulations of ATP7A and ATP7B have been tested in diverse AD models. For example, a significant reduction in APP concentration and a downregulation of its expression were observed after generating high intracellular copper levels by ATP7A overexpression in fibroblast [34]. An AD mouse model also showed that ATP7A is significantly expressed in microglial cells activated around Aβ plaques, suggesting that changes in copper homeostasis in AD may be strongly related to inflammatory processes.

## 7.7 Systemic View: Our Contribution

### *7.7.1 Iron*

In a recent AD study [35], we evaluated the activation of the main plasma antioxidant system: the Cp–Tf system. As a measure of the Cp–Tf system's functionality, we took the Cp/Tf ratio, which is known to reflect the combined antioxidant activity of ceruloplasmin-bound copper in the oxidized state ($Cu^{2+}$) and of apo-transferrin. Results showed a significant activation of the Cp–Tf system, suggesting the presence of extensive oxidative stress in AD, afflicting patients as a systemic condition [35], correlating with MMSE and the MTA [35].

In another recent AD study [33], we evaluated whether and how all the above listed iron markers are related to liver status. Results revealed that AD patients have lower albumin, longer prothrombin time, and higher transaminases ratios (aspartate/alanine transaminases, AST/ALT) than controls, clearly indicating a liver hypofunction. AD patients also showed decreased transferrin and increased ferritin levels. Healthy control carriers of a single copy of the *HFE* gene's *H63D* mutation showed a normal iron status, whereas AD carriers of the same mutation had higher levels of iron and lower levels of transferrin and ceruloplasmin. The latter panel resembles hemochromatosis and was not found in noncarrier patients, suggesting that *H63D* mutation is not itself sufficient to increase the risk of AD; rather, it is a synergy of at least a single copy H63D mutation together with liver dysfunction that increases the probability of developing AD.

### 7.7.2 Copper Bound to Ceruloplasmin and Free Copper

In the systemic view of copper metabolism and homeostasis, it is fundamental to distinguish between bound and non-ceruloplasmin or free copper. The key difference between them lies in the fact that both the limited size of the low molecular weight transporters and the labile nature of their binding allow free copper to easily cross the BBB. Consequently, free copper represents the bulk of copper transported into the brain [18], whereas ceruloplasmin-bound copper, which is up to 95 % of serum copper, represents less than 1 % of the copper present in the brain (see Sect. 7.2.2). We measured free copper concentrations up to 3.7 times higher in AD patients than in controls correlating with CSF AD markers [36], AD main cognitive deficits [28], neuroanatomical and electroencephalographic (EEG) changes [37], and known genetic risk factors, such as *APOE ε4* [28, 37]. Also, ceruloplasmin appeared fragmented, suggesting that the higher-than-normal levels of free copper might be due to some impairment in the incorporation of copper into that protein during biosynthesis [28]. Ceruloplasmin fragmentation could be the cause or the result of a disturbed hepatocyte function, possibly resulting in liver hypometabolism [38].

Since published reports on systemic copper dysfunctions are not univocal [25], we ran meta-analyses encompassing all data on AD serum, plasma, and CSF studies published between 1983 and 2012 [39]. Our analyses concluded that AD patients actually have significant higher levels of serum copper than healthy controls. Even though modest, the assessed copper increase is high enough to distinguish AD from healthy controls on the basis of the increase in the free copper fraction, which is a metabolic error of copper metabolism, and it is increased in a large percentage of AD patients [28]. The evidence of increased serum concentrations of free copper in AD has been reported by most of the studies published from 1983 to January 2013 (Table 7.3). Only the recent AIBL study [40] seems to be substantially different from the others. In this study the mean value of free (non-ceruloplasmin in [40]) copper for healthy controls was in the negative range, differently from all the other studies (Table 7.3). Twomey et al. [41] proposed 6.6 as the theoretical optimal mean value of the copper-to-ceruloplasmin (Cu:Cp) ratio in healthy subjects. This ratio is an index of the accordance of copper and ceruloplasmin pair for each serum measurement to be in a physiological range. As reported in Table 7.3, the Cu:Cp ratio for the Rembach et al. study [40] was lower than 6.6 and even lower than 6.0 which

**Table 7.3** Copper (Cu), ceruloplasmin (Cp), Cp-bound copper (Cp–Cu), non-Cp-bound copper (non-Cp–Cu), and Cu:Cp ratio as reported in all the studies published since 1983 until January 2013 on Cu and Cp measurements in control individuals (CTRL) and Alzheimer's disease (AD)

| | | CTRL | | | | | | | | AD | | | | | | | |
|---|---|---|---|---|---|---|---|---|---|---|---|---|---|---|---|---|---|
| Author | Cp | | | | Cu (μM) | Cp (mg/dL) | Cp–Cu (μM) | Non-Cp–Cu[a] | | | | | Cu (μM) | Cp (mg/dL) | Cp–Cu (μM) | Non-Cp–Cu[a] | |
| (year) | assay | N | F % | Mean age | Mean (SD) | Mean (SD) | Mean (SD) | Mean | Cu:Cp[b] | N | F % | Mean age | Mean (SD) | Mean (SD) | Mean (SD) | Mean | Cu:Cp[b] |
| Arnal (2010) [54] | Enz. | 79 | 48 | 77.8 (3.7) | 12.7 (0.9) | 26.5 (7) | 12.5 (3.3) | 0.2 | 6.34 | 110 | 58.2 | 74.7 (4.1) | 15.7 (2.7) | 25.2 (10.7) | 11.9 (5.0) | 3.79 | 8.21 |
| Brewer (2010) [55] | Neph | 29 | | 68.6 (56–80) | 18.4 (3.1) | 25.7 (3.7) | 12.1 (1.7) | 6.27 | 9.45 | 28 | | 76.2 (53–89) | 17.0 (2.4) | 24.4 (3.0) | 11.5 (1.4) | 5.46 | 9.19 |
| Agarwal (2008) [56] | Imm. | 50 | | 55.3 (10.88) | 21.1 (4.9) | 28.82 (8) | 13.6 (3.8) | 7.5 | 9.68 | 50 | | 60 (11.6) | 24.56 (4.7) | 41.60 (9.6) | 19.6 (4.5) | 4.92 | 7.79 |
| Sedighi (2006) [57] | Enz. | 50 | 50 | 67.8 | 20.83 (2.47) | 31.1 (5.40) | 14.7 (2.5) | 6.2 | 8.84 | 50 | 50 | 76.4 | 21.7 (3.1) | 27.7 (9.60) | 13.1 (4.5) | 8.59 | 10.32 |
| Snaedal (1998) [58] | Neph | 44 | 72.7 | 74.3 (53–89) | 19.4 (15–36.4) | 38.3 (22.2–65.5) | 18.1 | 1.30 | 6.69 | 44 | 72.7 | 74.3 (53–89) | 19.1 (11.4–30.0) | 38.2 (24.7–56.2) | 18.03 | 1.07 | 6.60 |
| Molaschi (1996) [59] | Neph | 421 | 100 | 77.6 (2.3) | 19.29 (3.8) | 40.6 (9.1) | 19.2 (4.3) | 0.14 | 6.27 | 31 | 100 | 77.2 (2.4) | 18.9 (3.4) | 38.8 (7.1) | 18.3 (3.3) | 0.54 | 6.41 |
| Squitti (2005) [60] | Imm. | 44 | 45.5 | 71.1 (11) | 12.6 (2.5) | 26.5 (4.7) | 12.5 (2.2) | 0.10 | 6.28 | 47 | 74.5 | 75.6 (7.7) | 17.2 (5.9) | 30.1 (5.9) | 14.2 (2.8) | 2.99 | 7.54 |
| Squitti (2006) [36] | Imm. | 25 | 68 | 71 (9.6) | 12.8 (2.3) | 24.6 (2.9) | 11.6 (1.4) | 1.20 | 6.87 | 28 | 71.4 | 71.4 (8.6) | 16.2 (3.2) | 27.9 (5.3) | 13.2 (2.5) | 3.03 | 7.66 |
| Squitti (2007) [38] | Imm. | 53 | 66 | 70 (10) | 13 (2.8) | 26.8 (5.3) | 12.6 (2.5) | 0.40 | 6.40 | 51 | 78.4 | 73 (8) | 16.1 (5.3) | 28.4 (4.7) | 13.4 (2.2) | 2.70 | 7.48 |

| | | | | | | | | | | | | | | | | | |
|---|---|---|---|---|---|---|---|---|---|---|---|---|---|---|---|---|---|
| Zappasodi (2008) [37] | Imm. | 20 | 65 | 71.55 (9.2) | 12.9 (3) | 27 (5.3) | 12.7 (1.4) | 0.20 | 6.31 | 54 | 81.5 | 73.7 (8.7) | 15.1 (3.4) | 27.6 (5.9) | 13.03 (2.78) | 2.07 | 7.22 |
| Squitti (2011) [61] | Imm. | 100 | 57 | 69 (9.7) | 14 (2.7) | 28 (5.6) | 13.2 (2.6) | 0.80 | 6.6 | 105 | 78 | 74 (8) | 16 (3.7) | 28 (5.4) | 13.2 (2.5) | 2.78 | 7.54 |
| Rembach (2013) [40] | Neph | 716 | 57.5 | 69 (6.8) | 14.32 (2.9) | 31,68 (7.9) | 14.9 (3.7) | −0.58 | 5.97 | 152 | 59.9 | 77 (7.9) | 13.9 (2.7) | 31,6 (7.9) | 14.9 (3.7) | −0.99 | 5.82 |
| Lopez (2013) [47] | Imm. | 33 | 63.60 | 74.0 (5.0) | 13.82 (3.7) | 25.82 (4.3) | 12.2 (2.0) | 1.6 | 7.06 | 36 | 55.5 | 77.7 (5.3) | 15.83 (2.9) | 25.9 (4.1) | 12.2 (1.9) | 3.61 | 8.07 |

In the table is reported also the assay employed to measure ceruloplasmin: enzymatic (Enz.), nephelometric (Neph), or immunoturbidimetric (Imm.)

Twomey et al. [41] proposed that 6.6 is the theoretical optimal mean value of this ratio in healthy subjects. In Table 7.3 the values consistently higher than 6.6 are reported in pale pink, and they represent an overestimation of copper with respect to Cp values. Values consistently lower than 6.6 are represented in pale blue and represent an underestimation of copper with respect to Cp values

[a]Data concerning Cp–Cu have been calculated following the Walshe equation [62]. Non-Cp–Cu was calculated subtracting Cp–Cu from Cu as reported by [62] and was expressed in terms of negative or positive values. The Rembach et al. study [40] is the only study reporting mean value <0

[b]Cu:Cp ratio is calculated as reported in Twomey et al. [41]. These authors provide the equation as follows: [copper (μM)] × [132,000 g/mol]/[Cp (mg/dL) × $10^4$]

indicates an underestimation of copper with respect to ceruloplasmin values, pointing out that in the AIBL study ceruloplasmin values were not in line with the other studies. Ethnicity may account for this result. In fact, as described by AIBL group, Australia is a geographically isolated country with a different ethnic mix and AIBL participants reflected this admixture [42]. Serum ceruloplasmin levels are affected by genetics [43], and several studies have demonstrated that genetic variation among human populations is associated with significant functional differences in health-related aspects [44]. Among human populations, Australians showed a very complex genetic structure [45], which, thereby, may explain the differences between AIBL results and those from the other studies.

We tested the hypothesis that free copper increase may allow distinguishing individuals affected by MCI from healthy individuals [35]. That study revealed that free copper and the *APOE* genotype differentiate MCI from healthy subjects, while free copper and *APOEε4* differentiate MCI individuals also from AD patients and associate to MMSE worsening, as do age and sex [28]. This is in agreement with recent data demonstrating that an increase in the ratio of serum copper to non-heme iron can predict MCI conversion to AD [46, 47].

An interesting insight into this issue came out of a study of ours on normal women over age 50 [28]. We measured their levels of free and ceruloplasmin-bound copper and studied the latter's relationship with MMSE scores and with a battery of 19 neuropsychological tests. We found that free (but not ceruloplasmin-bound) copper levels correlated inversely with MMSE scores and with all memory-related tests.

Furthermore, since homocysteine is an accepted risk factor for atherosclerosis and has been reported to be linked to AD risk, we studied its levels in patients with cognitive impairment both in relation to carotid intima-media thickness and C677T methylenetetrahydrofolate reductase (*MTHFR*) polymorphism [48] and to EEG rhythms in awake resting subjects [28]. We found homocysteine to be higher in cognitively disturbed subjects and to be associated with the typical slowing of EEG rhythms, which characterize AD. The increased levels of plasma homocysteine appeared to be promoted by the TT677 *MTHFR* genotype and to affect the thickening of the carotid intima-media. Homocysteine may disturb brain function through diverse pathways, including microvascular damage, and direct or indirect toxic interaction with copper. A direct interaction of homocysteine with copper has been already demonstrated in cell cultures where homocysteine selectively potentiated toxicity even at low copper concentration [49].

### 7.7.3 ATP7B as a New Potential Genetic Risk Factor for AD

Maintenance of body copper homeostasis requires copper transporters intensingly working in the liver to supply copper to cuproproteins. P-type ATPases, and in particular ATP7B, are copper efflux pumps that regulate the amount of copper leaving the hepatocyte via bile canaliculi and copper supply to nascent ceruloplasmin. Defects in copper packing into ceruloplasmin increase the amount of free copper released into blood circulation. We demonstrated increased apo-ceruloplasmin in the serum of AD patients [28] and signs of liver hypofunction ascribable to disturbing effects of free copper on hepatocytes [38]. Moreover, the free copper increase was tightly related to the AD clinical picture, revealing direct effects on brain functions [35, 37, 50], which imply increased free copper levels inside the brain [51]. We also found signs of free copper moving from serum to CSF through the BBB [28] and of free copper correlations with CSF markers of AD [28]. It is well known from WD studies that ATP7B failure causes serum free copper to increase beyond the normal reference range (>1.6 μmol/L). Thus we hypothesized that LoF variants in *ATP7B* gene may cause the increased free copper that we observed in AD previously, strongly supporting free copper as a causative risk factor for AD, accelerating pathogenic processes.

Our first study on this topic analyzed the ATP7B sequence changes in 2, 5, 8, 10, 14, and 16 exons of the *ATP7B* gene, where most of the Mediterranean mutations causing WD are located [17].

In a group of mild–moderate AD patients, we found high frequencies of the minor allele in two SNPs causing non-synonymous substitutions in the protein: the rs1801243 (c.1216T>G) associated with amino acid change serine to alanine in position 406 and the rs1061472 (c.2495A>G) that cause the amino acid substitution of lysine to arginine in position 832. Subsequently, we studied another non-synonymous change arginine in lysine in position 952 (rs732774).

In a larger study population, we confirmed the significant association of rs1061472 genotypes and AD risk and revealed an association for the rs772774 [52].

No other genetic association studies explored the *ATP7B* hypothesis, and no GWAS found significant SNPs in the chromosomal region where *ATP7B* is located. The GWAS outcome may be due to some limitations of this approach and to complex structure of *ATP7B* gene. The linear modeling framework often used for GWASs usually considers only one SNP at a time, thus ignoring the genomic and environmental context of each SNP.

Thus GWAS success depends critically on the assumptions made about disease complexity [53].

Furthermore, the GWASs hardly detect rare variants, which are instead supposed to account for the missing hereditability of complex diseases [53].

The genetic studies on *ATP7B* gene revealed a great heterogeneity in this locus, in which a number of common, uncommon, and rare variants are present [53].

On these items, we hypothesized that *ATP7B* gene may hide multiple rare variants associated with large effect sizes for AD [53], and we are currently working to verify this hypothesis and to test the interaction between these genetic variants and copper-related biochemical variables in AD patients and in elderly healthy controls.

## 7.8 Conclusions

The above-described work shows that a systemic metal disarrangement occurs in AD and that even though copper dysfunctions cannot be assumed as the cause of the disease, a causative rather than associative role can be claimed in terms of risk factor.

This conclusion is supported by solid clinical, epidemiological, experimental, meta-analytical, and genetic [28] data. However, further research is needed to understand the role of *ATP7B* as a potential harbor of rare variants which could account for a percentage of the missing hereditability of AD [53].

## References


1. Wyman S, Simpson RJ, McKie AT, Sharp PA. Dcytb (Cybrd1) functions as both a ferric and a cupric reductase in vitro. FEBS Lett. 2008;582(13):1901–6. doi:S0014-5793(08)00410-9 [pii] 10.1016/j.febslet.2008.05.010.
2. Sacher A, Cohen A, Nelson N. Properties of the mammalian and yeast metal-ion transporters DCT1 and Smf1p expressed in *Xenopus laevis* oocytes. J Exp Biol. 2001;204(Pt 6):1053–61.
3. Ganz T, Nemeth E. Hepcidin and disorders of iron metabolism. Annu Rev Med. 2011;62:347–60. doi:10.1146/annurev-med-050109-142444.
4. Nemeth E, Tuttle MS, Powelson J, Vaughn MB, Donovan A, Ward DM, Ganz T, Kaplan J. Hepcidin regulates cellular iron efflux by binding to ferroportin and inducing its internalization. Science. 2004;306(5704):2090–3. doi:1104742 [pii] 10.1126/science.1104742.
5. Linder MC, Hazegh-Azam M. Copper biochemistry and molecular biology. Am J Clin Nutr. 1996;63(5):797S–811.
6. Brewer GJ. Toxicity of copper in drinking water. J Toxicol Environ Health B Crit Rev. 2010;13 (6):449–52; author reply 453–49. doi:925758771 [pii] 10.1080/10937404.2010.499732
7. Kim BE, Nevitt T, Thiele DJ. Mechanisms for copper acquisition, distribution and regulation. Nat Chem Biol. 2008;4(3):176–85. doi:nchembio.72 [pii] 10.1038/nchembio.72.
8. Collins JF, Franck CA, Kowdley KV, Ghishan FK. Identification of differentially expressed genes in response to dietary iron deprivation in rat duodenum. Am J Physiol Gastrointest Liver Physiol. 2005;288(5):G964–71. doi:00489.2004 [pii] 10.1152/ajpgi.00489.2004.
9. Jacobs DS, Oxley DK, DeMott WR. Laboratory test handbook concise with disease index. 2nd ed. Hudson: Lexi-Comp Inc; 2002.
10. Wang J, Pantopoulos K. Regulation of cellular iron metabolism. Biochem J. 2011;434(3):365–81. doi:BJ20101825 [pii] 10.1042/BJ20101825.
11. Aisen P. Transferrin receptor 1. Int J Biochem Cell Biol. 2004;36(11):2137–43. doi:10.1016/j.biocel.2004.02.007 S1357272504000615 [pii].
12. Lill R. Function and biogenesis of iron-sulphur proteins. Nature. 2009;460(7257):831–8. doi:nature08301 [pii] 10.1038/nature08301.
13. Muckenthaler MU, Galy B, Hentze MW. Systemic iron homeostasis and the iron-responsive element/iron-regulatory protein (IRE/IRP) regulatory network. Annu Rev Nutr. 2008;28:197–213. doi:10.1146/annurev.nutr.28.061807.155521.
14. Moos T, Rosengren Nielsen T, Skjorringe T, Morgan EH. Iron trafficking inside the brain. J Neurochem. 2007;103(5):1730–40. doi:JNC4976 [pii] 10.1111/j.1471-4159.2007.04976.x.
15. Wu LJ, Leenders AG, Cooperman S, Meyron-Holtz E, Smith S, Land W, Tsai RY, Berger UV, Sheng ZH, Rouault TA. Expression of the iron transporter ferroportin in synaptic vesicles and the blood-brain barrier. Brain Res. 2004;1001(1–2):108–17. doi:10.1016/j.brainres.2003.10.066 S0006899303041234 [pii].
16. Michalke B, Nischwitz V. Review on metal speciation analysis in cerebrospinal fluid-current methods and results: a review. Anal Chim Acta. 2010;682(1–2):23–36. doi:S0003-2670(10)01237-7 [pii] 10.1016/j.aca.2010.09.054.
17. Bucossi S, Ventriglia M, Panetta V, Salustri C, Pasqualetti P, Mariani S, Siotto M, Rossini PM, Squitti R. Copper in Alzheimer's disease: a meta-analysis of serum, plasma, and cerebrospinal fluid studies. J Alzheimers Dis. 2011;24(1):175–85. doi:R474R375HXT67344 [pii] 10.3233/JAD-2010-101473.

18. Choi BS, Zheng W. Copper transport to the brain by the blood-brain barrier and blood-CSF barrier. Brain Res. 2009;1248:14–21. doi:S0006-8993(08)02634-6 [pii] 10.1016/j.brainres.2008.10.056.
19. Schlief ML, Gitlin JD. Copper homeostasis in the CNS: a novel link between the NMDA receptor and copper homeostasis in the hippocampus. Mol Neurobiol. 2006;33(2):81–90. doi:MN:33:2:81 [pii] 10.1385/MN:33:2:81.
20. Gaggelli E, Kozlowski H, Valensin D, Valensin G. Copper homeostasis and neurodegenerative disorders (Alzheimer's, prion, and Parkinson's diseases and amyotrophic lateral sclerosis). Chem Rev. 2006;106(6):1995–2044. doi:10.1021/cr040410w.
21. Bielli P, Calabrese L. Structure to function relationships in ceruloplasmin: a 'moonlighting' protein. Cell Mol Life Sci. 2002;59(9):1413–27.
22. Hoogenraad T. Wilson's disease. Amsterdam: Intermed Medical Publishers; 2001.
23. Madsen E, Gitlin JD. Copper and iron disorders of the brain. Annu Rev Neurosci. 2007;30:317–37. doi:10.1146/annurev.neuro.30.051606.094232.
24. Vassiliev V, Harris ZL, Zatta P. Ceruloplasmin in neurodegenerative diseases. Brain Res Brain Res Rev. 2005;49(3):633–40. doi:S0165-0173(05)00046-9 [pii] 10.1016/j.brainresrev.2005.03.003.
25. Schrag M, Mueller C, Oyoyo U, Smith MA, Kirsch WM. Iron, zinc and copper in the Alzheimer's disease brain: a quantitative meta-analysis. Some insight on the influence of citation bias on scientific opinion. Prog Neurobiol. 2011;94(3):296–306. doi:S0301-0082(11)00072-4 [pii] 10.1016/j.pneurobio.2011.05.001.
26. Bartzokis G, Tishler TA. MRI evaluation of basal ganglia ferritin iron and neurotoxicity in Alzheimer's and Huntington's disease. Cell Mol Biol (Noisy-le-Grand). 2000;46(4):821–33.
27. Bush AI, Tanzi RE. Therapeutics for Alzheimer's disease based on the metal hypothesis. Neurotherapeutics. 2008;5(3):421–32. doi:S1933-7213(08)00090-1 [pii] 10.1016/j.nurt.2008.05.001.
28. Squitti R. Metals in Alzheimer's disease: a systemic perspective. Front Biosci. 2012;17:451–72. doi:3938 [pii].
29. Atwood CS, Perry G, Zeng H, Kato Y, Jones WD, Ling KQ, Huang X, Moir RD, Wang D, Sayre LM, Smith MA, Chen SG, Bush AI. Copper mediates dityrosine cross-linking of Alzheimer's amyloid-beta. Biochemistry. 2004;43(2):560–8. doi:10.1021/bi0358824.
30. Roberts BR, Ryan TM, Bush AI, Masters CL, Duce JA. The role of metallobiology and amyloid-beta peptides in Alzheimer's disease. J Neurochem. 2012;120 Suppl 1:149–66. doi:10.1111/j.1471-4159.2011.07500.x.
31. Rogers JT, Lahiri DK. Metal and inflammatory targets for Alzheimer's disease. Curr Drug Targets. 2004;5(6):535–51.
32. Moalem S, Percy ME, Andrews DF, Kruck TP, Wong S, Dalton AJ, Mehta P, Fedor B, Warren AC. Are hereditary hemochromatosis mutations involved in Alzheimer disease? Am J Med Genet. 2000;93(1):58–66. doi:10.1002/1096-8628(20000703)93:1<58::AID-AJMG10>3.0.CO;2-L [pii].
33. Giambattistelli F, Bucossi S, Salustri C, Panetta V, Mariani S, Siotto M, Ventriglia M, Vernieri F, Dell'acqua ML, Cassetta E, Rossini PM, Squitti R. Effects of hemochromatosis and transferrin gene mutations on iron dyshomeostasis, liver dysfunction and on the risk of Alzheimer's disease. Neurobiol Aging. 2011. doi:S0197-4580(11)00062-5 [pii] 10.1016/j.neurobiolaging.2011.03.005.
34. Bellingham SA, Lahiri DK, Maloney B, La Fontaine S, Multhaup G, Camakaris J. Copper depletion down-regulates expression of the Alzheimer's disease amyloid-beta precursor protein gene. J Biol Chem. 2004;279(19):20378–86. doi:10.1074/jbc.M400805200 M400805200 [pii].
35. Squitti R, Salustri C, Siotto M, Ventriglia M, Vernieri F, Lupoi D, Cassetta E, Rossini PM. Ceruloplasmin/Transferrin ratio changes in Alzheimer's disease. Int J Alzheimers Dis. 2010;2011:231595. doi:10.4061/2011/231595.
36. Squitti R, Barbati G, Rossi L, Ventriglia M, Dal Forno G, Cesaretti S, Moffa F, Caridi I, Cassetta E, Pasqualetti P, Calabrese L, Lupoi D, Rossini PM. Excess of nonceruloplasmin serum copper in AD correlates with MMSE, CSF [beta]-amyloid, and h-tau. Neurology. 2006;67(1):76–82. doi:67/1/76 [pii] 10.1212/01.wnl.0000223343.82809.cf.
37. Zappasodi F, Salustri C, Babiloni C, Cassetta E, Del Percio C, Ercolani M, Rossini PM, Squitti R. An observational study on the influence of the APOE-epsilon4 allele on the correlation

between 'free' copper toxicosis and EEG activity in Alzheimer disease. Brain Res. 2008;1215:183–9. doi:S0006-8993(08)00820-2 [pii] 10.1016/j.brainres.2008.03.066.
38. Squitti R, Ventriglia M, Barbati G, Cassetta E, Ferreri F, Dal Forno G, Ramires S, Zappasodi F, Rossini PM. 'Free' copper in serum of Alzheimer's disease patients correlates with markers of liver function. J Neural Transm. 2007;114(12):1589–94. doi:10.1007/s00702-007-0777-6.
39. Ventriglia M, Bucossi S, Panetta V, Squitti R. Copper in Alzheimer's disease: a meta-analysis of serum, plasma, and cerebrospinal fluid studies. J Alzheimers Dis. 2012. doi:39R1H6VR87U10373 [pii] 10.3233/JAD-2012-120244.
40. Rembach A, Doecke JD, Roberts BR, Watt AD, Faux NG, Volitakis I, Pertile KK, Rumble RL, Trounson BO, Fowler CJ, Wilson W, Ellis KA, Martins RN, Rowe CC, Villemagne VL, Ames D, Masters CL, Bush AI. Longitudinal analysis of serum copper and ceruloplasmin in Alzheimer's disease. J Alzheimers Dis. 2012;2:8. doi:K107150570452374 [pii] 10.3233/JAD-121474.
41. Twomey PJ, Viljoen A, House IM, Reynolds TM, Wierzbicki AS. Copper:caeruloplasmin ratio. J Clin Pathol. 2007;60(4):441–2. doi:60/4/441 [pii] 10.1136/jcp.2006.041756.
42. Gardener S, Gu Y, Rainey-Smith SR, Keogh JB, Clifton PM, Mathieson SL, Taddei K, Mondal A, Ward VK, Scarmeas N, Barnes M, Ellis KA, Head R, Masters CL, Ames D, Macaulay SL, Rowe CC, Szoeke C, Martins RN. Adherence to a Mediterranean diet and Alzheimer's disease risk in an Australian population. Transl Psychiatry. 2012;2:e164. doi:tp201291 [pii] 10.1038/tp.2012.91.
43. Tang WH, Hartiala J, Fan Y, Wu Y, Stewart AF, Erdmann J, Kathiresan S, Roberts R, McPherson R, Allayee H, Hazen SL. Clinical and genetic association of serum paraoxonase and arylesterase activities with cardiovascular risk. Arterioscler Thromb Vasc Biol. 2012;32(11):2803–12. doi:ATVBAHA.112.253930 [pii] 10.1161/ATVBAHA.112.253930.
44. Polimanti R, Piacentini S, Manfellotto D, Fuciarelli M. Human genetic variation of CYP450 superfamily: analysis of functional diversity in worldwide populations. Pharmacogenomics. 2012;13(16):1951–60. doi:10.2217/pgs.12.163.
45. Ballantyne KN, van Oven M, Ralf A, Stoneking M, Mitchell RJ, van Oorschot RA, Kayser M. MtDNA SNP multiplexes for efficient inference of matrilineal genetic ancestry within Oceania. Forensic Sci Int Genet. 2012;6(4):425–36. doi:S1872-4973(11)00183-9 [pii] 10.1016/j.fsigen.2011.08.010.
46. Mueller C, Schrag M, Crofton A, Stolte J, Muckenthaler MU, Magaki S, Kirsch W. Altered serum iron and copper homeostasis predicts cognitive decline in mild cognitive impairment. J Alzheimers Dis. 2012;29(2):341–50. doi:N021303923584N63 [pii] 10.3233/JAD-2011-111841.
47. Lopez N, Tormo C, De Blas I, Llinares I, Alom J. Oxidative stress in Alzheimer's disease and mild cognitive impairment with high sensitivity and specificity. J Alzheimers Dis. 2013;33(3):823–9. doi:W27083817P732L45 [pii] 10.3233/JAD-2012-121528.
48. Gorgone G, Ursini F, Altamura C, Bressi F, Tombini M, Curcio G, Chiovenda P, Squitti R, Silvestrini M, Ientile R, Pisani F, Rossini PM, Vernieri F. Hyperhomocysteinemia, intima-media thickness and C677T MTHFR gene polymorphism: a correlation study in patients with cognitive impairment. Atherosclerosis. 2009;206(1):309–13. doi:S0021-9150(09)00155-5 [pii] 10.1016/j.atherosclerosis.2009.02.028.
49. White AR, Huang X, Jobling MF, Barrow CJ, Beyreuther K, Masters CL, Bush AI, Cappai R. Homocysteine potentiates copper- and amyloid beta peptide-mediated toxicity in primary neuronal cultures: possible risk factors in the Alzheimer's-type neurodegenerative pathways. J Neurochem. 2001;76(5):1509–20.
50. Squitti R, Bressi F, Pasqualetti P, Bonomini C, Ghidoni R, Binetti G, Cassetta E, Moffa F, Ventriglia M, Vernieri F, Rossini PM. Longitudinal prognostic value of serum "free" copper in patients with Alzheimer disease. Neurology. 2009;72(1):50–5. doi:72/1/50 [pii] 10.1212/01.wnl.0000338568.28960.3f.
51. James SA, Volitakis I, Adlard PA, Duce JA, Masters CL, Cherny RA, Bush AI. Elevated labile Cu is associated with oxidative pathology in Alzheimer disease. Free Radic Biol Med. 2012;52(2):298–302. doi:S0891-5849(11)01110-5 [pii] 10.1016/j.freeradbiomed.2011.10.446.

52. Bucossi S, Polimanti R, Mariani S, Ventriglia M, Bonvicini C, Migliore S, Manfellotto D, Salustri C, Vernieri F, Rossini PM, Squitti R. Association of K832R and R952K SNPs of Wilson's Disease Gene with Alzheimer's Disease. J Alzheimers Dis. 2012;29(4):913–9. doi:D729J362Q40W53R3 [pii] 10.3233/JAD-2012-111997.
53. Squitti R, Polimanti R. Copper Hypothesis in the Missing Hereditability of Sporadic Alzheimer's Disease: ATP7B Gene as Potential Harbor of Rare Variants. J Alzheimers Dis. 2012;29(3):493–501. doi:NM01V338M14607R1 [pii] 10.3233/JAD-2011-111991.
54. Arnal N, Cristalli DO, de Alaniz MJ, Marra CA. Clinical utility of copper, ceruloplasmin, and metallothionein plasma determinations in human neurodegenerative patients and their first-degree relatives. Brain Res. 2010;1319:118–30. doi:S0006-8993(09)02666-3 [pii] 10.1016/j.brainres.2009.11.085.
55. Brewer GJ, Kanzer SH, Zimmerman EA, Celmins DF, Heckman SM, Dick R. Copper and ceruloplasmin abnormalities in Alzheimer's disease. Am J Alzheimers Dis Other Demen. 2010;25(6):490–7. doi:1533317510375083 [pii] 10.1177/1533317510375083.
56. Agarwal R, Kushwaha SS, Tripathi CB, Singh N, Chhillar N. Serum copper in Alzheimer's disease and vascular dementia. Indian J Clin Biochem. 2008;23(4):369–74. doi:10.1007/s12291-008-0081-8 81 [pii].
57. Sedighi B, Shafa MA, Shariati M. A study of serum copper and ceruloplasmin in Alzheimer's disease in Kerman, Iran. Neurol Asia. 2006;11:107–9.
58. Snaedal J, Kristinsson J, Gunnarsdottir S, Olafsdottir BM, Johannesson T. Copper, ceruloplasmin and superoxide dismutase in patients with Alzheimer's disease. a case-control study. Dement Geriatr Cogn Disord. 1998;9(5):239–42. doi:dem09239 [pii].
59. Molaschi M, Ponzetto M, Bertacna B, Berrino E, Ferrario E. Determination of selected trace elements in patients affected by dementia. Arch Gerontol Geriatr. 1996;22 Suppl 1:39–42. doi:0167-4943(96)86910-X [pii] 10.1016/0167-4943(96)86910-X.
60. Squitti R, Pasqualetti P, Dal Forno G, Moffa F, Cassetta E, Lupoi D, Vernieri F, Rossi L, Baldassini M, Rossini PM. Excess of serum copper not related to ceruloplasmin in Alzheimer disease. Neurology. 2005;64(6):1040–6. doi:64/6/1040 [pii] 10.1212/01.WNL.0000154531.79362.23.
61. Squitti R, Ghidoni R, Scrascia F, Benussi L, Panetta V, Pasqualetti P, Moffa F, Bernardini S, Ventriglia M, Binetti G, Rossini PM. Free copper distinguishes mild cognitive impairment subjects from healthy elderly individuals. J Alzheimers Dis. 2011;23(2):239–48. doi:6716PK524717404N [pii] 10.3233/JAD-2010-101098.
62. Walshe JM. Wilson's disease: the importance of measuring serum caeruloplasmin non-immunologically. Ann Clin Biochem. 2003;40(Pt 2):115–21. doi:10.1258/000456303763046021.

# Chapter 8
# Brain Oxidative Stress in the Pathogenesis and Progression of Alzheimer's Disease

**Rukhsana Sultana, Aaron M. Swomley, and D. Allan Butterfield**

**Abstract** Alzheimer's disease is an age-related neurodegenerative disease and is characterized by the presence of senile plaques (SP), neurofibrillary tangles, and synapse loss. Amyloid-beta, one of the main components of SP, has been known to induce oxidative stress and is highly toxic. Using redox proteomics approaches a number of oxidatively modified proteins were identified in AD and mild cognitive impairment (MCI) brain that are consistent with the clinical presentation, pathology, and biochemistry. The identification of key proteins that are highly susceptible to amyloid-beta-mediated oxidation might serve as biomarkers for use in diagnosing and also in the identification of therapeutic targets to prevent or delay this devastating disorder.

## Abbreviation

| | |
|---|---|
| 3-NT | 3-nitro tyrosine |
| AD | Alzheimer's disease |
| ADDL | β-amyloid-derived-diffusible ligands |
| AGEs | Advance glycation end products |
| AICD | APPs intracellular c-terminal domain |
| APC | Anaphase promoting complex |
| *APOE 4* | *Apolipoprotein E* allele 4 |
| APP | Amyloid precursor protein |
| Aβ | Amyloid beta-peptide |
| AβPP | β-amyloid precursor protein |

R. Sultana, Ph.D. • A.M. Swomley, Ph.D. • D.A. Butterfield, Ph.D. (✉)
Department of Chemistry, Center of Membrane Sciences, The Sanders-Brown Center on Aging, University of Kentucky, Lexington, KY 40506-0055, USA
e-mail: dabcns@uky.edu

D. Praticò and P. Mecocci (eds.), *Studies on Alzheimer's Disease*, Oxidative Stress in Applied Basic Research and Clinical Practice, DOI 10.1007/978-1-62703-598-9_8,

| | |
|---|---|
| BR | Bilirubin-IX-alpha |
| BRCA 1 | Breast cancer type-1 susceptibility protein |
| BV | Biliverdin-IX-alpha |
| BVR-A | Biliverdin reductase-A |
| CBP | Creb response binding protein |
| CDKs | Cyclin-dependent kinases |
| CEL | *N*-carboxyethyl-lysine |
| CML | Nε-(carboxymethyl) lysine |
| c-MYC | Cellular-myelocytomatosis |
| CR | Carbonyl reductase |
| CSF | Cerebral spinal fluid |
| DMDMAH-1 | Dimethylarginine dimethylaminohydrolase 1 |
| DRP-2 | Dihydropyrimidinase-related protein 2 |
| EOAD | Early onset-AD |
| $F_2$-IsoP | $F_2$-isoprostane |
| $F_4$-NP | $F_4$-Neuroprostane |
| GAPDH | Glyceraldehyde-3-phosphate dehydrogenase |
| Glycer-AGE | Glyceraldehyde-derived AGEs |
| GPX | Glutathione peroxidase |
| GRD | Glutathione reductase |
| GRP | Glucose regulated protein precursor |
| GSH | Glutathione |
| GSK3β | Glycogen synthase kinase 3-β |
| GST M | Glutathione-S-transferase Mu |
| HNE | 4-hydroxy 2-trans-nonenal |
| HNE-GSH | HNE-glutathione |
| HO-1 | Heme oxygenase-1 |
| HO-2 | Heme oxygenase-2 |
| HSPA8 | Heat shock protein A8 |
| iASPP | Inhibitor of apoptosis-stimulating protein of p53 (iASPP) |
| IDE | Insulin degrading enzyme |
| IGF-1 | Insulin growth factor-1 (IGF-1) |
| IKKβ | IkB kinase type β |
| iNOS | Inducible nitric oxide synthase |
| IPL | Inferior parietal lobule |
| LDH | Lactate dehydrogenase |
| LOAD | Late onset-AD |
| MAPK | Mitogen-activated protein kinase |
| MCI/aMCI | Non-amnestic/amnestic mild-cognitive impairment |
| MDA | Malondialdehyde |
| MDH | Malate dehydrogenase 1 (MDH) |
| Mdm-2 | Murine double minute-2 |
| MMSE | Mini-mental state examination |
| MnSOD | Manganese superoxide dismutase |
| MRP3 | Multidrug resistant protein-3 |
| mtDNA | Mitochondrial DNA |

| | |
|---|---|
| nDNA | Nuclear DNA |
| NF-E2 | Nuclear factor-erythroid 2 |
| NFT | Neurofibrillary tangles |
| NF-κB | Nuclear factor kappa-B |
| NINCDS-ADRDA | National Institute of Neurological and Communicative Disorders and Stroke and the Alzheimer's Disease and Related Disorders Association |
| NMDA | *N*-methyl-D-aspartate |
| NO | Nitric oxide |
| NOS | Nitric oxide synthase |
| NPrG | 1-*N*2-propanodeoxyguanosine |
| Nrf-2 | Nuclear factor related factor 2 |
| PDI | Protein disulfide isomerase |
| PICALM | Phosphatidylinositol-binding clathrin assembly protein |
| Pin1 | Peptidyl prolyl cis-trans isomerase |
| PLK | Polo-like kinase |
| PP2A | Protein phosphatase 2-A |
| PPIase | Peptidyl-prolyl *cis/trans* isomerase |
| PRVI | Peroxiredoxin 6 |
| PS-1 | *Presenilin-1* |
| PS-2 | *presenilin-2* |
| Rb | Retinoblastoma protein |
| RNS | Reactive nitrogen species |
| ROS | Reactive oxygen species |
| rRNA | Ribosomal RNA |
| RSNO | S-nitrosothiols |
| SBP1 | Syntaxin binding protein I |
| $SCF^{skp2}$ | Skp cullin, F-box containing complex |
| SOD | Superoxide dismutase |
| SOD1/SOD2 | Superoxide dismutase1/2 |
| Thio-1 | Thioredoxin-1 |
| TNF-α | Tumor necrosis factor-α |
| TPI | Triose phosphate isomerase (TPI) |
| tRNA | transfer RNA |
| UCH L-1 | Ubiquitin carboxy-terminal hydrolase L-1 |
| UPR | Unfolded protein response |
| VDAC | Voltage-dependent anion channel |

## 8.1 Introduction

Alzheimer's disease (AD) is the sixth leading cause of death in the USA and is the most common form of dementia. This disease currently is estimated to affect 5.1 million people aged 65 and older in the USA. With approximately 80 million people

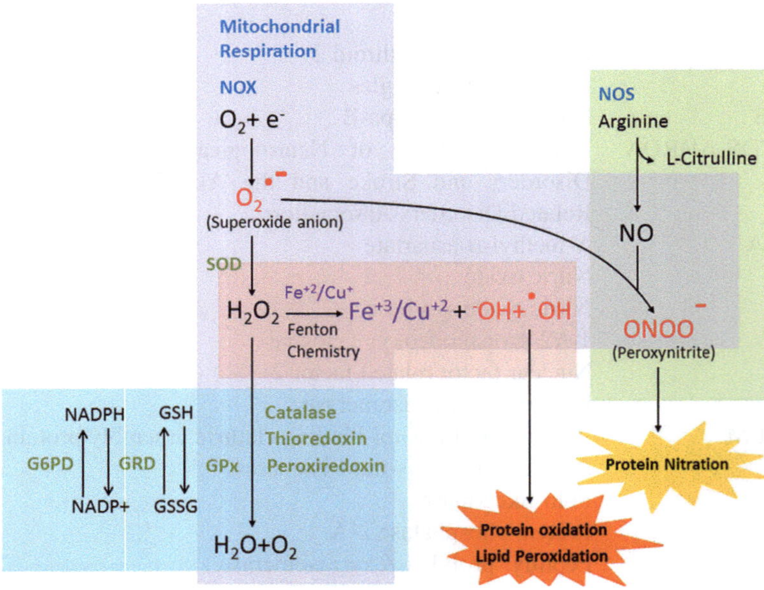


**Fig. 8.1** Presented above are important pathways in the production (names represented in *blue*) and breakdown (names of antioxidant enzymes in *green*) of ROS/RNS (compounds represented in *red*) that lead to damaging posttranslational modifications of biomolecules

in the "Baby Boom" population, it is expected that the number of AD patients will increase to as many as 20 million by 2050 unless means to delay the onset or progression of the disease are developed [41] (Fig. 8.1).

## 8.2 Alzheimer's Disease

AD is histopathologically characterized by the presence of abnormal protein deposits, including senile plaques (rich in amyloid β-peptide [Aβ]) and neurofibrillary tangles (NFT, rich in hyperphosphoryalted tau) [71], and synaptic loss. Structural MRI studies demonstrate early brain atrophy in AD that is predominant in hippocampus, precuneus, temporal and parietal lobes, parts of the frontal cortex and cingulate gyrus. In contrast, the cerebellum typically displays minor, if any, changes in AD brain compared to controls [37]. FDG-PET analysis demonstrates a characteristic pattern of decreased glucose metabolism in parietal-temporal association cortices [62].

Diagnosis of AD is established by criteria outlined by the National Institute of Neurological and Communicative Disorders and Stroke and the Alzheimer's Disease and Related Disorders Association (NINCDS-ADRDA) in which both possible and probable AD is diagnosed using cognitive examinations with the exclusion of other possible causes, while definite diagnosis is obtained postmortem or via biopsy. Within the realm of possible and probable AD, there are also subclassifications that have been recognized such as non-amnestic/amnestic mild-cognitive impairment (MCI/aMCI), early onset-AD (EOAD), and late onset-AD (LOAD).

## 8.3 Mild Cognitive Impairment

MCI is arguably the earliest form of AD. Patients who have developed aMCI have an increased probability of developing clinical AD [58], yet it remains a challenge to identify which patients with MCI will proceed to clinical AD and which patients will not [43]. In this stage of the disease, patients are found to have decreased hippocampal volume based on MRI measurements [42], decreased energy metabolism, increased number of NFTs and subsequent Braak staging (Stage II–III) [11], lowered MMSE scores, increased levels of amyloid deposition correlated with both non-amnestic MCI as well as amnestic MCI, as well as cell cycle protein oxidation and general cycle aberrations [35, 76].

The exact molecular mechanisms that lead to the loss of neurons and development of AD pathology are still unclear. Mutations of *presenilin-1* (PS-1), *presenilin-2* (PS-2), and *APP* genes have been reported to cause familial AD (FAD). In addition, other genes, such as *apolipoprotein E* allele 4 (*APOE 4*), *clusterin* (CLU aka APOJ), *phosphatidylinositol-binding clathrin assembly protein* (PICALM), *endothelial nitric oxide synthase-3*, and *alpha-2-macroglobulin,* have been suggested as risk factors for AD.

A number of hypotheses have been proposed to link the pathologic lesions, neuronal histopathology, biochemistry, and clinical symptoms of AD, including the amyloid cascade, excitotoxicity, oxidative stress, and inflammation hypotheses. All these hypotheses are based, to some extent, on the role of Aβ. The oxidative stress hypothesis for the pathogenesis and progression of AD [14, 18, 55] is based on the observations of increased cellular free radical production in AD. Such production is proposed to exceed the capacity to scavenge or otherwise neutralize these damaging moieties, leading to neuronal dysfunction and death that is mediated by protein oxidation, lipid peroxidation, and nucleic acid oxidation [20].

## 8.4 Amyloid Beta-Peptide (Aβ), Oxidative Stress, and AD Pathogenesis

The Aβ-induced oxidative stress hypothesis in AD [14, 55] is supported by Aβ-induced elevation of oxidative stress markers in brain, and subsequent neuronal degeneration [38]. Senile plaques are composed of a core of Aβ surrounded by degenerating neurites. Aβ is derived by the proteolytic cleavage of amyloid precursor protein (APP) by β- and γ-secretases. Both Aβ(1-40) and Aβ(1-42) peptides, among other Aβ peptides, are found in human brain and the former two are the most abundant forms of this peptide. Aβ exists in at least four aggregated states: monomers, oligomers, protofibrils, and fibrils. Recent data suggest that oligomeric Aβ is the toxic species of this peptide rather than Aβ fibrils themselves [39].

Previous studies [13, 18] have identified the single methionine residue at position 35 of Aβ(1-42) as a critical mediator of increased oxidative stress and neurotoxic properties of this peptide. Recent in vivo from our laboratory showed for the first the time methionine is critical to induce oxidative stress and consequently, a key player in amyloid β-induced oxidative stress and AD pathogenesis [81].

## 8.5 Evidence of Oxidative Stress in Brain of Subjects with AD and MCI

The atrophy, indicative of neuronal loss, observed in AD and MCI brain by MRI studies correlates well with increased levels of oxidative stress markers in AD and MCI brain. Further, the specific targets of oxidative damage reported to date are linked to the biochemical, clinical, and pathological development of MCI and AD, and are discussed further below.

### *8.5.1 Protein Carbonyls*

Brain protein carbonyls are increased in AD [28, 31, 74]. In the frontal cortex of Swedish APP670/671 FAD mutation, increased levels of protein carbonyls, diene conjugates, and lipid peroxides were found [8]. Further, the levels of carbonyl reductase (CR) are increased in brain of AD and Down syndrome subjects [3], suggesting enzyme induction due to increased levels of protein carbonyls. The authors of this report hypothesized a possible role of Aβ in this observed induction; however, the activity of CR was not measured, and the mechanism(s) accounting for increased CR levels were not identified. Since the 20S proteasome is known to degrade oxidized proteins normally, it is conceivable that oxidative inhibition of the 20S proteasome, similar to the case for the 26S proteasome in AD [47], may occur. Reduced degradation of protein carbonyls by the proteasome complex may contribute to the elevated protein carbonyls found in AD brain.

Beta-actin and creatine kinase BB have been identified as specifically oxidized proteins in AD brain using 2D electrophoresis and 2D Western blots [1]. These techniques form the basis of the methodology needed to further examine the role of oxidative modifications of specific brain proteins in AD pathogenesis and have led to the development and use of redox proteomic [32] techniques to identify carbonylated brain proteins in AD [23]. 2D gel electrophoresis coupled to mass spectrometry [32, 77] have allowed the discovery of increased carbonylation of creatine kinase BB, glutamine synthase, ubiquitin carboxy-terminal hydrolase L-1 (UCH L-1), dihydropyrimidinase-related protein 2 (DRP-2, also called CRMP2), alpha-enolase, and heat shock cognate 71 in AD inferior parietal lobule (IPL) compared to age-matched controls [23]. Subsequent studies of AD hippocampus demonstrated specific carbonylation of peptidyl prolyl *cis–trans* isomerase (Pin1), phosphoglycerate mutase 1, UCH L-1, DRP-2, carbonic anhydrase II, triose phosphate isomerase (TPI), alpha-enolase, and gamma-SNAP compared to age-matched controls [74]. Consistent with the notion that oxidative modification of proteins leads to dysfunction of normal cellular processes in AD, the activities of Pin1, enolase, and carbonic anhydrase II were significantly lower in AD hippocampus compared to matched tissue samples from control subjects [74]. Others [49] using redox proteomics showed significant decreased protein carbonyls in malate dehydrogenase 1 (MDH), glutamate dehydrogenase, 14-3-3 protein zeta/delta, aldolases A and C and increased oxidation of carbonic anhydrase 1. The sample processing in this study did not use detergents, and may have led to identification of fewer oxidized proteins than that seen in other studies as a result of decreased exposure of protein carbonyls. In the IPL of FAD subjects, increased carbonylation of UCH-L1, gamma-enolase, actin, and dimethylarginine dimethylaminohydrolase 1 (DMDMAH-1) have been reported [17]. Others also reported oxidation and accumulation of proteins like UCH L1, ATP synthase, and Cu,Zn-superoxide dismutase in AD brain [28, 82], confirming our prior results.

MCI brain also demonstrates increased levels of protein carbonyls [17, 48]. Redox proteomics studies in MCI hippocampus led to the identification of alpha-enolase, glutamine synthetase, pyruvate kinase M2, and Pin1 as specifically carbonylated proteins recapitulating many of the findings seen in fulminate AD brain tissue [17]. Recent reports have also identified increased specific carbonylation of carbonic anhydrase II, heat shock protein 70, mitogen activated protein kinase I, and syntaxin binding protein I (SBP1) in MCI [78].

### 8.5.2 Protein Nitration

Increased protein nitration has been reported in AD brain [72], and correlates with increased nitric oxide synthase (NOS) levels, suggesting a role of nitration in AD pathophysiology [36]. Redox proteomics studies identified a large number of proteins that are specifically nitrated in AD hippocampal and IPL compared to control brain, including alpha- and gamma-enolase, lactate dehydrogenase (LDH), neuropolypeptide h3, TPI, and alpha-actin in AD IPL [24], and alpha-enolase, glyceraldehyde-3-phosphate dehydrogenase (GAPDH), ATP synthase alpha-chain, carbonic anhydrase-II, and voltage-dependent anion channel (VDAC) protein in AD hippocampus [79]. These nitrated proteins are involved in various cellular functions such as energy metabolism, structural maintenance, pH regulation, and mitochondrial function. Oxidative modification (i.e., nitration, carbonylation, etc.) may alter protein functionality [79]. Excess nitration of TPI was recently confirmed [40] in hippocampus and frontal cortex of AD subjects, suggesting a link among decreased glucose metabolism via an impaired glycolytic pathway, nitrosylation of TPI, and the formation of Aβ and paired helical filaments. However, it is not clear why, in spite of oxidative modification, its activity remains unchanged in AD brain. It is possible that this nitrosylation is a nonspecific marker of increased oxidative stress rather than a direct contributor to the development of AD. In contrast, Reyes et al. demonstrated nitration of Tyr 18 followed by Tyr 29 of tau, which is mostly associated with or in close proximity to amyloid plaques [64]. Hence, nitration of proteins may reflect underlying posttranslational modification of proteins in AD.

Consistent with this notion, increased levels of 3-NT in MCI hippocampus and IPL using immunochemistry were reported [80]. There is also evidence for AD-specific nitration of MDH, α enolase, glucose regulated protein precursor (GRP), aldolase, glutathione-S-transferase Mu (GST M), multidrug resistant protein-3 (MRP3), and 14-3-3 protein gamma in MCI IPL [80]. In MCI hippocampus, α enolase, MDH, peroxiredoxin 6 (PR VI), DRP-2, fascin 1, and heat shock protein A8 (HSPA8) were identified as specifically nitrated compared to age-matched controls [80]. A recent study [65] reported S-nitrosyl-cysteine modification of DRP-2, alpha-internexin, glutamate dehydrogenase 1, alpha-enolase, GFAP, MDH, ProSAAS precursor protein, proopiomelanocortin, proenkephalin, and septin in the entorhinal cortex of AD, and suggested that Aβ activation of glial cells surrounding the SP might have led to increased nitrosylation of GFAP contributing to the pathogenesis of AD. Protein disulfide isomerase (PDI), an enzyme that catalyzes thiol–disulphide exchange, has been reported to be *S*-nitrosylated in AD brain [84]. Increased nitrosylation and decreased activity of this protein in AD may lead to alteration in its ability to facilitate disulfide bond formation and rearrangement reactions, increased accumulation of polyubiquitinated proteins, and activation of the ER-resident unfolded protein response (UPR). Recently Cho et al. [27] reported increased levels of S-nitrosylation of dynamin-related protein 1 in brains of subjects with AD and suggested that S-nitrosylation of this protein may trigger mitochondrial fission, consequently adding to known mitochondrial damage in AD, which could contribute to synapse loss and neuronal damage in this disorder.

### 8.5.3 Lipid Peroxidation

Increased lipid peroxidation in AD brain has been identified by measuring elevated levels of free and protein-bound HNE, acrolein, $F_2$-isoprostane ($F_2$-IsoP), $F_4$-neuroprostane ($F_4$-NP), and isoprostane 8,12-iso-iP$F_2\alpha$-VI [56, 86]. Further, the increased levels of another marker of lipid peroxidation, malondialdehyde (MDA), in AD brain have been correlated with the decreased activity of superoxide dismutase (SOD) [22].

Increased levels of adducts of HNE and glutathione (HNE-GSH) were found in AD [85]. In normal cells the HNE-GSH adducts are removed by the combined action of GST, GSH, and MRP-1. However, in AD brain all these detoxification components were targets of HNE modification themselves, leading to decreased clearance of HNE and subsequent accumulation of HNE protein adducts [50, 75]. Further, the proteasome, which removes damaged proteins from the cells, has elevated HNE- and neuroprostane-conjugation in brain in both MCI and AD [25]. Further, increased levels of lipid peroxidation markers such as thiobarbituric acid reactive substances, MDA, $F_2$-IsoP, $F_4$-NP, and protein-bound HNE also were reported in subjects with MCI [19, 48, 56].

Proteomics studies identified regionally specific HNE modification of proteins, i.e., ATP synthase, GS, MnSOD, DRP-2 in AD hippocampus and α-enolase, aconitase, aldolase, peroxiredoxin 6, and α-tubulin in AD cortex [60]. Some of these proteins were previously found to either nitrated or carbonylated in AD [24, 74, 79]. The appearance of different oxidative modifications in common target proteins supports the role of oxidative stress in AD and is consistent with the notion that these specific proteins may be involved in AD.

In MCI hippocampus and cortex, increased levels of protein-bound HNE in neuropolypeptide h3, carbonyl reductase (NADPH), α-enolase, lactate dehydrogenase B, phosphoglycerate kinase, heat shock protein 70, ATP synthase alpha chain, pyruvate kinase, actin, elongation factor Tu, and translation initiation factor alpha were identified by proteomics [63]. Increased lipid peroxidation in AD and MCI brain and a role for Aβ(1-42) in this process were further supported by studies that showed loss of apoptosis-related phospholipid asymmetry in AD and MCI [2]. Noting that the high reactivity of free radicals requires that the initiator of lipid peroxidation must reside in the lipids, the findings above suggest that, in AD and MCI brain, oligomeric and hydrophobic Aβ (1-42) inserts into the membrane of brain cells to cause lipid peroxidation and that such changes are an early event in the pathogenesis and progression of AD.

The proteomics studies suggest that oxidation of proteins is an integral part of the progression and pathophysiology of AD [57]. The appearance of common targets of oxidation of proteins between MCI and AD implies their important roles in loss of cellular energetics, alterations in neurotransmission and cell signaling pathways, as well as SP and NFT formation. In the following section we discussed about some the proteins that critical in the progression and pathogenesis of AD.

## 8.6 Glucose Metabolism and AD

Enolase, an oxidatively modified protein in AD and MCI brain, is important for regulating glucose metabolism. However, a number of recent studies showed that enolase also plays important roles in cell signaling, Aβ clearance, and activation of cell survival pathways [16]. Oxidative dysfunction of one protein may alter several cellular pathways implicated in the pathogenesis of AD. This point is further illustrated by GAPDH, which is also selectively oxidized in AD [79]. GAPDH is a key enzyme in the glycolytic pathway; oxidation of GAPDH increases the levels of the glycolytic intermediates that are upstream to GAPDH such as glyceraldehyde-3-phosphate, which can activate the glycation pathway leading to increased formation of methylglyoxal that can react with the biomolecules, causing further damage and altered cellular function. Further, the increased levels of glyceraldehyde-3-phosphate can also lead to activation of protein kinase C pathway [21]. In addition, to up regulation of other pathways the decreased activity of GADPH increases influx of glucose through polyol pathway during which NADPH is consumed, which leads to reduced availability of GSH [29].

GAPDH has recently been shown to play key roles in transcription regulation, cell signaling, and vesicular transportation in addition to binding to other small molecules such as nitric oxide (NO), glutathione (GSH), tumor necrosis factor (TNF)-α. GAPDH also interacts with β-amyloid precursor protein (AβPP) [70]. Hence, oxidative dysfunction of enolase and GAPDH can lead to multiple changes consistent with pathology, biochemistry, and clinical presentations of AD and MCI [15]. Modulation of the cellular pathways altered by the selective oxidation of both GADPH and enolase could prove to be fertile ground for the development of novel therapeutic agents for AD [15, 16].

Recent longitudinal studies have shown a relationship between AD and glucose metabolism disorders [10]. One of the key proteins in regulation of glucose levels is insulin. The regulation of brain levels of insulin is important for proper cognitive function. For example, insulin is known to regulate the expression of *N*-methyl-D-aspartate (NMDA) receptors, one of the calcium ($Ca^{2+}$) regulating protein [33], that regulates the functions of other proteins that are important in learning and memory process. Further, the levels of insulin also influences the acetylcholine transferase, an enzyme involved in the synthesis of acetylcholine which has been reported to be altered in AD brain and is consequently a key in the etiology of AD [66].

Interestingly, a recent study showed that insulin protects the neuron against β-amyloid-derived-diffusible ligands (ADDL), important for synaptic deterioration, induced oxidative stress [10]. Insulin protects the neuron not by simply binding to ADDL but rather via an insulin dependent signaling mechanism. A pilot study showed that intranasal insulin administration improves memory and attention in patients AD patients without affecting the glucose levels [30]. Interestingly, the enzyme that regulates the insulin levels i.e., insulin degrading enzyme (IDE), is also important in the degradation of Aβ [61]. The reduced levels of insulin may also induce GSK-3beta activity which might lead to increased phosphorylation of tau protein and consequently NFT formations [69]. Hence, the decrease in insulin as reported in diabetes might lead to increased accumulation of Aβ in brain, cognitive impairment, and eventually AD pathogenesis.

In addition to decreasing the IDE levels, decreased activity of the enzymes involved in glucose metabolism by oxidation leads to increased glucose accumulation, which may have toxic effects on neurons through osmotic insults and oxidative stress subsequent to AGE formation. In addition to the effect on neurons, AGEs can also activate the microglia in the CNS, which can promote more free radicals and other inflammatory markers. The role of microglia has been proposed to be one of the underlying mechanism(s) of AD pathogenesis.

## 8.7 Pin1 and Canonical AD

Proline is an amino acid that may adopt one of two spatial conformations within a peptide. These two conformations are referred to as *cis* and *trans* orientations, of which the isomerization between the two is known to be a slow step in protein folding. Pin1, or Protein Interacting with NIMA (never in mitosis-A), is a PPIase (peptidyl-prolyl *cis/trans* isomerase) of the parvulin subfamily that specifically acts upon protein substrates with the motif of phosphorylated Ser/Thr on the amino-terminal side of an adjacent Pro (pSer/Thr-Pro) [51].

We have recently demonstrated that not only was Pin1 oxidatively modified in age matched control AD hippocampus, but that the activity as well as the protein expression of Pin1 were both found to be significantly reduced [73]. The implications of this finding are vast when considering the large amount of literature that links Pin1 as a crucial regulator of intracellular processes. When considering AD, Pin1 has been shown to regulate both directly and indirectly amyloid precursor protein (APP) and Tau, the proteins of which constitute the primary components of the neuropathological hallmarks β-amyloid plaques and NFTs, respectively. Pin1 has been shown to induce dephosphorylation of Tau at key sites via regulation of protein phosphatase 2-A (PP2A) and the knock down of Pin1 leads to hyperphosphorylation and aggregation of Tau positive NFTs [52].

Interestingly, Pin1 has been shown to act on the intracellular c-terminal domain (AICD) pT668-P of APP to direct processing of this protein toward the non-amyloidogenic pathway [59]. More recently, it has been found that Pin1 interacts with and inhibits the kinase activity of glycogen synthase kinase 3-β (GSK3β) which has been implicated in both Tau hyperphosphorylation as well as directing APP toward the amyloidogenic pathway [52]. The discovery that Pin1 both directly associates with the protein precursors to the classical AD pathological hallmarks as well as the protein kinase systems that contribute to this pathology, illustrates not only the complex and integral control that Pin1 exerts over this system, but also the potential ramifications that an oxidized and dysfunctional Pin1 could have.

## 8.8 Implications of an Oxidized Pin1 in Context of the Cell Cycle in MCI/AD

In normal non-neuronal cells, the means in which cells divide is through the induction of the cell cycle creating two daughter cells from one parental cell. This process may be broken down into four distinct segments ($G_1$, S, $G_2$, M); two growth phases $G_1$ and $G_2$ which themselves are built around an S phase during which DNA is replicated, followed by the M phase, or mitosis, which consists of chromosomal as well as cellular separation into two distinct daughter cells. In an effort to control abnormal proliferation of cells through the different phases of the cell cycle, regulatory points exist that the cell must fulfill for division to proceed. These checkpoints exist between the phases and are maintained by numerous control proteins. Neurons in a typical non-disease brain of an adult remain in what is called a quiescent state (the $G_o$ phase) wherein the cell cycle has been halted. Neurodegenerative diseases in which the protein Tau is key demonstrate aberrant cell cycle reentry of quiescent neurons and has been shown to proceed through multiple cell cycle check points, yet ultimately failing to complete mitosis and proceeds to apoptosis.

To date, a large number of control proteins important to the cell cycle have been identified as having aberrant expression, localization, or posttranslational modifications in AD [9, 46]. Moreover, cell cycle proteins have been found to have irregular expression in earlier stages of AD. We have reported that levels of CDK2, CDK5, and cyclin G1 were elevated in both the hippocampus IPL in brains of subjects with aMCI, demonstrating that cell cycle changes may appear prior to the classical AD diagnosis [45, 76].

While induction of neurons into the cell cycle by ambiguous means has been shown to be characteristic of AD, the progression and ultimate failure of the neuron to complete the cell cycle remains to be clarified in its own right, possibly due to the vast complexity of cell cycle control. Interestingly, not only does Pin1 interact with and promote cell cycle entry, but Pin1 has also been found to be associated with a number of proteins fundamental to the progression and completion of the cell cycle; associations that could have disastrous effects in vivo due to deficient expression or oxidized Pin1.

Pin1 has been found to regulate the tumor suppressor p53, a protein of significance in controlling cell proliferation, through multiple mechanisms including increasing p53 stability through the interaction of p53 with the E3 ubiquitin ligase murine double minute-2 (Mdm-2), facilitating the DNA binding of p53 to a number of pro-apoptotic promoter sites under conditions of cellular stress, enhancing p53 acetylation by p300 or creb response binding protein (CBP) acetylases, and the dissociation of p53 from the inhibitor of apoptosis-stimulating protein of p53 (iASPP) which is a key inhibitor of p53 [54]. In addition to p53, Pin1 has been implicated as a crucial player in the regulation of phosphorylated retinoblastoma protein (pRb), a key mediator of the G1/S phase transition. Evidence demonstrates that Pin1 plays a crucial role in the association of pRb with CDK/Cyclin complexes and release of E2F transcription factors important for entry into the S-phase [67]. Pin1 has been shown to directly affect the stability of the cell cycle inhibitor $p27^{kip1}$ via interaction with the $SCF^{skp2}$ (skp, cullin, F-box containing complex) E3 ubiquitin ligase complex [87].

The finding of an oxidized and dysfunctional Pin1 makes a significant contribution to the two-hit hypothesis proposed by Zhu et al., providing a direct link in which increased oxidative stress in conjunction with cell cycle abnormalities play vital roles in AD pathogenesis [88].

### *8.8.1 Heme Oxygenase-1/Biliverdin Reductase-A*

HO is an enzyme that oxidizes the heme moiety of heme-associated proteins and exists in two major isoforms arising from different genes: constitutively expressed heme oxygenase-2 (HO-2) and inducible HO-1, also known as heat shock protein-32

(Hsp-32). Using NADPH and oxygen, HO-1 produces ferrous iron, carbon monoxide, and biliverdin-IX-alpha (BV) through the hydrolysis of the tetrapyrrolic ring of heme moieties. Working with BVR-A, HO-1 provides the substrate, BV, for the production of the antioxidant bilirubin-IX-alpha (BR), which has been hypothesized to redox cycle between BR and BV in the presence of BVR-A and ROS, thereby protecting the cell from cytotoxicity [4, 53]. Notably, BVR has also been shown to function as a dual-specificity Ser/Thr and Tyr kinase plugging into the insulin growth factor-1 (IGF-1) and MAPK pathways as well as regulates the expression of various oxidative stress adaptive responsive genes [44, 83]. Research conducted by our laboratory revealed a significant elevation in HO-1, the inducible isoform, in AD hippocampus as well as increased posttranslational modifications including global protein phosphorylation and protein/lipid oxidation in both AD and MCI hippocampus and the classically unaffected AD region of the brain, the cerebellum [5].

We also found that BVR-A expression is significantly increased, has a significant reduction of phosphorylation on Ser/Thr as well as Tyr residues, and displays a significant reduction in both AD and MCI hippocampus [6]. In addition to these findings, it was demonstrated that in AD and MCI hippocampus, BVR-A had significant nitrosative modifications in the form of 3-NT, while simultaneously displaying a marked decrease in both protein bound carbonyls and HNE. Protein bound 3-NT levels correlated with a significant increase in inducible nitric oxide synthase (iNOS), which produces the radical NO, in both AD and MCI hippocampus and which, in the presence of superoxide, produces the peroxynitrite anion shown to be a precursor to protein-bound 3-NT [6]. Because BVR-A has been shown to plug into several important metabolic pathways, including upstream insulin signaling, the effects of impaired or aberrant BVR-A activity could prove disastrous for the neuron in AD.

### *8.8.2 Altered p53 and Cellular Redox Status*

p53 is a multifunctional protein that plays a major role in monitoring the state of cell proliferation and stress response through the transcription of key proteins in an effort to prevent damaged DNA from progressing to the next cellular generation [7]. The activity and stability of p53 as a transcription factor is regulated by many independent mechanisms [34, 54]. It is not clear exactly how p53 influences or is influenced by the redox state of the cell, research has shown that p53 plays a dual role, either increasing or decreasing ROS/RNS generation dependent on the cell type involved [7]. It has been hypothesized that p53 activity is dependent upon a threshold level of oxidative stress, having a number of key residues that are redox sensitive, as well as a possible altered or "mutant" 3D conformation upon posttranslational modification that may change the type of genes transactivated by p53. Importantly, we showed that such oxidative/nitrosative posttranslational modifications of p53 do occur in both MCI and AD inferior parietal lobule [26].

Buizza et al. immortalized lymphocytes collected from patients of controls, early onset AD (EOAD), and AD to determine if progression of the disease correlated with increased oxidative stress, decreased oxidative stress responsive enzyme expression or activity, and altered conformation of p53 [12]. The researchers found that protein bound HNE and 3-NT were significantly elevated ($p < 0.05$) in lymphocytes of AD patients, while the difference in oxidative parameters measured in EOAD compared to controls were elevated but not statistically significant. To determine the state of cellular machinery designed to regulate proper cellular redox status, the key proteins superoxide dismutase1/2 (SOD1/SOD2), glutathione peroxidase (GPX), glutathione reductase (GR) had their expression levels and activity measured. It was found that while the expression of these enzymes was not statistically altered, the activity of SOD and GR were found to be significantly decreased while GPX activity was found to be simply decreased. Most interestingly, using conformationally specific anti-p53 antibodies, it was found that an unfolded or altered p53 conformation correlated with an increasingly oxidized/nitrated p53 and diminished SOD activity indicating that the redox status of the cell influences the structural composition of p53 in lymphocytes. This finding gives rise to the potential of a peripheral diagnostic marker for the progression of AD [12].

As discussed, p53 has been found to play a dual role in both increasing and decreasing ROS/RNS dependent on the cell type, as demonstrated in non-neuronal cells by the ability of p53 to induce expression of antioxidant genes beneficial to cellular homeostasis in the absence of exogenous stress; while an increase in oxidative stress may lead to a pro-oxidant and pro-apoptotic p53 [68]. In neurons the threshold level discussed appears to be lower than in non-neuronal cells leading to a pro-oxidant p53 under basal conditions. For this reason, we used p53 knockout [$p53^{(-/-)}$] mice to determine the basal oxidative and nitrosative stress levels as well as antioxidant defense system activation in various neuronal compartments in the absence of p53 [7]. It was found that the oxidative/nitrosative stress was statistically decreased in the p53 null mice, the most drastic changes occurring in the nucleus and mitochondria. The expression levels of the redox protective entities thioredoxin-1 (Thio-1), BVR-A, manganese superoxide dismutase (MnSOD), IkB kinase type β (IKKβ), and nuclear factor kappa-B (NF-κB) were all found to have significantly increased levels, while a significant decrease in nuclear factor-erythroid 2 (NF-E2) related factor 2 (Nrf-2) was observed in the nuclear fraction [7]. The explanation given by the authors was that a loss of p53 appears to perturb the normal cellular homeostasis, resulting in the activation of protective machinery to avoid cellular damage or death. While the overall findings support their conclusions, it is evident that the intricacies of the mechanisms induced upon loss of p53 require further evaluation.

## 8.9 Conclusion

The oxidative stress elevation at different stages of AD and the appearance of specific protein targets of oxidation such as enolase and Pin1 suggest that specific proteins might be key players in AD pathogenesis (Fig. 8.2). As discussed above, most of the targets of protein oxidation have multiple functions, and suggest a link between metabolic disorders and the initiation and progression of AD (Fig. 8.2).

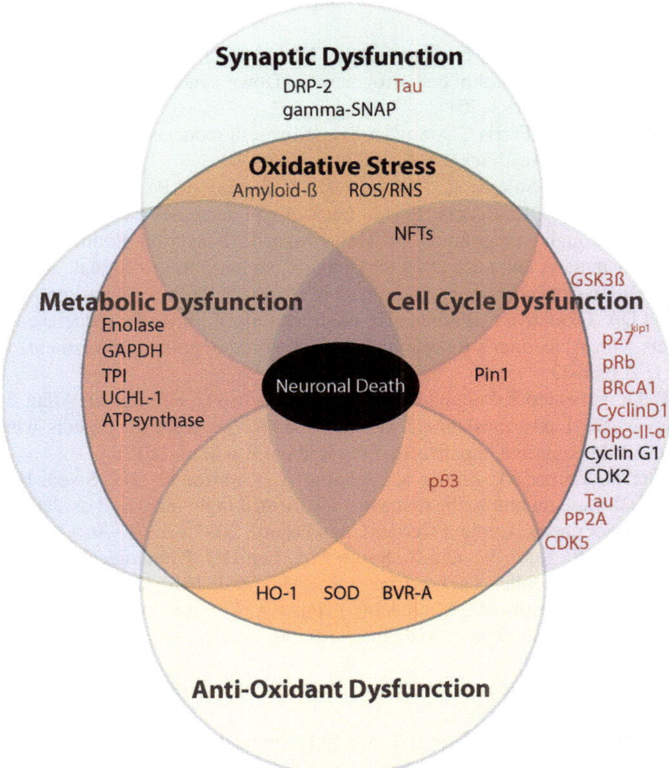


**Fig. 8.2** Venn diagram representing selected key proteins implicated in the different dysfunctional systems found within Alzheimer and MCI diseased brains/model systems. Protein names colored *red* represent a direct or indirect regulation by Pin1

There have been many studies tasked with finding ways to use biomarkers in the cerebral spinal fluid (CSF), oxidative stress biomarkers (OSBs) in blood and plasma [89], and protein and lipid oxidation of mitochondria [90] in order to correlate pathology with cognition obtained from examinations such as MMSE. The importance of identifying biomarkers for use in diagnosing the stages of progressing AD is underscored by the fact that known pathological markers such as β-amyloid deposition occur prior to cognitive deficits which could allow for earlier and potentially more effective treatments.

## References

1. Aksenov MY, Aksenova MV, Butterfield DA, Geddes JW, Markesbery WR. Protein oxidation in the brain in Alzheimer's disease. Neuroscience. 2001;103:373–83.
2. Bader Lange ML, Cenini G, Piroddi M, Abdul HM, Sultana R, Galli F, Memo M, Butterfield DA. Loss of phospholipid asymmetry and elevated brain apoptotic protein levels in subjects

with amnestic mild cognitive impairment and Alzheimer disease. Neurobiol Dis. 2008;29: 456–64.
3. Balcz B, Kirchner L, Cairns N, Fountoulakis M, Lubec G. Increased brain protein levels of carbonyl reductase and alcohol dehydrogenase in Down syndrome and Alzheimer's disease. J Neural Transm. 2001(193–201).
4. Baranano DE, Rao M, Ferris CD, Snyder SH. Biliverdin reductase: a major physiologic cytoprotectant. Proc Natl Acad Sci U S A. 2002;99:16093–8.
5. Barone E, Cenini G, Sultana R, Di Domenico F, Fiorini A, Perluigi M, Noel T, Wang C, Mancuso C, St Clair DK, Butterfield DA. Lack of p53 decreases basal oxidative stress levels in the brain through upregulation of Thioredoxin-1, Biliverdin Reductase-A, manganese superoxide dismutase, and nuclear factor kappa-b. Antioxid Redox Signal. 2012;16:1407–20.
6. Barone E, Di Domenico F, Cenini G, Sultana R, Coccia R, Preziosi P, Perluigi M, Mancuso C, Butterfield DA. Oxidative and nitrosative modifications of biliverdin reductase-A in the brain of subjects with Alzheimer's disease and amnestic mild cognitive impairment. J Alzheimers Dis. 2011;25:623–33.
7. Barone E, Di domenico F, Sultana R, Coccia R, Mancuso C, Perluigi M, Allan Butterfield D. Heme oxygenase-1 posttranslational modifications in the brain of subjects with Alzheimer disease and mild cognitive impairment. Free Radical Bio Med. 2012b.
8. Bogdanovic N, Zilmer M, Zilmer K, Rehema A, Karelson E. The Swedish APP670/671 Alzheimer's disease mutation: the first evidence for strikingly increased oxidative injury in the temporal inferior cortex. Dement Geriatr Cogn Disord. 2001;12:364–70.
9. Bonda DJ, Lee HP, Kudo W, Zhu X, Smith MA, Lee HG. Pathological implications of cell cycle re-entry in Alzheimer disease. Expert Rev Mol Med. 2010;12:e19.
10. Bosco D, Fava A, Plastino M, Montalcini T, Pujia A. Possible implications of insulin resistance and glucose metabolism in Alzheimer's disease pathogenesis. J Cell Mol Med. 2011;15: 1807–21.
11. Braak H, Braak E. Neuropathological stageing of Alzheimer-related changes. Acta Neuropathol. 1991;82:239–59.
12. Buizza L, Cenini G, Lanni C, Ferrari-Toninelli G, Prandelli C, Govoni S, Buoso E, Racchi M, Barcikowska M, Styczynska M, Szybinska A, Butterfield DA, Memo M, Uberti D. Conformational altered p53 as an early marker of oxidative stress in Alzheimer's disease. PLoS One. 2012;7:e29789.
13. Butterfield DA, Boyd-Kimball D. The critical role of methionine 35 in Alzheimer's amyloid beta-peptide (1-42)-induced oxidative stress and neurotoxicity. Biochim Biophys Acta. 2005;1703:149–56.
14. Butterfield DA, Drake J, Pocernich C, Castegna A. Evidence of oxidative damage in Alzheimer's disease brain: central role for amyloid beta-peptide. Trends Mol Med. 2001;7:548–54.
15. Butterfield DA, Hardas SS, Lange ML. Oxidatively modified glyceraldehyde-3-phosphate dehydrogenase (GAPDH) and Alzheimer disease: many pathways to neurodegeneration. J Alzheimers Dis. 2009;20(2):369–93.
16. Butterfield DA, Lange ML. Multifunctional roles of enolase in Alzheimer's disease brain: beyond altered glucose metabolism. J Neurochem. 2009;111:915–33.
17. Butterfield DA, Poon HF, St Clair D, Keller JN, Pierce WM, Klein JB, Markesbery WR. Redox proteomics identification of oxidatively modified hippocampal proteins in mild cognitive impairment: insights into the development of Alzheimer's disease. Neurobiol Dis. 2006; 22:223–32.
18. Butterfield DA, Reed T, Newman SF, Sultana R. Roles of amyloid beta-peptide-associated oxidative stress and brain protein modifications in the pathogenesis of Alzheimer's disease and mild cognitive impairment. Free Radic Biol Med. 2007;43:658–77.
19. Butterfield DA, Reed T, Perluigi M, De Marco C, Coccia R, Cini C, Sultana R. Elevated protein-bound levels of the lipid peroxidation product, 4-hydroxy-2-nonenal, in brain from persons with mild cognitive impairment. Neurosci Lett. 2006;397:170–3.
20. Butterfield DA, Stadtman ER. Protein oxidation processes in aging brain. Adv Cell Aging Gerontol. 1997;2:161–91.

21. Cantero AV, Portero-Otin M, Ayala V, Auge N, Sanson M, Elbaz M, Thiers JC, Pamplona R, Salvayre R, Negre-Salvayre A. Methylglyoxal induces advanced glycation end product (AGEs) formation and dysfunction of PDGF receptor-beta: implications for diabetic atherosclerosis. FASEB J. 2007;21:3096–106.
22. Casado A, Encarnacion Lopez-Fernandez M, Concepcion Casado M, de La Torre R. Lipid peroxidation and antioxidant enzyme activities in vascular and Alzheimer dementias. Neurochem Res. 2008;33:450–8.
23. Castegna A, Aksenov M, Thongboonkerd V, Klein JB, Pierce WM, Booze R, Markesbery WR, Butterfield DA. Proteomic identification of oxidatively modified proteins in Alzheimer's disease brain. Part II: dihydropyrimidinase-related protein 2, alpha-enolase and heat shock cognate 71. J Neurochem. 2002;82:1524–32.
24. Castegna A, Thongboonkerd V, Klein JB, Lynn B, Markesbery WR, Butterfield DA. Proteomic identification of nitrated proteins in Alzheimer's disease brain. J Neurochem. 2003;85:1394–401.
25. Cecarini V, Ding Q, Keller JN. Oxidative inactivation of the proteasome in Alzheimer's disease. Free Radic Res. 2007;41:673–80.
26. Cenini G, Sultana R, Memo M, Butterfield DA. Effects of oxidative and nitrosative stress in brain on p53 proapoptotic protein in amnestic mild cognitive impairment and Alzheimer disease. Free Radic Biol Med. 2008;45:81–5.
27. Cho DH, Nakamura T, Fang J, Cieplak P, Godzik A, Gu Z, Lipton SA. S-nitrosylation of Drp1 mediates beta-amyloid-related mitochondrial fission and neuronal injury. Science. 2009; 324:102–5.
28. Choi J, Levey AI, Weintraub ST, Rees HD, Gearing M, Chin LS, Li L. Oxidative modifications and down-regulation of ubiquitin carboxyl-terminal hydrolase L1 associated with idiopathic Parkinson's and Alzheimer's diseases. J Biol Chem. 2004;279:13256–64.
29. Chung SS, Ho EC, Lam KS, Chung SK. Contribution of polyol pathway to diabetes-induced oxidative stress. J Am Soc Nephrol. 2003;14:S233–6.
30. Craft S, Baker LD, Montine TJ, Minoshima S, Watson GS, Claxton A, Arbuckle M, Callaghan M, Tsai E, Plymate SR, Green PS, Leverenz J, Cross D, Gerton B. Intranasal insulin therapy for Alzheimer disease and amnestic mild cognitive impairment: a pilot clinical trial. Arch Neurol. 2012;69:29–38.
31. Dalle-Donne I, Aldini G, Carini M, Colombo R, Rossi R, Milzani A. Protein carbonylation, cellular dysfunction, and disease progression. J Cell Mol Med. 2006;10:389–406.
32. Dalle-Donne I, Scaloni A, Butterfield DA. Redox proteomics: from protein modifications to cellular dysfunction and diseases. Hoboken, NJ: John Wiley and Sons; 2006.
33. De Felice FG, Vieira MN, Bomfim TR, Decker H, Velasco PT, Lambert MP, Viola KL, Zhao WQ, Ferreira ST, Klein WL. Protection of synapses against Alzheimer's-linked toxins: insulin signaling prevents the pathogenic binding of Abeta oligomers. Proc Natl Acad Sci U S A. 2009;106:1971–6.
34. Di Domenico F, Cenini G, Sultana R, Perluigi M, Uberti D, Memo M, Butterfield DA. Glutathionylation of the pro-apoptotic protein p53 in Alzheimer's disease brain: implications for AD pathogenesis. Neurochem Res. 2009;34:727–33.
35. Evans TA, Raina AK, Delacourte A, Aprelikova O, Lee HG, Zhu X, Perry G, Smith MA. BRCA1 may modulate neuronal cell cycle re-entry in Alzheimer disease. Int J Med Sci. 2007;4:140–5.
36. Fernandez-Vizarra P, Fernandez AP, Castro-Blanco S, Encinas JM, Serrano J, Bentura ML, Munoz P, Martinez-Murillo R, Rodrigo J. Expression of nitric oxide system in clinically evaluated cases of Alzheimer's disease. Neurobiol Dis. 2004;15:287–305.
37. Fox NC, Crum WR, Scahill RI, Stevens JM, Janssen JC, Rossor MN. Imaging of onset and progression of Alzheimer's disease with voxel-compression mapping of serial magnetic resonance images. Lancet. 2001;358:201–5.
38. Frautschy SA, Baird A, Cole GM. Effects of injected Alzheimer beta-amyloid cores in rat brain. Proc Natl Acad Sci U S A. 1991;88:8362–6.
39. Glabe CC. Amyloid accumulation and pathogensis of Alzheimer's disease: significance of monomeric, oligomeric and fibrillar Abeta. Subcell Biochem. 2005;38:167–77.

40. Guix FX, Ill-Raga G, Bravo R, Nakaya T, de Fabritiis G, Coma M, Miscione GP, Villa-Freixa J, Suzuki T, Fernandez-Busquets X, Valverde MA, de Strooper B, Munoz FJ. Amyloid-dependent triosephosphate isomerase nitrotyrosination induces glycation and tau fibrillation. Brain. 2009;132:1335–45.
41. Hebert LE, Scherr PA, Bienias JL, Bennett DA, Evans DA. Alzheimer disease in the US population: prevalence estimates using the 2000 census. Arch Neurol. 2003;60:1119–22.
42. Jack Jr CR, Petersen RC, Xu YC, O'Brien PC, Smith GE, Ivnik RJ, Boeve BF, Waring SC, Tangalos EG, Kokmen E. Prediction of AD with MRI-based hippocampal volume in mild cognitive impairment. Neurology. 1999;52:1397–403.
43. Jicha GA, Parisi JE, Dickson DW, Johnson K, Cha R, Ivnik RJ, Tangalos EG, Boeve BF, Knopman DS, Braak H, Petersen RC. Neuropathologic outcome of mild cognitive impairment following progression to clinical dementia. Arch Neurol. 2006;63:674–81.
44. Kapitulnik J, Maines MD. Pleiotropic functions of biliverdin reductase: cellular signaling and generation of cytoprotective and cytotoxic bilirubin. Trends Pharmacol Sci. 2009;30:129–37.
45. Keeney JT, Swomley AM, Harris JL, Fiorini A, Mitov MI, Perluigi M, Sultana R, Butterfield DA. Cell cycle proteins in brain in mild cognitive impairment: insights into progression to Alzheimer disease. Neurotox Res. 2011.
46. Keeney JT, Swomley AM, Harris JL, Fiorini A, Mitov MI, Perluigi M, Sultana R, Butterfield DA. Cell cycle proteins in brain in mild cognitive impairment: insights into progression to Alzheimer disease. Neurotox Res. 2012;22:220–30.
47. Keller JN, Hanni KB, Markesbery WR. Impaired proteasome function in Alzheimer's disease. J Neurochem. 2000;75:436–9.
48. Keller JN, Schmitt FA, Scheff SW, Ding Q, Chen Q, Butterfield DA, Markesbery WR. Evidence of increased oxidative damage in subjects with mild cognitive impairment. Neurology. 2005;64:1152–6.
49. Korolainen MA, Goldsteins G, Nyman TA, Alafuzoff I, Koistinaho J, Pirttila T. Oxidative modification of proteins in the frontal cortex of Alzheimer's disease brain. Neurobiol Aging. 2006;27:42–53.
50. Lovell MA, Xie C, Markesbery WR. Decreased glutathione transferase activity in brain and ventricular fluid in Alzheimer's disease. Neurology. 1998;51:1562–6.
51. Lu KP, Hanes SD, Hunter T. A human peptidyl-prolyl isomerase essential for regulation of mitosis. Nature. 1996;380:544–7.
52. Ma SL, Pastorino L, Zhou XZ, Lu KP. Prolyl isomerase Pin1 promotes amyloid precursor protein (APP) turnover by inhibiting glycogen synthase kinase-3beta (GSK3beta) activity: novel mechanism for Pin1 to protect against Alzheimer disease. J Biol Chem. 2012;287:6969–73.
53. Mancuso C. Heme oxygenase and its products in the nervous system. Antioxid Redox Signal. 2004;6:878–87.
54. Mantovani F, Tocco F, Girardini J, Smith P, Gasco M, Lu X, Crook T, Del Sal G. The prolyl isomerase Pin1 orchestrates p53 acetylation and dissociation from the apoptosis inhibitor iASPP. Nat Struct Mol Biol. 2007;14:912–20.
55. Markesbery WR. Oxidative stress hypothesis in Alzheimer's disease. Free Radic Biol Med. 1997;23:134–47.
56. Markesbery WR, Kryscio RJ, Lovell MA, Morrow JD. Lipid peroxidation is an early event in the brain in amnestic mild cognitive impairment. Ann Neurol. 2005;58:730–5.
57. Martinez A, Portero-Otin M, Pamplona R, Ferrer I. Protein targets of oxidative damage in human neurodegenerative diseases with abnormal protein aggregates. Brain pathol. 2009; 10.1111/j.1750-3639.2009.00326.
58. Morris JC, Storandt M, Miller JP, McKeel DW, Price JL, Rubin EH, Berg L. Mild cognitive impairment represents early-stage Alzheimer disease. Arch Neurol. 2001;58:397–405.
59. Pastorino L, Sun A, Lu PJ, Zhou XZ, Balastik M, Finn G, Wulf G, Lim J, Li SH, Li X, Xia W, Nicholson LK, Lu KP. The prolyl isomerase Pin1 regulates amyloid precursor protein processing and amyloid-beta production. Nature. 2006;440:528–34.
60. Perluigi M, Sultana R, Cenini G, Di Domenico F, Memo M, Pierce WM, Coccia R, Butterfield DA. Redox proteomics identification of HNE-modified brain proteins in Alzheimer's disease:

role of lipid peroxidation in Alzheimer's disease pathogenesis. Proteomics Clin Appl. 2009; 3:682–93.
61. Qiu WQ, Folstein MF. Insulin, insulin-degrading enzyme and amyloid-beta peptide in Alzheimer's disease: review and hypothesis. Neurobiol Aging. 2006;27:190–8.
62. Rapoport SI. In vivo PET imaging and postmortem studies suggest potentially reversible and irreversible stages of brain metabolic failure in Alzheimer's disease. Eur Arch Psychiatry Clin Neurosci. 1999;249 Suppl 3:46–55.
63. Reed T, Perluigi M, Sultana R, Pierce WM, Klein JB, Turner DM, Coccia R, Markesbery WR, Butterfield DA. Redox proteomic identification of 4-hydroxy-2-nonenal-modified brain proteins in amnestic mild cognitive impairment: insight into the role of lipid peroxidation in the progression and pathogenesis of Alzheimer's disease. Neurobiol Dis. 2008;30:107–20.
64. Reyes JF, Reynolds MR, Horowitz PM, Fu Y, Guillozet-Bongaarts AL, Berry R, Binder LI. A possible link between astrocyte activation and tau nitration in Alzheimer's disease. Neurobiol Dis. 2008;31:198–208.
65. Riederer IM, Schiffrin M, Kovari E, Bouras C, Riederer BM. Ubiquitination and cysteine nitrosylation during aging and Alzheimer's disease. Brain Res Bull. 2009;80(4–5):233–41.
66. Rivera EJ, Goldin A, Fulmer N, Tavares R, Wands JR, de la Monte SM. Insulin and insulin-like growth factor expression and function deteriorate with progression of Alzheimer's disease: link to brain reductions in acetylcholine. J Alzheimers Dis. 2005;8:247–68.
67. Rizzolio F, Lucchetti C, Caligiuri I, Marchesi I, Caputo M, Klein-Szanto AJ, Bagella L, Castronovo M, Giordano A. Retinoblastoma tumor-suppressor protein phosphorylation and inactivation depend on direct interaction with Pin1. Cell Death Differ. 2012;19(7):1152–61.
68. Sablina AA, Budanov AV, Ilyinskaya GV, Agapova LS, Kravchenko JE, Chumakov PM. The antioxidant function of the p53 tumor suppressor. Nat Med. 2005;11:1306–13.
69. Salkovic-Petrisic M, Tribl F, Schmidt M, Hoyer S, Riederer P. Alzheimer-like changes in protein kinase B and glycogen synthase kinase-3 in rat frontal cortex and hippocampus after damage to the insulin signalling pathway. J Neurochem. 2006;96:1005–15.
70. Schulze H, Schuler A, Stuber D, Dobeli H, Langen H, Huber G. Rat brain glyceraldehyde-3-phosphate dehydrogenase interacts with the recombinant cytoplasmic domain of Alzheimer's beta-amyloid precursor protein. J Neurochem. 1993;60:1915–22.
71. Selkoe DJ. Alzheimer's disease: genes, proteins, and therapy. Physiol Rev. 2001;81:741–66.
72. Smith MA, Richey Harris PL, Sayre LM, Beckman JS, Perry G. Widespread peroxynitrite-mediated damage in Alzheimer's disease. J Neurosci. 1997;17:2653–7.
73. Sultana R, Boyd-Kimball D, Poon HF, Cai J, Pierce WM, Klein JB, Markesbery WR, Zhou XZ, Lu KP, Butterfield DA. Oxidative modification and down-regulation of Pin1 in Alzheimer's disease hippocampus: A redox proteomics analysis. Neurobiol Aging. 2006;27:918–25.
74. Sultana R, Boyd-Kimball D, Poon HF, Cai J, Pierce WM, Klein JB, Merchant M, Markesbery WR, Butterfield DA. Redox proteomics identification of oxidized proteins in Alzheimer's disease hippocampus and cerebellum: an approach to understand pathological and biochemical alterations in AD. Neurobiol Aging. 2006;27:1564–76.
75. Sultana R, Butterfield DA. Oxidatively modified GST and MRP1 in Alzheimer's disease brain: implications for accumulation of reactive lipid peroxidation products. Neurochem Res. 2004;29:2215–20.
76. Sultana R, Butterfield DA. Regional expression of key cell cycle proteins in brain from subjects with amnestic mild cognitive impairment. Neurochem Res. 2007;32:655–62.
77. Sultana R, Perluigi M, Butterfield DA. Redox proteomics identification of oxidatively modified proteins in Alzheimer's disease brain and in vivo and in vitro models of AD centered around Abeta(1-42). J Chromatogr. 2006;833:3–11.
78. Sultana R, Perluigi M, Butterfield DA. Oxidatively modified proteins in Alzheimer's disease (AD), mild cognitive impairment and animal models of AD: role of Abeta in pathogenesis. Acta Neuropathol. 2009;118:131–50.
79. Sultana R, Poon HF, Cai J, Pierce WM, Merchant M, Klein JB, Markesbery WR, Butterfield DA. Identification of nitrated proteins in Alzheimer's disease brain using a redox proteomics approach. Neurobiol Dis. 2006;22:76–87.

80. Sultana R, Reed T, Perluigi M, Coccia R, Pierce WM, Butterfield DA. Proteomic identification of nitrated brain proteins in amnestic mild cognitive impairment: a regional study. J Cell Mol Med. 2007;11:839–51.
81. Sultana R, Robinson RA, Bader Lange M, Fiorini A, Galvan V, Fombonne J, Baker A, Gorostiza O, Zhang J, Cai J, Pierce WM, Bredesen DE, Butterfield A. Do proteomics analyses provide insights into reduced oxidative stress in brain of an Alzheimer disease transgenic mouse model with a M631L APP substitution and thereby the importance of Abeta-resident methionine 35 in AD pathogenesis? Antioxid Redox Signal. 2012;17(11):1507–14.
82. Terni B, Boada J, Portero-Otin M, Pamplona R, Ferrer I. Mitochondrial ATP-synthase in the entorhinal cortex is a target of oxidative stress at stages I/II of Alzheimer's disease pathology. Brain pathol. 2009; 10.1111/j.1750-3639.2009.00266.x.
83. Tudor C, Lerner-Marmarosh N, Engelborghs Y, Gibbs PE, Maines MD. Biliverdin reductase is a transporter of haem into the nucleus and is essential for regulation of HO-1 gene expression by haematin. Biochem J. 2008;413:405–16.
84. Uehara T, Nakamura T, Yao D, Shi ZQ, Gu Z, Ma Y, Masliah E, Nomura Y, Lipton SA. S-nitrosylated protein-disulphide isomerase links protein misfolding to neurodegeneration. Nature. 2006;441:513–7.
85. Volkel W, Sicilia T, Pahler A, Gsell W, Tatschner T, Jellinger K, Leblhuber F, Riederer P, Lutz WK, Gotz ME. Increased brain levels of 4-hydroxy-2-nonenal glutathione conjugates in severe Alzheimer's disease. Neurochem Int. 2006;48:679–86.
86. Yao Y, Zhukareva V, Sung S, Clark CM, Rokach J, Lee VM, Trojanowski JQ, Pratico D. Enhanced brain levels of 8,12-iso-iPF2alpha-VI differentiate AD from frontotemporal dementia. Neurology. 2003;61:475–8.
87. Zhou W, Yang Q, Low CB, Karthik BC, Wang Y, Ryo A, Yao SQ, Yang D, Liou YC. Pin1 catalyzes conformational changes of Thr-187 in p27Kip1 and mediates its stability through a polyubiquitination process. J Biol Chem. 2009;284:23980–8.
88. Zhu X, Raina AK, Perry G, Smith MA. Alzheimer's disease: the two-hit hypothesis. Lancet Neurol. 2004;3:219–26.
89. Cristalli D, Arnal N, Maria F, de Alaniz M, Marra C. Peripheral markers in neurodegenerative patients and their first-degree relatives. J Neural Sci. 2011;314(1–2):48–56.
90. Sultana R, Perlvigi M, Newman S, Pierce W, Cini C, Coccia R, Butterfield DA. Redox proteomic analysis of carbonylated brain proteins in mild cognitive impairment and early Alzheimer's disease. Antioxid Redox Signaling. 2010;12:327–36.

# Chapter 9
# Cholesterol Metabolism and Oxidative Stress in Alzheimer's Disease

**Luigi Iuliano and Valerio Leoni**

**Abstract** Cholesterol has a prominent role in cell structure and function, in the brain, including signal transduction, neurotransmitter release, synaptogenesis, and membrane trafficking. Perturbation of cholesterol trafficking in the brain is potentially linked to the pathogenesis of Alzheimer's disease (AD). Cholesterol is unable to pass the blood–brain barrier, and its level in the brain depends exclusively on de novo synthesis and elimination, which rely on local transcription of ApoE and cassette transporters (ABCA1, ABCG1, and ABCG4) involved in the lipid transfer across membranes. These pathways are controlled by oxysterols. In order to maintain homeostasis, cholesterol is converted into 24-hydroxycholesterol by the neuronal specific cholesterol 24-hydroxylase, which is located in the endoplasmic reticulum. The putative role of upregulated oxidative stress in AD has raised interest in nonenzymatic oxysterols, which are generated by free radical species, such as those arising from the superoxide/hydrogen peroxide/hydroxyl radical system and by non-radical highly reactive oxygen species such as singlet oxygen, HOCl, and ozone. The analysis of oxysterols is a valuable tool to noninvasively investigating the role of cholesterol metabolism in the pathogenesis of neurodegeneration.

L. Iuliano, M.D. (✉)
Laboratory of Vascular Medicine and Mass Spectrometry, Department of Medico-Surgical Sciences and Biotechnology, Sapienza University of Rome, Corso della Republica 79, Rome, Italy
e-mail: luigi.iuliano@uniroma1.it

V. Leoni, M.D., Ph.D.
Laboratory of Clinical Pathology and Medical Genetics, Foundation IRCCS Institute of Neurology Carlo Besta, Milan, Italy

D. Praticò and P. Mecocci (eds.), *Studies on Alzheimer's Disease*, Oxidative Stress in Applied Basic Research and Clinical Practice, DOI 10.1007/978-1-62703-598-9_9,

## 9.1 Body Cholesterol Metabolism

Cholesterol, b-hydroxycholest-5-ene, is a structural element of the eukaryotic membranes regarded to be the most prominent "fluidity buffer" supporting the structure and function of lipid bilayers, is involved in varied cellular processes. Cholesterol metabolism—which is tightly regulated and depends on absorption, synthesis and biotransformation—is involved in a variety of metabolic pathways (Fig. 9.1), including those of bile acids, hormones, vitamin D, and oxysterols, which are compounds generated by the addition of oxygen into the cholesterol structure. Our coverage concentrates on cholesterol and oxysterols in the attempt to give a summary of their action in brain, while bile acids, hormones, and vitamin D are out of the scope of this review.

### *9.1.1 Cholesterol Absorption and Synthesis*

Cholesterol pool in the body results from dietary sources and de-novo synthesis; except for the brain that has no access to cholesterol of the blood compartment due to blood–brain barrier constraints. Under normal conditions, about 60 % of body's cholesterol is synthesized (about 700 mg/day), and the remaining is of dietary origin.

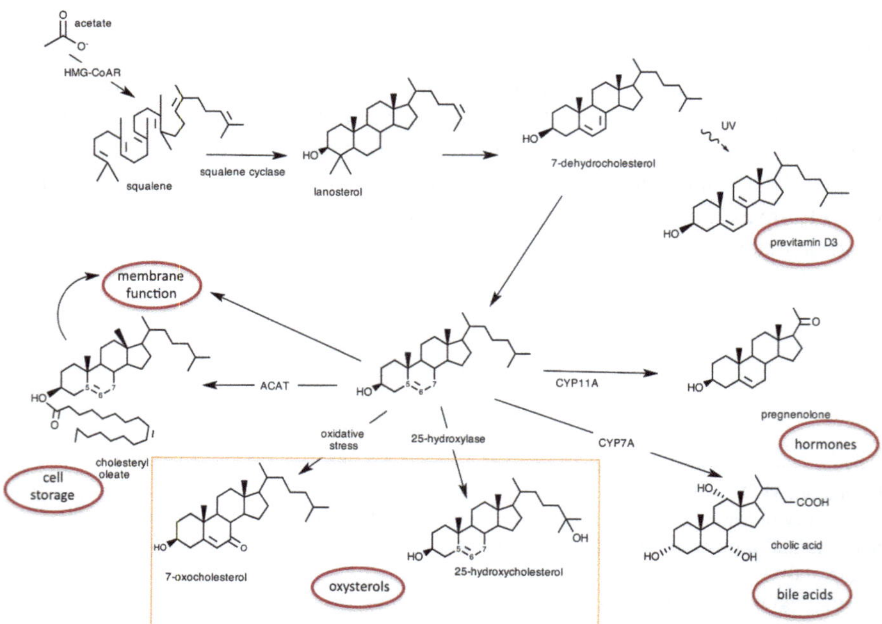


**Fig. 9.1** -->Main cholesterol functions. Synthetic representation of cholesterol biosynthesis and pathways of transformation. Prototypical oxysterols are represented in the *box*

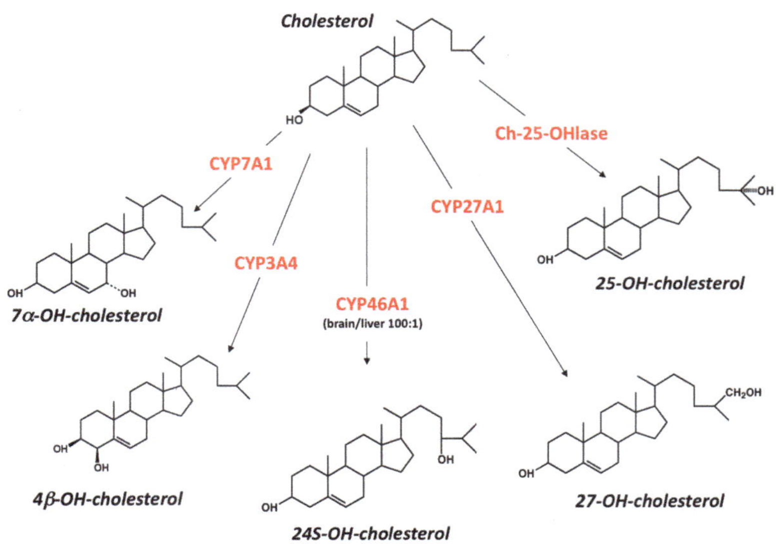


**Fig. 9.2** Main enzymatic pathways of oxysterol production from cholesterol

Cholesterol is absorbed from the intestinal lumen and is transported by chylomicrons to the liver where it can be esterified, converted into bile acids or loaded into very low density lipoproteins (VLDL) to be delivered to extrahepatic tissues. In the circulation, VLDL are remodelled by lipoprotein lipase to form low-density lipoproteins LDL. Cells fulfill their needs of cholesterol by de-novo synthesis or by taking up cholesterol from the LDL that circulate in the blood. LDL uptake, except in macrophages, is strictly regulated via the LDL receptors that bind Apolipoprotein B [1]. The influx of cholesterol inhibits HMG-CoA reductase and LDL receptor expression, and stimulates cholesterol esterification by acylCoA:cholesterol acyltransferase. Eventually, cholesterol is extruded by the cell to be transported back to the liver by HDL that acquire cholesterol directly from the plasma membrane through the action of proteins of the ABC-transporter cassette family. About 1 g of cholesterol is eliminated into the feces every day as it is or after conversion into bile acids, which have a decisive role in the solubilization and absorption of lipids, vitamins, and drugs.

### 9.1.2 Cholesterol Biotransformation into Oxysterols

Enzymatic and nonenzymatic addition of oxygen to the cholesterol molecule generate a class of derivatives, namely, oxysterols.

There are numerous cholesterol metabolizing enzymes (Fig. 9.2), many of which belong to the cytochrome P450 family [2], involved in the production of oxysterols. Cholesterol 7α-hydroxylase (CYP7A1) is a hepatic enzyme that converts cholesterol

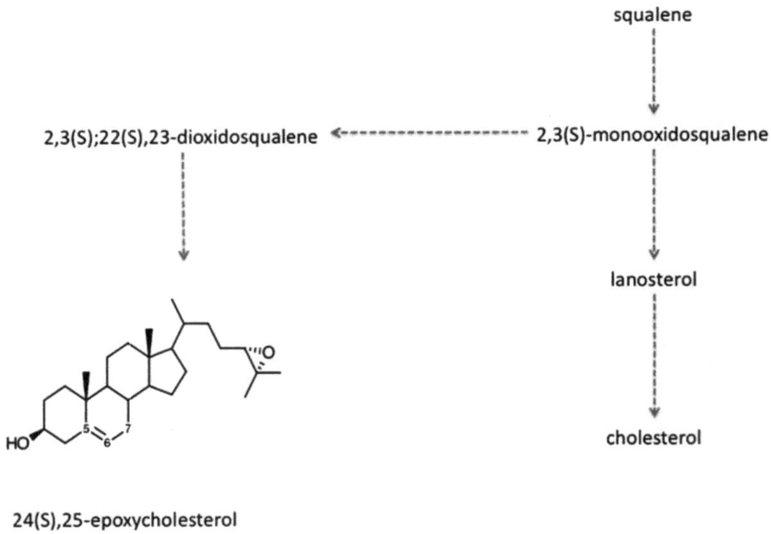


**Fig. 9.3** Formation of 24(S),25-epoxycholesterol during cholesterol synthesis

into 7α-hydroxycholesterol (7α-OHC), a starting intermediate in the biosynthesis of bile acids [3]; 7α-OHC is next converted into 7alpha-hydroxy-4-cholesten-3-one, by a microsomal 3β-hydroxy-$\Delta^5$-$C_{27}$-steroid oxidoreductase (C27 3β-HSD) that can be used as a marker of bile acid synthesis [4]. A portion of 7α-OHC formed in the liver leaks out into the circulatory system: the concentration of 7α-OHC in blood plasma correlates with the activity of CYP7A1 [5]. The drug-metabolizing enzyme CYP3A4 has been shown to convert cholesterol into 4β-hydroxycholesterol [6], an oxysterol that can be used as an endogenous marker of CYP3A4 and CYP3A5 activity [7]. 24(S),25-epoxycholesterol (24,25EC) is synthesized in a shunt in the mevalonate pathway, parallel to cholesterol and utilizing the same enzymes, and is not derived from preformed cholesterol [8] (Fig. 9.3).

Oxysterols can be formed in the absence of enzymatic catalysis, in a pathway termed "autoxidation" produced by free radical species, i.e., the superoxide/hydrogen peroxide/hydroxyl radical system, and by non-radical reactive oxygen species such as singlet oxygen, HOCl, and ozone [9] (Fig. 9.4). Cholesterol autoxidation falls into the field of lipid peroxidation, a field that can be traced back to the 1930s when studies related to the deterioration of foods, mostly due to susceptibility to oxidation of polyunsaturated fatty acids (PUFA), were carried out [10]. Lipid peroxidation causes alterations in fluidity and permeability, and the damage can be transferred a distance away from the initial site through reactive intermediates to damage other cellular constituents, such as proteins, sugars and nucleic acids, accounting for the putative pathophysiological role of lipid peroxidation in several conditions such as cardiovascular diseases, neurological diseases, aging and cancer. The unique double bond at position 5,6 makes cholesterol susceptible to nonenzymatic oxidation via

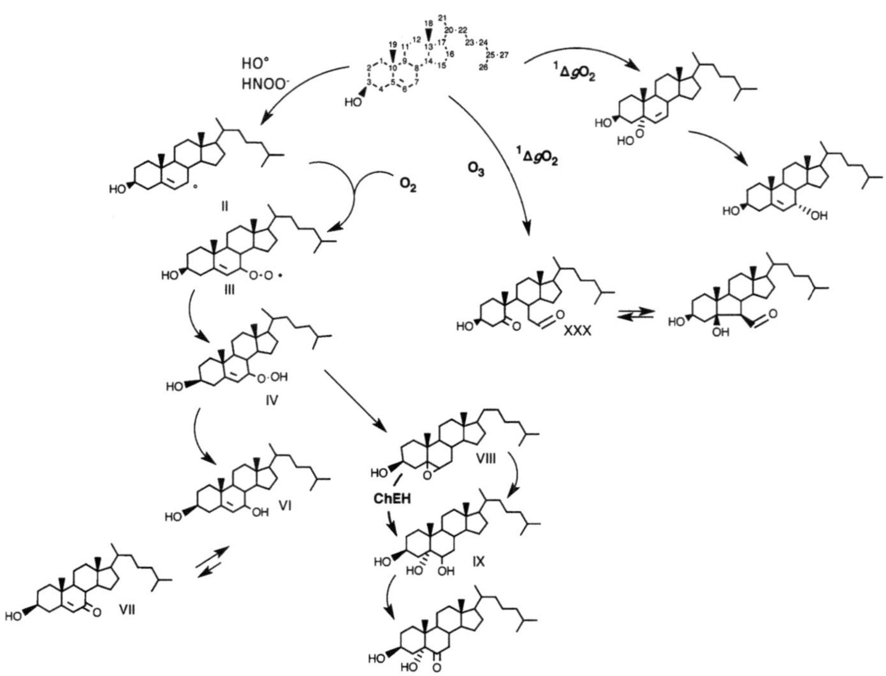


**Fig. 9.4** Schematic representation of cholesterol autoxidation by free radical (HO• and HNOO- are given as example of initiators of cholesterol autoxidation) and high reactive oxygen species (ozone and singlet oxygen)

free radical and non-free radical reactions, which lead to an array of oxysterols [9]. The preferential site of oxidation of cholesterol by free radicals is at carbon-7 where the carbon–hydrogen bond is relatively weak; the dissociation energy of this bond is 88 kcal/mol [11] The resulting 7β-hydroxycholesterol (7β-OCH), for which there is no specific enzyme detected in vivo [12] is relatively stable and suitable for analytical purposes as surrogate marker of oxidative stress [13].

Oxysterols are known to exert a multitude of biological effects of potential pathophysiological relevance, which are mediated by biophysical effects on membranes and/or stereospecific interactions with proteins. Oxysterols are part of the cellular machinery governing the metabolic function and integrity of the cell acting as signals or translational/posttranslational-gene expression regulators. Key proteins in the control of metabolism reportedly targets for oxysterols include: receptors transcription factors such as are the liver X receptors (LXRs) [14], which exert a key role in lipid metabolism, the sterol regulatory element binding proteins (SBREPs) [15], or the estrogen receptor (ER) and other proteins involved in cholesterol homeostasis such as the HMGCoA reductase, the oxysterol-binding proteins (OSBP), the estrogen receptor, the Nieman Pick disease proteins (NCP1, NCP2) [16], and the Hedgehog system [17]. The discovery in oxysterols area is expanding dramatically and reveals more and more oxysterols of valuable potential importance

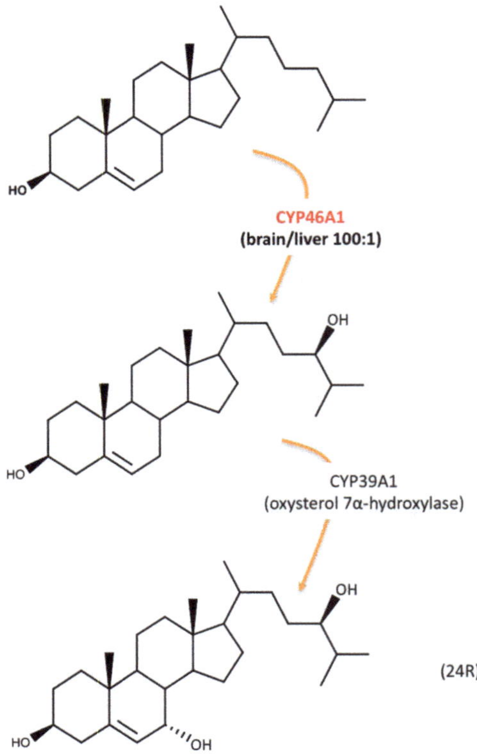


**Fig. 9.5** Metabolism of 24(S)-hydroxycholesterol

in biological systems. For instance, 5,6-secosterol and its carboxaldehyde are likely biomarkers in inflammation-oxidative stress processes [18]; 24-cholest-5-ene-3β,7α-triol (a metabolic transformation product of 24-OHC [19]) (Fig. 9.5), and 7α-hydroxy-3-oxo-4-cholestanoic acid 3β-hydroxy-5-cholestanoic acid (transformation products of 27-OH-cholesterol) (Fig. 9.6)[20] impact on brain control of cholesterol metabolism with potential relevance in neurodegenerative diseases, including in aging.

## 9.2 Brain Cholesterol

In the brain, cholesterol is about tenfold higher than in any other organ—about 25 % of the total body cholesterol is located in the brain—and has a remarkably long half-life, turning over some 250–300 times slower than that in the circulation. Cholesterol is involved in synaptogenesis, turnover, maintenance, stabilization and restoring of synapses and is a limiting factor for outgrowth of neuritis being involved in vesicle transport and exocytosis at synaptic levels (for reviews see [21]).

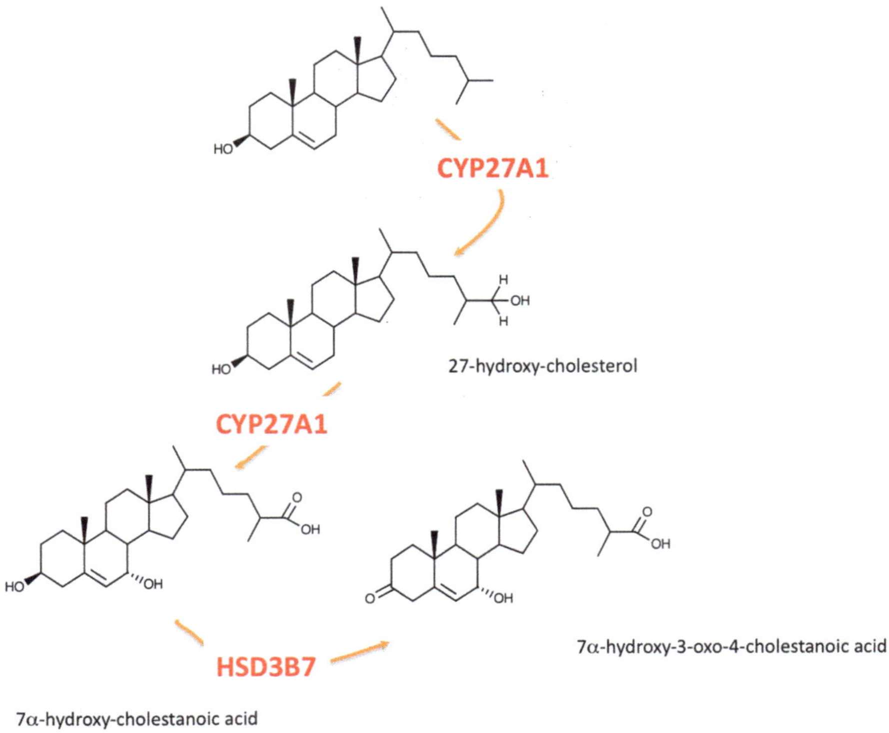


**Fig. 9.6** Formation of 27-hydroxycholesteol and downstream metabolism into its acids

Within the brain, about 70 % of cholesterol is present in myelin, 20 % in glial cells, and 10 % in neurons. According to various in vitro studies with cultured cells, astrocytes synthesize at least five- to tenfold more cholesterol than neurons, while oligodendrocytes have an even higher capacity for cholesterol synthesis, at least during periods of active myelination [22]. After brain maturation, i.e., in the adulthood, neurons down-regulate their cholesterol synthesis and use at least partly cholesterol from astrocytes (the "outsourcing" hypothesis of brain cholesterol metabolism [21]). ApoE is the main lipid carrier protein in the CNS and is released by astrocytes in order to supply neurons and synaptogenesis with lipids and cholesterol [23–25]. Due to the efficiency of blood–brain barrier, there is no passage of lipoprotein-bound cholesterol from the circulation into the brain [22]. Almost all (at least 99 %) cholesterol in the nervous system is unesterified.

## 9.3 Oxysterols and Brain Cholesterol Homeostasis

In the adult brain, most of the synthesis of cholesterol is balanced by the formation of 24OHC, which is able to cross the blood–brain barrier and enter the circulation [22, 26]. The addition of hydrophilic moieties allows oxysterols to transfer between membranes with several orders of magnitude faster than cholesterol. About 6–8 mg/24 h of cholesterol is released as 24OHC by the brain into the circulation [26]. In parallel, ApoE mediates a small efflux of cholesterol from the brain into the cerebrospinal fluid [27].

Under normal conditions, cholesterol 24-hydroxylase (CYP46A1), the enzyme forming 24OHC, is only expressed in neuronal cells associated with grey matter, mainly in cerebral cortex, hippocampus, dentate gyrus, amygdala, putamen, and thalamus [28].

CYP46A1 knockout mice show alteration in synaptic maturation associated with severe deficiencies in spatial, associative and motor learning associated with a delay of long lasting potential, but a modest reduction of HMCoA-reductase activity and cholesterol synthesis rate unaffecting the total brain cholesterol content [29].

Biotransformation of 24OHChol should also be considered. CYP39A1, which is highly expressed in the liver and specifically transform 24-hydroxycholesterol into (24R)-cholest-5-ene 3-b,7-α,24-triol [19] (Fig. 9.5) to be eliminated through the bile acid system, is also expressed in the brain [30]. Such a biotransformation of 24OHC may play an additional role in cholesterol trafficking in the brain.

In most cells outside the brain, cholesterol is eliminated by side chain oxidation of carbon-27 by the enzyme sterol 27-hydroxylase (CYP27A1), which is located in the inner membrane of the mitochondria and is abundantly expressed in macrophages. 27OHC is able to pass the blood–brain barrier and its levels in the cerebrospinal fluid (CSF) are significantly correlated with the corresponding levels in the circulation [31]. A putative highly efficient mechanism for eliminating 27OHC from the brain, is the metabolic transformation of 27OHC to its metabolites 7-α-hydroxy-cholestanoic acid and 7α-hydroxy-3-oxo-4-cholestanoic acid, which are able to rapidly cross the blood–brain barrier [20] (Fig. 9.6).

Recently, Wong et al. [32] reported that 24,25-epoxycholesterol is synthesized in primary human brain cells, specifically in astrocytes and neurons, and is able to modulate the expression of LXR and SREBP target genes.

## 9.4 Cholesterol Metabolism and Alzheimer's Disease

### *9.4.1 AD Pathophysiology*

Cholesterol influences a number of processes involved in the generation of the neuritic plaques and neurofibrillary tangles [33].

AD begins with abnormal processing of amyloid precursor protein (APP) that leads to excess of production and/or reduction in clearance of Aβ in the brain. APP, a transmembrane protein with a large extracellular domain, can go through an initial α- or β-cleavage, followed by a further γ-cleavage, to yield αAPP and P3 peptides or the βAPP and Aβ peptides, respectively. The increased amount of Aβ oligomers turns in Aβ aggregates and fibrils that progressively increase the number and dimension of plaques. Thus, several mechanisms (i.e., toxic effect of β oligomers, perturbation of brain metabolism, presence of ApoE *ε4* allele, oxidative stress, inflammatory response, and impaired cholesterol metabolism) contribute to abnormal tau aggregation, synaptic dysfunction, cell death, and brain atrophy and shrinkage [34].

Aβ plaques and NFT are associated with neurodegeneration characterized by atrophy, neuron loss and gliosis. Neurodegeneration and NFT deposition are neuronal processes with a similar topographic distribution; Aβ plaques are extracellular and occur with a different but somehow overlapping distribution to NFT and neurodegenerative pathological changes. The cognitive impairment and other clinical symptoms are related to neurodegeneration and especially to loss of synapses. Cerebral atrophy and not Aβ or NFT burden is the pathological substrate of cognitive impairment [34].

### *9.4.2 Cholesterol and Neurodegeneration*

High plasma cholesterol at midlife [35] and the *ε4* allele of Apolipoprotein E (ApoE) [36] are significant risk factors for sporadic AD. Cholesterol influences a number of processes involved in the generation of the neuritic plaques and neurofibrillary tangles [37]. Increased membrane cholesterol favors the activity of the β-secretase pathway, accumulation of Aβ1-40 and Aβ1-42 peptides, and increase in amyloid deposits [38]. Conflicting results are, however, reported in the literature by other studies [39]. In AD patients, but not in MCI or control individuals, lower plasma total cholesterol and LDL-C were found associated to lower brain volumes/higher CSF volumes [40]. A large 21-year follow-up study presented an association between serum total cholesterol changes from midlife to late-life and late-life cognitive status. A moderate decrease in serum cholesterol is associated with increased risk of a most impaired late-life cognitive status [40]. It is likely that, while high midlife cholesterol is a risk factor for AD, decreased cholesterol later in life instead reflects an ongoing pathological process in the brain and should be considered as a frailty marker, predictive of worse cognitive functioning [41].

The study of oxysterols may help to explain the apparently contrasting observations about higher levels of cholesterol as a risk factor, and impaired cholesterol metabolism in patients with neurodegenerative diseases.

### *9.4.3 Oxysterols and Neurodegeneration*

Under in vitro conditions, 24OHC is an efficient inhibitor of the formation of Aβ [25, 42], while 27OHC shows a much lower capacity to inhibit this reaction. 24OHC was shown reduced and 27OHC increased in brain samples from AD patient [43]. 27-OHC-challenged SH-SY5Y cells release high amounts of Aβ42, beta-amyloid precursor protein (APP), and β-secretase (BACE1). These effects are not observed with 24OHC that, in contrast, increases the levels of sAPPalpha, suggesting that 24-OHC favors the processing of APP towards a non-amyloidogenic pathway [42]. It is likely that the reduced levels of 24OHC may accelerate the progress of the disease and the increased levels of 27OHC may not be able to compensate for this [44]. In hypercholesterolemia plasma levels of 27OHC are higher compared to normal individuals, and may account for an accelerated deleterious brain uptake that antagonize the protective effect of 24OHC [44].

The biochemical function of APP and the mechanism of the toxicity of β-amyloid are still unclear. It has been show that β-amyloid can oxidize cholesterol to form 7β-hydroxycholesterol (7β-OHC) in a highly efficient mechanism [45]. The rate of reaction between cholesterol and β-amyloid was comparable to the rates of cholesterol metabolizing enzymes (kcat $\approx$ 0.211 $min^{-1}$). In turn, 7β-OHC was able to inhibit secretion of soluble APP from cultured rat hippocampal H19–7/IGF-IR neuronal cells, as well as to inhibit tumor necrosis factor-α-converting enzyme β-secretase activity. 7β-OHC causes rearrangement of the liquid-ordered phase [46], which results in the creation of lipid rafts, and is a potent inhibitor of α-protein kinase C, an enzyme critical for memory consolidation and synaptic plasticity and implicated in AD [45].

Recently, cholesterol ozonization has attracted the attention for the putative formation of ozone within the body. In a series of studies, Wentworth provided evidence that ozone may be produced inside the body at sites of inflammatory reactions by the superoxide-hydrogen peroxide/HOCl system and of a newly discovered catalytic domain on the antibody molecules [47]. Concomitantly, the putative specific cholesterol ozonization product 5,6-secosterol was described within the atherosclerotic plaques [48] and in brain tissue extracts obtained from patients with AD [49].

On the one hand, the relevance of these reports is limited by some controversy concerning the identification of the products and the mechanism of formation of ozone [50] along with the pathway of cholesterol autoxidation that does not support the fingerprint molecules of cholesterol ozonation. On the other hand, the isolation of 5,6-secosterol, from pathologic tissue specimens of blood vessels and from the brain, is of great importance to further explore the role of oxysterols in disease mechanism. Cholesterol ozonolysis products, including 5,6-secosterol and its carboxaldehyde, may participate in neurodegenerative disease mechanisms at the amyloid peptide level. From a chemical perspective, this group of oxysterols is highly reactive in that they contain an aldehyde group, which in Aβ may condense with amine at the N-terminus and with ε-amino groups of Lys-16 and Lys-28. This modification would increase the hydrophobicity of the peptide, possibly increasing its propensity to misfold and assemble into the quaternary structures associated with AD. Zhang et al. [49] demonstrated that 5,6-secosterol and its carboxaldehyde covalently modify Aβ, dramatically accelerating its amyloidogenesis in vitro. These results, validated by the detection of ozonolysis products in the brain [49], provide a possible chemical link between hypercholesterolemia, inflammation, atherosclerosis, and sporadic AD.

7-dehydrocholesterol, a precursor of cholesterol and vitamin D, is extremely reactive and has been found to have the highest propagation rate constant known for any lipid toward free radical chain oxidation [51]. The rate constant for autoxidation of 7-dehydrocholesterol ($kp \approx 2{,}200\ M^{-1}\ s^{-1}$) is ten times higher than that of arachidonic acid ($kp \approx 200\ M^{-1}\ s^{-1}$)[51]. Elevated levels of 7-DHC and reduced levels of cholesterol are observed in tissues and fluids of patients with Smith–Lemli–Opitz syndrome (SLOS), which is a metabolic disorder resulting from mutations in the gene encoding 7-dehydrocholesterol reductase (*DHCR7*), the enzyme that catalyzes the reduction of 7-dehydrocholesterol (7-DHC) to cholesterol [52]. SLOS is characterized by a broad spectrum of phenotypes including multiple congenital malformations and mental retardation. Oxysterols derivatives of 7-DHC oxidation are biologically active towards Neuro2a cells [53], and at sub-μM to μM concentrations trigger changes in expression of genes involved in lipid biosynthesis and cell proliferation [53]. Among these oxysterols, the compound 3β,5α,dihydroxycholest-7-ene-6-one has been identified in fibroblasts from patients with Smith–Lemli–Optiz syndrome [51], with a proposed mechanism of formation based on transformation of 7DHC into the intermediates 5α,6α-epoxycholest-7-en-3β-ol and 7-cholesten-3β,5α,6β-triol. These oxysterols, derived from autoxidation processes, may be useful biomarkers to investigate the pathological mechanisms in SLOS and could be exported to other clinical settings of neurodegenerative disorders.

## 9.5 Plasma Oxysterols in Alzheimer's Disease

Almost all the plasma 24OHC in humans seems to have a cerebral origin, and to reflect the number of metabolically active neuronal cells in the brain. In line with the hypothesis that the number of metabolically active neuronal cells are decreased in the brain of patients with neurodegenerative diseases, plasma levels of 24OHC have been reported to be decreased in AD, Vascular Dementia, Multiple Sclerosis and Huntington's Disease in relation to the disease burden and the loss of metabolically active neuronal cells [54–61]. The progressive, extensive, atrophy in AD would be expected to be associated to a progressive reduction of plasma 24OHC [62–64]. In support of this concept, a significant correlation between 24OHC and hippocampal volume [59] was found in aged individuals referring episodes of cognitive impairment [60]. Higher plasma 24OHC was described in AD patients in agreement with the hypothesis that neuronal damage could be associated to a higher turnover of neuronal membranes, which provides higher levels of cholesterol to be converted into 24OHC [65]. Recently, plasma 24OHC was found significantly increased in AD patients [63]. However, in about 50 % of controls evidence of brain atrophy was detected and CSF biomarker definition was missing. Levels of 24OHC were found to correlate with C reactive protein, providing a link between 24OHC metabolism and systemic inflammation [63]. Difference in patients' selection might partially explain these conflicting data. Other studies reported that plasma levels of

24OHC and 27-OCH in AD patients were not significantly different than in control subjects [66]. In this study, a high positive correlation was observed between 24S-OHC and 27-OHC in the whole patient group, suggesting that the two oxysterols share the mechanism contributing to their plasma level [66].

Plasma levels of oxidative stress-driven oxysterols, including 7β-OCH and 7-oxoC, and α-tocopherol were investigated in AD patients and two amnestic MCI subtypes, amnestic single-domain MCI and multi-domain MCI, compared to healthy control subjects [66]. 7β-OHC and 7-oxoC, which are validated markers of lipid peroxidation, have been described as elevated in several conditions associated with increased oxidative stress, such as the vascular disorders [67]. Unlike previous studies in other disease settings [12, 68–70], this study did not detect any significant increase in blood 7β-OCH or 7-oxoC levels of patients, compared to healthy controls. This neutral result could limit the usefulness of these cholesterol oxidation markers in this specific disease. However, an upregulated oxidative stress was reported in this study inasmuch it was found a progressive reduction of the α-tocopherol/cholesterol ratio from HC, throughout MCI, to AD patients. Consumption of vitamin E in the presence of unaltered concentrations of plasma 7β-OHC and 7-oxoC is consistent with an increased flux of free radicals in the brain of dementia patients with consumption of the antioxidant, which is reflected back into the plasmatic vitamin E pool.

Taken together, these results suggest that plasma 24OHC should not be used to discriminate aging from AD patients [31, 66, 71]. A bias in patient selection or in the analytical methods used by different laboratories could account for the discrepancies in blood oxysterol levels in these studies. Because these potential biases limit the usefulness of oxysterols as molecular probes of disease, the standardization of analytical methods is mandatory.

## 9.6 Oxysterols in Cerebrospinal Fluid

About 99 % of the flux of 24OHC from the brain goes directly into the blood-stream and less than 1 % goes first to the CSF. The small fraction in CSF seems to reflect the rate of neuronal degeneration rather than the mass of metabolically active neuronal cells [72]. Also 27OHC could be detected in CSF and its levels were found to correlate with the function of the blood–brain barrier and the albumin ratio [72]. Increased CSF levels of side-chain oxysterols were described in some neurodegenerative disorders such as MS and AD, with a relative higher 24OHC/27OHC ratio [31, 72–75]. In active demyelinating processes, significantly higher levels of both oxysterols were found [31, 72].

Hereditary spastic paresis is a genetically, and clinically, heterogeneous group of degenerative diseases that involve the upper motor neuron, and is characterized by selective axonal loss in the corticospinal tracts and dorsal columns. In Hereditary spastic paresis, in which a SPG5 mutation in CYP7B1 is described, high levels of 27OHC are reported, both in plasma (up to 5–9-fold) and CSF (up to 30-fold) [76].

## 9.7 Conclusions

There is increasing need to understand mechanisms of neurodegenerative disorders, and to identify reliable biomarkers that may allow for less invasive and more accurate diagnosis, as well as serve as a predictor of disease progression and treatment response. Cholesterol and its oxygenated products, oxysterols, play prominent roles in brain functions and in the pathophysiology of amyloidogenesis and neurodegeneration and are subjects of intense investigation. The increasing number of oxysterols, formed by enzymatic catalysis and/or oxidative stress mechanisms, have potential biological importance and require further characterization at the chemical, biological and clinical level. Cholesterol metabolism in the brain may provide novel therapeutic target for management of neurodegenerative disorders.

**Acknowledgments** The authors wish to gratefully acknowledge the collaboration along the years of Prof. I. Björkhem, Prof. U. Diczfalusy, Dr. A. Salomon, Prof. M. Kivipelto, Dr. M. Shafaati at Karolinska Insitutet, Stockholm, Sweden. The authors are grateful for discussion with all members of the European Network on Oxysterols Research (ENOR)(www.oxysterols.com). This work was supported by grants from Ministero dell' Università, Ricerca Scientifica e Tecnologica (PRIN 2007L7BHK8) and Sapienza University of Rome (to L.I.); and Italian Ministry of Health (Fondi per giovani Ricercatori 2008) (to V.L.).

## References

1. Iuliano L. The oxidant stress hypothesis of atherogenesis. Lipids. 2001;36(Suppl):S41–4.
2. Pikuleva IA. Cholesterol-metabolizing cytochromes P450: implications for cholesterol lowering. Expert Opin Drug Metab Toxicol. 2008;4:1403–14.
3. Björkhem I, Eggertsen G. Genes involved in initial steps of bile acid synthesis. Curr Opin Lipidol. 2001;12:97–103.
4. Heverin M, Meaney S, Brafman A, Shafir M, Olin M, Shafaati M, von Bahr S, Larsson L, Lovgren-Sandblom A, Diczfalusy U, Parini P, Feinstein E, Bjorkhem I. Studies on the cholesterol-free mouse: strong activation of LXR-regulated hepatic genes when replacing cholesterol with desmosterol. Arterioscler Thromb Vasc Biol. 2007;27:2191–7.
5. Björkhem I, Reihner E, Angelin B, Ewerth S, Akerlund JE, Einarsson K. On the possible use of the serum level of 7 alpha-hydroxycholesterol as a marker for increased activity of the cholesterol 7 alpha-hydroxylase in humans. J Lipid Res. 1987;28:889–94.
6. Bodin K, Bretillon L, Aden Y, Bertilsson L, Broome U, Einarsson C, Diczfalusy U. Antiepileptic drugs increase plasma levels of 4beta-hydroxycholesterol in humans: evidence for involvement of cytochrome p450 3A4. J Biol Chem. 2001;276:38685–9.
7. Diczfalusy U, Miura J, Roh HK, Mirghani RA, Sayi J, Larsson H, Bodin KG, Allqvist A, Jande M, Kim JW, Aklillu E, Gustafsson LL, Bertilsson L. 4Beta-hydroxycholesterol is a new endogenous CYP3A marker: relationship to CYP3A5 genotype, quinine 3-hydroxylation and sex in Koreans, Swedes and Tanzanians. Pharmacogenet Genomics. 2008;18:201–8.
8. Nelson JA, Steckbeck SR, Spencer TA. Biosynthesis of 24,25-epoxycholesterol from squalene 2,3;22,23-dioxide. J Biol Chem. 1981;256:1067–8.
9. Iuliano L. Pathways of cholesterol oxidation via non-enzymatic mechanisms. Chem Phys Lipids. 2011;164:457–68.
10. Niki E. Lipid peroxidation: physiological levels and dual biological effects. Free Radic Biol Med. 2009;47:469–84.

11. Gardner HW. Oxygen radical chemistry of polyunsaturated fatty acids. Free Radic Biol Med. 1989;7:65–86.
12. Iuliano L, Micheletta F, Natoli S, Ginanni Corradini S, Iappelli M, Elisei W, Giovannelli L, Violi F, Diczfalusy U. Measurement of oxysterols and alpha-tocopherol in plasma and tissue samples as indices of oxidant stress status. Anal Biochem. 2003;312:217–23.
13. Larsson H, Bottiger Y, Iuliano L, Diczfalusy U. In vivo interconversion of 7beta-hydroxycholesterol and 7-ketocholesterol, potential surrogate markers for oxidative stress. Free Radic Biol Med. 2007;43:695–701.
14. Janowski BA, Willy PJ, Devi TR, Falck JR, Mangelsdorf DJ. An oxysterol signalling pathway mediated by the nuclear receptor LXR alpha. Nature. 1996;383:728–31.
15. Murphy C, Murray AM, Meaney S, Gafvels M. Regulation by SREBP-2 defines a potential link between isoprenoid and adenosylcobalamin metabolism. Biochem Biophys Res Commun. 2007;355:359–64.
16. Olkkonen VM, Hynynen R. Interactions of oxysterols with membranes and proteins. Mol Aspects Med. 2009;30:123–33.
17. Mann RK, Beachy PA. Novel lipid modifications of secreted protein signals. Annu Rev Biochem. 2004;73:891–923.
18. Murphy RC, Johnson KM. Cholesterol, reactive oxygen species, and the formation of biologically active mediators. J Biol Chem. 2008;283:15521–5.
19. Li-Hawkins J, Lund EG, Bronson AD, Russell DW. Expression cloning of an oxysterol 7alpha-hydroxylase selective for 24-hydroxycholesterol. J Biol Chem. 2000;275:16543–9.
20. Meaney S, Heverin M, Panzenboeck U, Ekstrom L, Axelsson M, Andersson U, Diczfalusy U, Pikuleva I, Wahren J, Sattler W, Bjorkhem I. Novel route for elimination of brain oxysterols across the blood–brain barrier: conversion into 7alpha-hydroxy-3-oxo-4-cholestenoic acid. J Lipid Res. 2007;48:944–51.
21. Pfrieger FW, Ungerer N. Cholesterol metabolism in neurons and astrocytes. Prog Lipid Res. 2011;50:357–71.
22. Björkhem I, Meaney S. Brain cholesterol: long secret life behind a barrier. Arterioscler Thromb Vasc Biol. 2004;24:806–15.
23. Björkhem I, Leoni V, Meaney S. Genetic connections between neurological disorders and cholesterol metabolism. J Lipid Res. 2010;51:2489–503.
24. de Chaves EP, Narayanaswami V. Apolipoprotein E and cholesterol in aging and disease in the brain. Future Lipidol. 2008;3:505–30.
25. Bu G. Apolipoprotein E and its receptors in Alzheimer's disease: pathways, pathogenesis and therapy. Nat Rev Neurosci. 2009;10:333–44.
26. Lutjohann D, Breuer O, Ahlborg G, Nennesmo I, Siden A, Diczfalusy U, Bjorkhem I. Cholesterol homeostasis in human brain: evidence for an age-dependent flux of 24S-hydroxycholesterol from the brain into the circulation. Proc Natl Acad Sci USA. 1996; 93:9799–804.
27. Xie C, Lund EG, Turley SD, Russell DW, Dietschy JM. Quantitation of two pathways for cholesterol excretion from the brain in normal mice and mice with neurodegeneration. J Lipid Res. 2003;44:1780–9.
28. Lund EG, Guileyardo JM, Russell DW. cDNA cloning of cholesterol 24-hydroxylase, a mediator of cholesterol homeostasis in the brain. Proc Natl Acad Sci USA. 1999;96: 7238–43.
29. Kotti TJ, Ramirez DM, Pfeiffer BE, Huber KM, Russell DW. Brain cholesterol turnover required for geranylgeraniol production and learning in mice. Proc Natl Acad Sci USA. 2006;103:3869–74.
30. Blalock EM, Geddes JW, Chen KC, Porter NM, Markesbery WR, Landfield PW. Incipient Alzheimer's disease: microarray correlation analyses reveal major transcriptional and tumor suppressor responses. Proc Natl Acad Sci USA. 2004;101:2173–8.
31. Leoni V, Masterman T, Patel P, Meaney S, Diczfalusy U, Bjorkhem I. Side chain oxidized oxysterols in cerebrospinal fluid and the integrity of blood–brain and blood-cerebrospinal fluid barriers. J Lipid Res. 2003;44:793–9.

32. Wong J, Quinn CM, Guillemin G, Brown AJ. Primary human astrocytes produce 24(S), 25-epoxycholesterol with implications for brain cholesterol homeostasis. J Neurochem. 2007;103:1764–73.
33. Di Paolo G, Kim TW. Linking lipids to Alzheimer's disease: cholesterol and beyond. Nat Rev Neurosci. 2011;12:284–96.
34. Jack Jr CR, Knopman DS, Jagust WJ, Shaw LM, Aisen PS, Weiner MW, Petersen RC, Trojanowski JQ. Hypothetical model of dynamic biomarkers of the Alzheimer's pathological cascade. Lancet Neurol. 2010;9:119–28.
35. Kivipelto M, Solomon A. Cholesterol as a risk factor for Alzheimer's disease – epidemiological evidence. Acta Neurol Scand Suppl. 2006;185:50–7.
36. Saunders AM, Strittmatter WJ, Schmechel D, George-Hyslop PH, Pericak-Vance MA, Joo SH, Rosi BL, Gusella JF, Crapper-MacLachlan DR, Alberts MJ, et al. Association of apolipoprotein E allele epsilon 4 with late-onset familial and sporadic Alzheimer's disease. Neurology. 1993;43:1467–72.
37. Frears ER, Stephens DJ, Walters CE, Davies H, Austen BM. The role of cholesterol in the biosynthesis of beta-amyloid. Neuroreport. 1999;10:1699–705.
38. Howland DS, Trusko SP, Savage MJ, Reaume AG, Lang DM, Hirsch JD, Maeda N, Siman R, Greenberg BD, Scott RW, Flood DG. Modulation of secreted beta-amyloid precursor protein and amyloid beta-peptide in brain by cholesterol. J Biol Chem. 1998;273:16576–82.
39. van den Kommer TN, Dik MG, Comijs HC, Lutjohann D, Lips P, Jonker C, Deeg DJ. The role of extracerebral cholesterol homeostasis and ApoE e4 in cognitive decline. Neurobiol Aging. 2012;33(622):e617–28.
40. Solomon A, Kareholt I, Ngandu T, Winblad B, Nissinen A, Tuomilehto J, Soininen H, Kivipelto M. Serum cholesterol changes after midlife and late-life cognition: twenty-one-year follow-up study. Neurology. 2007;68:751–6.
41. Stewart R, White LR, Xue QL, Launer LJ. Twenty-six-year change in total cholesterol levels and incident dementia: the Honolulu-Asia Aging Study. Arch Neurol. 2007;64:103–7.
42. Prasanthi JR, Huls A, Thomasson S, Thompson A, Schommer E, Ghribi O. Differential effects of 24-hydroxycholesterol and 27-hydroxycholesterol on beta-amyloid precursor protein levels and processing in human neuroblastoma SH-SY5Y cells. Mol Neurodegener. 2009;4:1.
43. Heverin M, Bogdanovic N, Lutjohann D, Bayer T, Pikuleva I, Bretillon L, Diczfalusy U, Winblad B, Bjorkhem I. Changes in the levels of cerebral and extracerebral sterols in the brain of patients with Alzheimer's disease. J Lipid Res. 2004;45:186–93.
44. Bjorkhem I, Cedazo-Minguez A, Leoni V, Meaney S. Oxysterols and neurodegenerative diseases. Mol Aspects Med. 2009;30:171–9.
45. Nelson TJ, Alkon DL. Oxidation of cholesterol by amyloid precursor protein and beta-amyloid peptide. J Biol Chem. 2005;280:7377–87.
46. Wang N, Yvan-Charvet L, Lutjohann D, Mulder M, Vanmierlo T, Kim TW, Tall AR. ATP-binding cassette transporters G1 and G4 mediate cholesterol and desmosterol efflux to HDL and regulate sterol accumulation in the brain. FASEB J. 2008;22:1073–82.
47. Kettle AJ, Clark BM, Winterbourn CC. Superoxide converts indigo carmine to isatin sulfonic acid: implications for the hypothesis that neutrophils produce ozone. J Biol Chem. 2004;279:18521–5.
48. Babior BM, Takeuchi C, Ruedi J, Gutierrez A, Wentworth Jr P. Investigating antibody-catalyzed ozone generation by human neutrophils. Proc Natl Acad Sci USA. 2003;100: 3031–4.
49. Zhang Q, Powers E, Nieva J, Huff M, Dendle M, Bieschke J, Glabe C, Eschenmoser A, Wentworth P, Lerner R. Metabolite-initiated protein misfolding may trigger Alzheimer's disease. Proc Natl Acad Sci. 2004;101:4752.
50. Kettle AJ, Winterbourn CC. Do neutrophils produce ozone? An appraisal of current evidence. Biofactors. 2005;24:41–5.
51. Honda A, Yamashita K, Hara T, Ikegami T, Miyazaki T, Shirai M, Xu G, Numazawa M, Matsuzaki Y. Highly sensitive quantification of key regulatory oxysterols in biological samples by LC-ESI-MS/MS. J Lipid Res. 2009;50:350–7.

52. Waterham HR. Inherited disorders of cholesterol biosynthesis. Clin Genet. 2002;61:393–403.
53. Korade Z, Xu L, Shelton R, Porter NA. Biological activities of 7-dehydrocholesterol-derived oxysterols: implications for Smith-Lemli-Opitz syndrome. J Lipid Res. 2010;51:3259–69.
54. Karrenbauer VD, Leoni V, Lim ET, Giovannoni G, Ingle GT, Sastre-Garriga J, Thompson AJ, Rashid W, Davies G, Miller DH, Bjorkhem I, Masterman T. Plasma cerebrosterol and magnetic resonance imaging measures in multiple sclerosis. Clin Neurol Neurosurg. 2006;108:456–60.
55. Leoni V, Strittmatter L, Zorzi G, Zibordi F, Dusi S, Garavaglia B, Venco P, Caccia C, Souza AL, Deik A, Clish CB, Rimoldi M, Ciusani E, Bertini E, Nardocci N, Mootha VK, Tiranti V. Metabolic consequences of mitochondrial coenzyme A deficiency in patients with PANK2 mutations. Mol Genet Metab. 2012;105:463–71.
56. Leoni V, Caccia C. Oxysterols as biomarkers in neurodegenerative diseases. Chem Phys Lipids. 2011;164:515–24.
57. Leoni V, Mariotti C, Tabrizi SJ, Valenza M, Wild EJ, Henley SM, Hobbs NZ, Mandelli ML, Grisoli M, Bjorkhem I, Cattaneo E, Di Donato S. Plasma 24S-hydroxycholesterol and caudate MRI in pre-manifest and early Huntington's disease. Brain. 2008;131:2851–9.
58. Leoni V, Masterman T, Diczfalusy U, De Luca G, Hillert J, Bjorkhem I. Changes in human plasma levels of the brain specific oxysterol 24S-hydroxycholesterol during progression of multiple sclerosis. Neurosci Lett. 2002;331:163–6.
59. Koschack J, Lutjohann D, Schmidt-Samoa C, Irle E. Serum 24S-hydroxycholesterol and hippocampal size in middle-aged normal individuals. Neurobiol Aging. 2009;30:898–902.
60. Solomon A, Kivipelto M, Wolozin B, Zhou J, Whitmer RA. Midlife serum cholesterol and increased risk of Alzheimer's and vascular dementia three decades later. Dement Geriatr Cogn Disord. 2009;28:75–80.
61. Kolsch H, Heun R, Kerksiek A, Bergmann KV, Maier W, Lutjohann D. Altered levels of plasma 24S- and 27-hydroxycholesterol in demented patients. Neurosci Lett. 2004;368:303–8.
62. Besga A, Cedazo-Minguez A, Kareholt I, Solomon A, Bjorkhem I, Winblad B, Leoni V, Hooshmand B, Spulber G, Gonzalez-Pinto A, Kivipelto M, Wahlund LO. Differences in brain cholesterol metabolism and insulin in two subgroups of patients with different CSF biomarkers but similar white matter lesions suggest different pathogenic mechanisms. Neurosci Lett. 2012;510:121–6.
63. Zuliani G, Donnorso MP, Bosi C, Passaro A, Dalla Nora E, Zurlo A, Bonetti F, Mozzi AF, Cortese C. Plasma 24S-hydroxycholesterol levels in elderly subjects with late onset Alzheimer's disease or vascular dementia: a case–control study. BMC Neurol. 2011;11:121.
64. Solomon A, Leoni V, Kivipelto M, Besga A, Oksengard AR, Julin P, Svensson L, Wahlund LO, Andreasen N, Winblad B, Soininen H, Bjorkhem I. Plasma levels of 24S-hydroxycholesterol reflect brain volumes in patients without objective cognitive impairment but not in those with Alzheimer's disease. Neurosci Lett. 2009;462:89–93.
65. Lutjohann D, Papassotiropoulos A, Bjorkhem I, Locatelli S, Bagli M, Oehring RD, Schlegel U, Jessen F, Rao ML, von Bergmann K, Heun R. Plasma 24S-hydroxycholesterol (cerebrosterol) is increased in Alzheimer and vascular demented patients. J Lipid Res. 2000;41:195–8.
66. Iuliano L, Monticolo R, Straface G, Spoletini I, Gianni W, Caltagirone C, Bossu P, Spalletta G. Vitamin E and enzymatic/oxidative stress-driven oxysterols in amnestic mild cognitive impairment subtypes and Alzheimer's disease. J Alzheimers Dis. 2010;21:1383–92.
67. Arca M, Natoli S, Micheletta F, Riggi S, Di Angelantonio E, Montali A, Antonini TM, Antonini R, Diczfalusy U, Iuliano L. Increased plasma levels of oxysterols, in vivo markers of oxidative stress, in patients with familial combined hyperlipidemia: reduction during atorvastatin and fenofibrate therapy. Free Radic Biol Med. 2007;42:698–705.
68. Iuliano L, Monticolo R, Straface G, Zullo S, Galli F, Boaz M, Quattrucci S. Association of cholesterol oxidation and abnormalities in fatty acid metabolism in cystic fibrosis. Am J Clin Nutr. 2009;90:477–84.
69. Micheletta F, Natoli S, Misuraca M, Sbarigia E, Diczfalusy U, Iuliano L. Vitamin E supplementation in patients with carotid atherosclerosis: reversal of altered oxidative stress status in plasma but not in plaque. Arterioscler Thromb Vasc Biol. 2004;24:136–40.

70. Corradini SG, Micheletta F, Natoli S, Iappelli M, Di Angelantonio E, De Marco R, Elisei W, Siciliano M, Rossi M, Berloco P, Attili AF, Diczfalusy U, Iuliano L. High preoperative recipient plasma 7beta-hydroxycholesterol is associated with initial poor graft function after liver transplantation. Liver Transpl. 2005;11:1494–504.
71. Teunissen CE, Dijkstra CD, Polman CH, Hoogervorst EL, von Bergmann K, Lutjohann D. Decreased levels of the brain specific 24S-hydroxycholesterol and cholesterol precursors in serum of multiple sclerosis patients. Neurosci Lett. 2003;347:159–62.
72. Leoni V, Masterman T, Mousavi FS, Wretlind B, Wahlund LO, Diczfalusy U, Hillert J, Bjorkhem I. Diagnostic use of cerebral and extracerebral oxysterols. Clin Chem Lab Med. 2004;42:186–91.
73. Shafaati M, Solomon A, Kivipelto M, Bjorkhem I, Leoni V. Levels of ApoE in cerebrospinal fluid are correlated with Tau and 24S-hydroxycholesterol in patients with cognitive disorders. Neurosci Lett. 2007;425:78–82.
74. Leoni V, Shafaati M, Salomon A, Kivipelto M, Bjorkhem I, Wahlund LO. Are the CSF levels of 24S-hydroxycholesterol a sensitive biomarker for mild cognitive impairment? Neurosci Lett. 2006;397:83–7.
75. Schonknecht P, Lutjohann D, Pantel J, Bardenheuer H, Hartmann T, von Bergmann K, Beyreuther K, Schroder J. Cerebrospinal fluid 24S-hydroxycholesterol is increased in patients with Alzheimer's disease compared to healthy controls. Neurosci Lett. 2002;324:83–5.
76. Schule R, Siddique T, Deng HX, Yang Y, Donkervoort S, Hansson M, Madrid RE, Siddique N, Schols L, Bjorkhem I. Marked accumulation of 27-hydroxycholesterol in SPG5 patients with hereditary spastic paresis. J Lipid Res. 2009;51:819–23.

# Chapter 10
# Brain Hypometabolism, Oxidative Stress, Maternal Transmission, and Risk of Late-Onset Alzheimer's Disease

**Lisa Mosconi, John Murray, Pauline McHugh, and Mony de Leon**

**Abstract** Alzheimer's disease (AD) is an age-dependent neurodegenerative disorder associated with progressive loss of cognitive function. Positron emission tomography (PET) imaging with 2-[$^{18}$F]fluoro-2-deoxy-D-glucose (FDG) as the tracer has long been used to measure resting-state cerebral metabolic rates of glucose, a proxy for neuronal activity. Several FDG-PET studies have shown that metabolic reductions occur decades before onset of AD symptoms, suggesting that metabolic deficits may be an upstream event in at least some late-onset AD cases. This review explores this possibility, initially by discussing the link between AD pathology and neurodegeneration, with a focus on the metabolic pathways involved in neuronal function and bioenergetics, and the relationship between glucose metabolism, oxidative stress, and AD. This will be followed by a summary of the FDG-PET method, ranging from physics to kinetic modeling, and PET findings in AD. We will then discuss recent findings of progressive FDG-PET hypometabolism in adult children of mothers, but not fathers, affected by late-onset AD. Given the connection between glucose metabolism and mitochondria and the fact that mitochondrial DNA is maternally inherited in humans, it will be argued that altered bioenergetics may be an upstream event in individuals with a maternal family history of AD. Biomarkers of AD have great potential for identifying AD endophenotypes in cognitively intact individuals at risk for AD, which may help direct investigation of potential susceptibility genes.

L. Mosconi, Ph.D. (✉) • J. Murray, M.S. • P. McHugh, M.D. • M. de Leon, D.Ed.
Department of Psychiatry, Center for Brain Health, NYU School of Medicine,
145 East 32nd Street, 2nd floor, New York, NY 10016, USA
e-mail: lisa.mosconi@nyumc.org

D. Praticò and P. Mecocci (eds.), *Studies on Alzheimer's Disease*, Oxidative Stress in Applied Basic Research and Clinical Practice, DOI 10.1007/978-1-62703-598-9_10,

## 10.1 Alzheimer's Disease: Clinical Features, Prevalence, and Pathology

Alzheimer's disease (AD), the leading cause of dementia in the elderly, is a neurodegenerative disorder with insidious onset and progressive declines in memory, attention, and language. At present, AD affects approximately 10 % of individuals 65 years of age, with the prevalence doubling every 5 years up to age 80, above which the prevalence exceeds 40 % [1].

Longitudinal studies of normal individuals who go on to develop AD show that there is a somewhat abrupt transition in cognitive symptom decline between the preclinical stage and the early stage of AD. Gradual cognitive decline in the preclinical stage reaches an inflection point that gives way to a comparatively steep loss of cognitive abilities, which is the hallmark of clinical AD. Typically, the relatively rapid loss of cognitive abilities following the inflection point is what leads family members or caregivers to bring patients in for evaluation. However, this kind of "tipping point" suggests that by the time patients come in for diagnosis, too much irreversible brain damage may have already occurred for any treatments to be effective. Interventions, once they are developed, ideally would be implemented long before symptoms occur. While risk factors such as apolipoprotein E (ApoE) ε4 genotype and family history have been identified, their predictive value remains to be established, and its presence may not be enough to justify the potential risks of medical interventions (as they become available) in non-symptomatic individuals. Therefore, another major goal in AD research is the identification of diagnostic markers, especially for the preclinical stages of disease when symptoms are not yet apparent.

Currently, a definite diagnosis of AD can only be made by neuropathology, which is regarded as the gold standard, and is based on postmortem detection of specific pathological lesions: amyloid-beta (Aβ) plaques in the extracellular space and blood vessels, intracellular neurofibrillary tangles (NFT), and neuronal and synaptic loss in specific brain regions [2–4]. Changes in brain histopathology, and consequently in brain structure and function, are known to precede the signs and symptoms of disease by many years. Neurodegeneration in AD is estimated to begin 20–30 years before any clinical manifestations of disease become evident [2–4].

Some decades ago the largely accepted paradigm was that people with AD pathology had dementia and people without AD pathology did not. This simple division started blurring when pathology studies revealed plaques and tangles in a sizable fraction of elders who had died with their cognition intact and that the number, density, and location of amyloid plaques were not particularly correlated with either AD symptoms or severity [4, 5]. Biomarker and longitudinal aging studies of the past 20 years swept aside this old binary view in favor of a more complex and dynamic picture. At present, the majority view holds that both pathologic and clinical changes occur gradually over time, and that, *while there can be no clinical symptoms without AD pathology, there may be AD pathology without clinical symptoms*. Presence of neurodegeneration, rather than plaques and tangles, is best associated with the onset of cognitive deficits in AD [6]. It has become increasingly clear that

AD is a polygenic and multifactorial disease, which may include different pathogenetic mechanisms that nonetheless result in similar clinical outcomes. The following paragraphs explore this possibility, initially by discussing the link between AD pathology and neurodegeneration, with a focus on the metabolic pathways involved in neuronal function and bioenergetics. We will then discuss the relationship between glucose metabolism, oxidative stress, and AD. This will be followed by a summary of positron emission tomography (PET) imaging of brain glucose metabolism, ranging from physics to kinetic modeling, and PET findings in AD. We will then discuss recent findings of selective reductions of PET brain glucose metabolism in adult children of mothers, but not fathers, affected by late-onset AD (LOAD). These data led to discovery of a maternally inherited AD endophenotype characterized by progressive brain hypometabolism, increased amyloid load, atrophy, and oxidative stress. Finally, given the connection between glucose metabolism and mitochondria, and the fact that mitochondrial DNA is maternally inherited in humans, it will be argued that altered bioenergetics may be an upstream event in individuals with a maternal family history of AD.

## 10.2 Alzheimer's Disease and Bioenergetics

### *10.2.1 Highly Metabolic Neurons Are Affected Early in Alzheimer's Disease*

The brain regions experiencing the earliest changes in AD are the medial temporal lobes (entorhinal cortex, hippocampus) [2, 3, 7]. In the neocortex, the pyramidal cells anatomically connected to the entorhinal cortex and the CA1 and subiculum regions of the hippocampus are particularly prone to NFT formation and degeneration, whereas primary sensorimotor and occipital areas and cerebellum exhibit minimal pathology and neuronal loss [2, 3, 7]. Disruption of the pyramidal neurons in the perforant path is thought to disconnect the hippocampus from the rest of the cortex, strongly contributing to the decline in memory observed in early AD [6]. Despite a predilection for the neocortex early on, Aβ depositions are also found in the medial temporal lobes at later stages of disease [7]. Soluble Aβ oligomers disrupt long-term potentiation and reduce synaptic plasticity in the hippocampus early in the course of the disease [8]. The medial temporal regions and functionally connected neocortex are most affected in AD, but the pathology is not uniform nor does it affect all cell types. Despite recent advances in our understanding of the molecular basis and pathophysiology of neurodegenerative disorders, the problem of *selective neuronal vulnerability* has proved difficult to solve. Why is the hippocampus initially affected in AD? Why and how does neuronal death spread to other brain regions as the disease progresses?

Recent progress has begun to show how cellular and molecular changes that occur during normal aging render neurons vulnerable to degeneration, and how disease specific genetic and environmental factors determine which neurons succumb and which are resistant to advancing pathology. The physical and molecular characteristics of neurons, their functional properties, and their location in neural circuits are all likely to influence their fate during aging (for review see [6, 9]). Large projection neurons with relatively long axons are most damaged in AD. These neurons have three main characteristics:

1. *Have high-energy requirements.* These neurons have high metabolic rates. As such, their functionality is directly dependent on glucose availability and utilization.
2. *Rely primarily on axonal transport* for functional support (anterograde and retrograde). The axons of cortico-cortical projection neurons travel long distances, which makes these neurons more prone to receiving multiple insults and more sensitive to cytoskeletal dysfunction [6, 9].
3. *Have a large cell surface area.* This increases exposure of the cells to toxic environmental conditions.

Normal synapse function requires a multitude of coordinated mechanisms, including generation of gene products responsible for formation and maintenance of membrane complexes, synthesis and delivery of mRNAs, proteins, and transmitters, regulation of vesicle trafficking, release and reuptake, and many more. It is implied that, for all these actions to be performed efficiently, *sufficient energetic substrates must be supplied and utilized.* Under normal physiological conditions, the brain relies almost exclusively on glucose as its main substrate for energy production. As there are only minor glycogen stores in brain, a permanent supply of glucose from the blood stream is necessary to maintain brain function [10, 11]. In the brain, the free energy necessary to drive most cellular reactions is derived from phosphorylation of ATP, which is mostly produced in the mitochondria from the oxidation of glucose under aerobic conditions. This also dictates that any interruption of glucose supply causes an almost immediate failure of brain function, as can be seen from the rapid loss of consciousness after interruption of blood supply to brain. Disruptions in glucose metabolism are a direct determinant of synaptic dysfunction [10, 11]. In keeping with this observation, recent evidence suggests that altered neuronal energy metabolism is a very early change and a strong correlate of clinical impairment in AD, as reviewed below.

## 10.2.2 Brain Glucose Metabolism

### 10.2.2.1 Molecular Basis of Glucose Metabolism

Brain glucose metabolism is comprised of a series of processes by which glucose is converted into ATP to be used for cellular energy. This is accomplished in two steps: glycolysis in the cytoplasm and Krebs cycle and oxidative phosphorylation in the mitochondria. As shown in Fig. 10.1, under aerobic conditions, glucose is transported from blood to tissue by glucose transporters where it is phosphorylated into glucose-6-phosphate (G-6-P) by the hexokinase and glucokinase enzymes and is

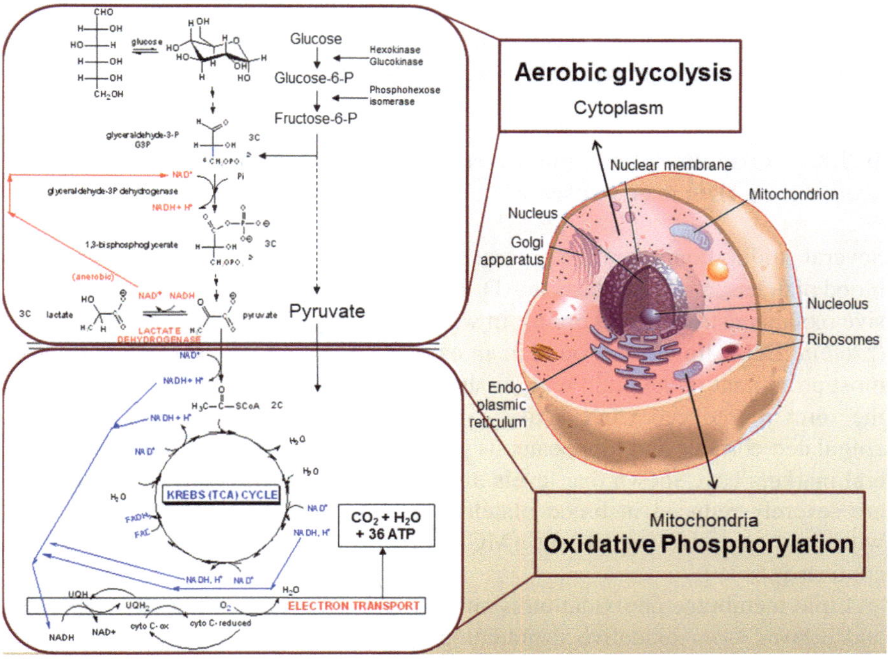


**Fig. 10.1** Aerobic cell respiration: glucose metabolism and the electron transport chain in mitochondria. During respiration, glucose is oxidized to pyruvate down the glycolytic pathway in the cytoplasm. Pyruvate enters the mitochondria and enters the Krebs cycle. Each time the fuel molecule is oxidized, the released electrons are collected by electron carriers and then delivered to the electron transport chain (ETC). The electrons are transferred from protein to protein down the ETC chain (oxidative phosphorylation), which results in production of $CO_2$, water and ATP: $C_6H_{12}O_6 + 6O_2 \Rightarrow 6CO_2 + 6H_2O + 30ATP$

then further metabolized down the glycolytic pathway until it is reduced to pyruvate. Pyruvate then enters the mitochondria to undergo further enzymatic degradation under oxidative phosphorylation (OXPHOS). OXPHOS is the process that accounts for the high ATP yield by the cells and is essential to the maintenance of cellular homeostasis and function. Pyruvate combines with oxygen to form an acetyl group, releasing carbon dioxide. The acetyl group ($CH_3CO$) is then combined with coenzyme A as acetyl coenzyme A and enters the Krebs cycle. During this series of reactions, each acetyl group is oxidized to form two molecules of carbon dioxide, and the energy released is transferred to four electron carrier molecules that form the electron transport chain (ETC, i.e., mitochondrial respiration). Although electron transport occurs with great efficiency, a small percentage of electrons are prematurely leaked to oxygen at the ETC complex I and III steps, resulting in the formation of reactive oxygen species (ROS), including free radicals and peroxides. Oxidative stress is caused by an imbalance between ROS production and the cell's ability to detoxify them [12]. Due to their reactions with macromolecules, the generation of ROS can lead to immediate oxidative damage of cell lipids, proteins, and DNA, which eventually results in neuronal damage and death [9, 12, 13]. Therefore, reduced activity of one or more of the ETC enzymes could compromise ATP synthesis and increase production of ROS, with damage to neurons.

#### 10.2.2.2 Oxidative Stress and Altered Glucose Metabolism in Alzheimer's Disease

Several studies confirmed the presence of severe oxidative stress in brain tissue, blood platelets, and fibroblasts in AD. At postmortem, there is evidence for extensive oxidative stress in AD brains, in which basically all cellular macromolecules (protein, DNA, lipids) are found in an oxidized form [14–16]. Oxidative stress is most prominent in the brain regions showing degeneration in AD, particularly in the form of reduced COX activity in parietotemporal, frontal, and posterior cingulated cortices and hippocampus [14–16]. Several in vivo studies of peripheral markers have shown that levels of ETC enzymes, particularly COX activity, are severely reduced in blood platelets and fibroblasts in AD and in patients with mild cognitive impairment (MCI), often a prodrome to AD, compared to controls [16–19].

Lipid membrane peroxidation is another early event following ROS increases and derives from oxidative degradation of polyunsaturated fatty acids in cell membranes. Lipid peroxidation products of arachidonic acid and DHA are most prominent in hippocampus and association cortex in AD. Increases in isoprostane levels, a marker of lipid membrane peroxidation, are found in brain regions vulnerable to NFT formation. Isoprostanes correlate with the extent of neurodegeneration in AD, and increases in this biomarker precede Aβ deposition in transgenic animals [20].

Overall, these studies provide evidence for peripheral oxidative stress, such as in blood platelets and fibroblasts that are non-degenerating tissue and should not be affected by CNS pathology. This suggests that COX reductions may not be simply a secondary, or epiphenomenal, consequence of neurodegeneration but may instead represent a systemic damage in AD. These findings support an alternative to the long held amyloid cascade hypothesis that has dominated AD research. The central role for Aβ in AD is strongly supported by studies of the rare early-onset (<60 years) forms of familial AD (EOFAD), which are caused by mutations in the amyloid precursor protein (APP), presenilin 1 and presenilin 2 genes, and are found in large multigenerational families with an autosomal dominant pattern of disease inheritance. However, it is less clear whether Aβ dysmetabolism is an upstream event also for the more common, late-onset form of AD, which accounts for almost 99 % of all AD cases. It is plausible that mitochondria and cell bioenergetics regulate Aβ, and changes in mitochondrial function and cell bioenergetics occur upstream to Aβ changes in sporadic AD, at least in some cases [21]. Investigations of oxidative stress and brain glucose metabolism are particularly suited to test this hypothesis in vivo, particularly by means of neuroimaging techniques.

## 10.3 Positron Emission Tomography Imaging of Brain Glucose Metabolism

Imaging techniques are invaluable in diagnosing AD, as they are generally noninvasive and relatively easy for patients to tolerate. Among these, positron emission tomography (PET) has long been used in the diagnosis of neurodegenerative diseases. PET is a nuclear medicine technique that produces a three-dimensional image or picture of functional processes in the body. The system detects pairs of gamma rays emitted by a positron-emitting radionuclide, which is introduced into the body on a biologically active molecule (i.e., tracer). PET uses unique isotopes (particularly those of carbon, nitrogen, oxygen, and fluorine) to study processes in the body that would be hardly possible to demonstrate by other means. In particular, the development of 2-[$^{18}$F]fluoro-2-deoxy-D-glucose (FDG) in the late 1970s was a major factor in expanding the scope of PET imaging [22]. The FDG tracer is an analog of glucose that couples $^{18}$fluorine (110 min half-life) to glucose. Metabolically active tissues within the body, such as tumors, the brain, and heart, are easily visualized with FDG as these take up more glucose than other tissues. The concentrations of imaged tracer reflect tissue metabolic activity, in terms of regional glucose uptake. FDG-PET offers the unique capability to both visualize and quantitate resting-state cerebral metabolic rates of glucose (CMRglc), which is a proxy for neuronal activity [10, 23] and a direct index of synaptic function and density [24, 25]. The FDG-PET method was developed based on Sokoloff's landmark autoradiographic studies over 25 years ago [22, 23, 26] and was validated for human brain by Michael Phelps in 1979 [27]. FDG-PET has since been used to track AD-related changes in brain metabolism. Oftentimes, PET researchers are asked by molecular and cell biologists what exactly is reflected in the FDG signal. This is a very important question, as the term "glucose metabolism" is used to describe somehow different substrates in nuclear medicine and biology. In biology, glucose metabolism is the process by which the carbohydrate is broken down via multiple enzymatic pathways to produce ATP during oxidative phosphorylation or lactate during anaerobic metabolism. FDG shares a similar beginning but not the same fate.

FDG-PET imaging begins with the intravenous injection of FDG into the subject's arm (Fig. 10.2). Within minutes, the isotope accumulates in the brain, where glucose is used as the primary source of energy. As the radioisotope undergoes positron emission decay (i.e., positive beta decay), it emits positrons, which are antiparticles of the electrons with opposite charge. The emitted positron travels in tissue for a short distance (less than 1 mm), while losing kinetic energy until it decelerates to a point where it collides with a free electron. The collision annihilates both electron and positron, resulting in the conversion of matter to energy in the form of two 511 keV photons (i.e., gamma rays) that are shot at almost 180° to each other. The photons are detected by an array of detectors which surround the patient when they reach the scintillator in the scanning device. As the gamma rays are emitted in opposite directions, it is possible to localize their source along a straight line of coincidence that connects the detectors (i.e., line of response, LOR).

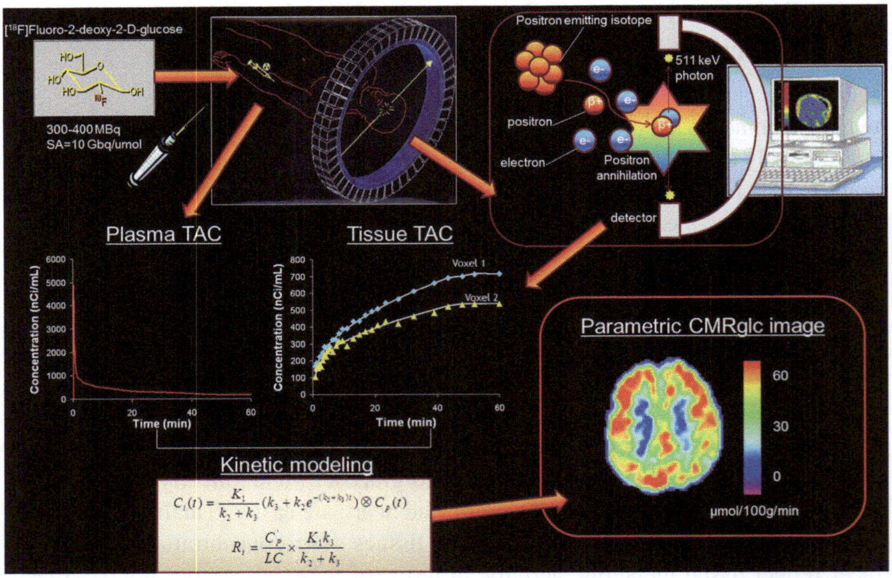


**Fig. 10.2** Schematic representation of FDG-PET imaging: acquisition and analysis

During the scan, a record of tissue concentration is made as the tracer decays. Using statistics collected from tens of thousands of coincidence events, tomographic reconstruction procedures are used to compute the total activity along many LORs and to derive a map of radioactivity concentration in tissue. As some photons are scattered and the signal from tissues deep inside the brain will be attenuated as compared to that from brain structures closer to the detectors, algorithms are applied to correct the images for scatter, attenuation, and decay. Image reconstruction transforms the corrected data back to spatial representation in order to create a visual image. Emission scans are often reconstructed as a two-dimensional image. Three-dimensional reconstructions can also be done using 2D projections from multiple angles. The FDG-PET signal is modeled using either absolute quantitation or by means of "relative" measurements. Absolute quantitation of CMRglc is based on measurement of the accumulation of phosphorylated 2-deoxyglucose in brain relative to the rate of exposure (the input function) while taking into account the effect of the radiolabeled derivative. The method was first developed with carbon-14-labeled deoxyglucose ($^{14C\text{-}DG}$) and was designed specifically to take advantage of the localization made possible by quantitative autoradiography [23, 28]. FDG is an $^{18}$F-labeled analog of glucose. Its molecular composition is similar to that of glucose except for a missing oxygen at the second carbon position which is replaced by $^{18}$F (Fig. 10.3). After a bolus injection into the blood, FDG is transported bidirectionally by the same carriers that transport glucose across the blood–brain barrier (BBB), enters brain tissue, and is phosphorylated into FDG-6-phosphate (FDG-6-P) by the hexokinase enzyme. However, because of the missing oxygen, the following enzyme, glucose-P-isomerase, does not recognize the molecule, and FDG-6-P cannot be converted to fructose-6-phosphate. FDG-6-phospate, once formed,

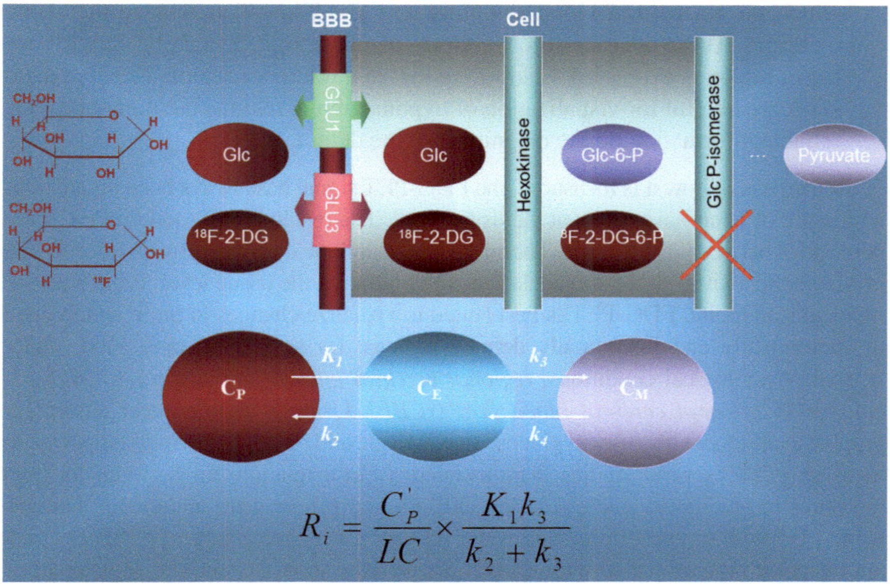


**Fig. 10.3** Theoretical basis of FDG method for measurement of local cerebral glucose utilization using a compartmental model. FDG and glucose (Glc) compete for the carriers that transport both between plasma and tissue (GLU1 and GLU3) across the blood–brain barrier (BBB) and for the hexokinase enzyme that phosphorylates them to their respective hexose 6-phosphates (Glc-6-P and $^{18}$F-2DG-6-P). The *dashed arrow* represents the possibility of glucose-6-phosphate (Glc-6-P) hydrolysis by glucose-6-idromerase activity, in which case the molecule proceeds down the glycolytic pathway to pyruvate. $^{18}$F-2DG-6-P is not further oxidized and remains trapped in tissue. In the theoretical model, $C_P$ represents the total concentration of glucose in arterial plasma, $C_E$ represents glucose concentration in the precursor tissue pool that serves as substrate for hexokinase, and $C_M$ represents glucose concentration in tissue. The constants $K_1$, $k_2$, and $k_3$ represent the rate constants for carrier-mediated transport of glucose from plasma to tissue, for carrier-mediated transport of glucose back from tissue to plasma, and for phosphorylation by hexokinase, respectively. The same model is applied to FDG. The metabolic rate of glucose ($R_i$) is estimated as a function of FDG concentration in plasma ($C_P'$), the lumped constant (LC), and the metabolic rate of FDG $[(K_1 \times k_3)/k_2 + k_3)]$

remains essentially "trapped" in tissue. There is no direct information on whether glucose is eventually converted to ATP in mitochondria, turned into glycogen or lactate, or taken up in the pentose shunt.

Image data can be acquired in different ways, and finding the best model to describe the tracer kinetics continues to be a lively area of research. The two principal FDG-PET measurements are:

(a) Quantitative measurements. These measurements are called "quantitative" or "absolute" as they are expressed as glucose metabolism units, i.e., micromoles of glucose taken up by 100 g of tissue per minute (μmol/100 g/min). Quantitative methods yield reliable estimates of CMRglc, as defined above.
(b) Semiquantitative and nonquantitative measurements. As an alternative to invasive kinetic modeling, simplified reference-tissue techniques have been developed

which do not require blood sampling and yield so-called relative metabolic measures. These measures are "relative" in nature since metabolic activity in a target region is expressed as a ratio to that in a reference region. Relative measures are helpful for visualization of hypometabolic regions but do not provide reliable information on the magnitude of the deficit.

Regardless of how it is measured, the FDG-PET signal has always been regarded as a measure of neuronal activity at the tissue level. The resolution of modern PET scanners varies between 2 and 7 mm. Therefore, the technique does not provide information of metabolic activity at the molecular level but rather at the tissue level. From a physiological standpoint, FDG-PET is best suited to visualize changes in brain metabolism in anatomically and functionally defined regions, for several reasons. First, under normal physiological conditions, *regional cerebral glucose utilization (rCMRglc) is coupled to regional cerebral blood flow (rCBF)*. Using a double-tracer autoradiographic technique with 2-deoxyglucose to rCMRglc and radiolabeled iodoantipyrine to measure rCBF, several studies demonstrated a close relationship between the two measures in preclinical studies [11, 29, 30]. Similar results were obtained in humans using PET with oxygen-15-labeled water ($H_2O_{15}$) for measurement of rCBF. Second, any increase in neuronal activity leads to an increase in rCBF, and the adjustment of rCBF to changes in neuronal activity occurs within a few seconds [31, 32]. Changes in rCBF are thought to be neurogenically mediated vascular responses in preparation for an altered metabolic demand. Third, a multitude of activation studies have shown that rCBF and rCMRglc increase in response to external stimulation in the brain regions responsible for the execution of a specific cognitive process. For example, rCBF and rCMRglc increase in the auditory cortex of individuals listening to music or sounds and in the visual cortex after exposure to visual stimuli [33, 34].

Finally, although it is widely accepted that CNS energy production is based almost exclusively upon the oxidation of glucose, there is a debate over exactly which cells, neurons, and glia are consuming the carbohydrate and what work the energy is being used to support [25, 35, 36]. While glial cells outnumber neurons ten to one, most ATP is probably used by neurons to reverse ion fluxes associated with action potentials and excitatory postsynaptic currents. It has been estimated that as much as 70 % of the energy derived from glucose oxidation is required to convert glutamate, the major neurotransmitter in the brain, to glutamine during glutamatergic transmission [25]. Therefore, glucose utilization as measured with FDG-PET reflects mostly excitatory neuronal activity in brain tissue.

## 10.4 The Search for Endophenotypes of AD

### *10.4.1 FDG-PET Studies of AD*

FDG-PET studies proved quite successful in diagnosing AD. On FDG-PET, AD is characterized by a specific regional pattern of CMRglc reductions in the parietotemporal areas, posterior cingulate cortex, and medial temporal regions [37–40] (Fig. 10.4).

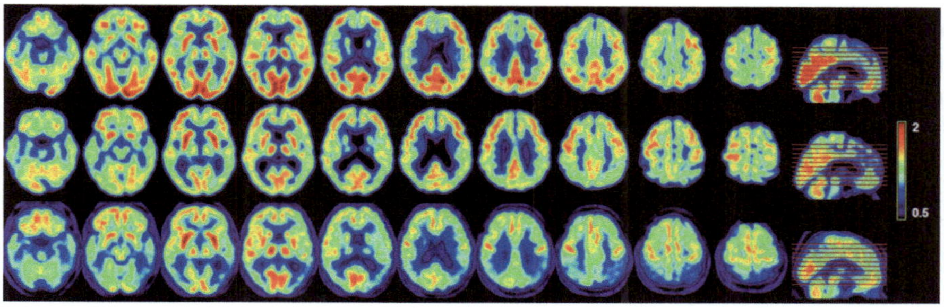


**Fig. 10.4** FDG-PET scans in normal aging and AD. Three representative cases are shown: a 70-year-old cognitively normal individual (*top*), a 70-year-old AD patient with mild dementia (*middle*), and a 75-year-old AD patient with severe AD (*bottom*). FDG-PET scans are standardized uptake value ratios (SUVR) to whole brain metabolism (range 0.5–2). Darker colors reflect lower metabolism. Regional hypometabolism is evident in the parietotemporal regions of both AD patients compared to the control subject and is more pronounced in the severe AD than in the mild AD case

As the disease progresses, frontal association cortices become involved, while cerebellum, striatum, basal ganglia, and primary visual and sensorimotor cortices remain preserved [37, 41]. This in vivo pattern of hypometabolism is found in the vast majority of clinically diagnosed AD patients and in over 85 % pathologically confirmed AD cases [39]. An interesting finding is that clinical AD symptoms essentially never occur without CMRglc decreases, the extent of which is related to the severity of cognitive impairment [42, 43]. Moreover, despite some overlap, the characteristic AD pattern of CMRglc reductions yields high sensitivity in distinguishing AD from controls and from other neurodegenerative dementias [39, 40, 44]. The method yields high discrimination accuracy in patients with mild dementia as well as moderate-to-severe dementia [39, 40, 44].

Additionally, FDG-PET has been used to detect regional hypometabolism consistent with AD in non-demented individuals who would go on to develop AD. A few longitudinal FDG-PET studies of cognitively normal (NL) elderly monitored the progression of some to AD and to mild cognitive impairment (MCI), often a prodrome to AD [45] compared with those who remained normal. These studies showed that CMRglc reductions in hippocampal and parietal regions precede the onset of dementia by many years [46–50] and predict decline from normal cognition to MCI and AD with over 80 % accuracy [47, 48]. Progressive CMRglc reductions were observed years in advance of clinical symptoms in a clinicopathological series of subjects followed with longitudinal in vivo FDG-PET scans from normal cognition to the clinical diagnosis and to postmortem confirmation of AD [49]. More work is needed to establish how early FDG-PET deficits become detectable in the course of disease. Nonetheless, published studies support the use of FDG-PET in the early detection of AD, and, thanks to the method's high sensitivity, regional hypometabolism in AD-vulnerable brain regions has been proposed as an endophenotype of AD. As discussed by Swerdlow [21], when a particular trait or biomarker typically found in conjunction with a disease is detected in persons who do not have the disease, the presence of that trait or biomarker is said to constitute an endophenotype. An endophenotype state does not indicate a carrier will develop the full-blown disease, although it infers that compared to persons without the endophenotype, those with the endophenotype carry an increased risk [21].

### 10.4.2 FDG-PET Endophenotypes and Maternal Transmission of AD

FDG-PET studies led to identification of an AD endophenotype in yet asymptomatic individuals at risk for developing the disease by virtue of having an AD-affected parent, especially those whose mothers had the disease. Evidence for a maternal link to LOAD further promotes the notion that mitochondria-induced oxidative stress may be a primary promoter of neurodegeneration in at least some forms of LOAD, as reviewed below.

Having a first-degree family history of LOAD is a major risk factor for developing the disease [51, 52]. However, the biological mechanisms through which a family history of LOAD confers risk to offspring are largely unknown. Clinical and epidemiology studies show that maternally inherited AD may account for over 20 % of all LOAD cases, and children of AD mothers are at higher risk for developing AD than children of AD fathers [53]. Recent FDG-PET studies revealed that adult children of AD-affected mothers are more likely to express an AD endophenotype than adult children of AD-affected fathers. As parent gender effects on brain function had not been explored prior to publication of FDG-PET reports, these findings set the stage for the exploration of a newly identified risk factor for AD (Fig. 10.5). Main results are summarized in Table 10.1.

The first FDG-PET study to show a "maternal effect" in AD used state-of-the-art full dynamic modeling with arterial input functions and showed that CMRglc was significantly lower in NL elderly with a maternal family history of AD (FHm) compared to those with a paternal family history (FHp) and to controls with negative family history of AD (FH−) [54]. CMRglc reductions in FHm individuals involved the brain regions typically affected in clinical AD patients and ranged from 8 % in frontal cortex to 27 % in PCC. No CMRglc differences were found between FHp and FH− groups [54]. A subsequent 2-year longitudinal FDG-PET study showed that CMRglc reductions over time are more severe in NL FHm than in FHp and FH− [55].

A further FDG-PET study investigated whether CMRglc deficits in NL FHm are associated with advancing maternal age at birth [56]. The study included a large sample of 96 NL individuals (36 FHm, 24 FHp, and 36 FH−) with quantitative FDG-PET scans. After replicating previous findings of selectively reduced CMRglc in NL FHm, the study showed that CMRglc in AD-vulnerable regions was negatively correlated with maternal age at birth only in the FHm group [56]. In contrast, CMRglc was not influenced by maternal age at birth for FHp and FH− subjects. Advanced paternal age was not significantly associated with metabolism in any group [56]. Importantly, in all above studies, no differences were observed between FHm males and females as well as in ApoE ε4 carriers and noncarriers [54–56].

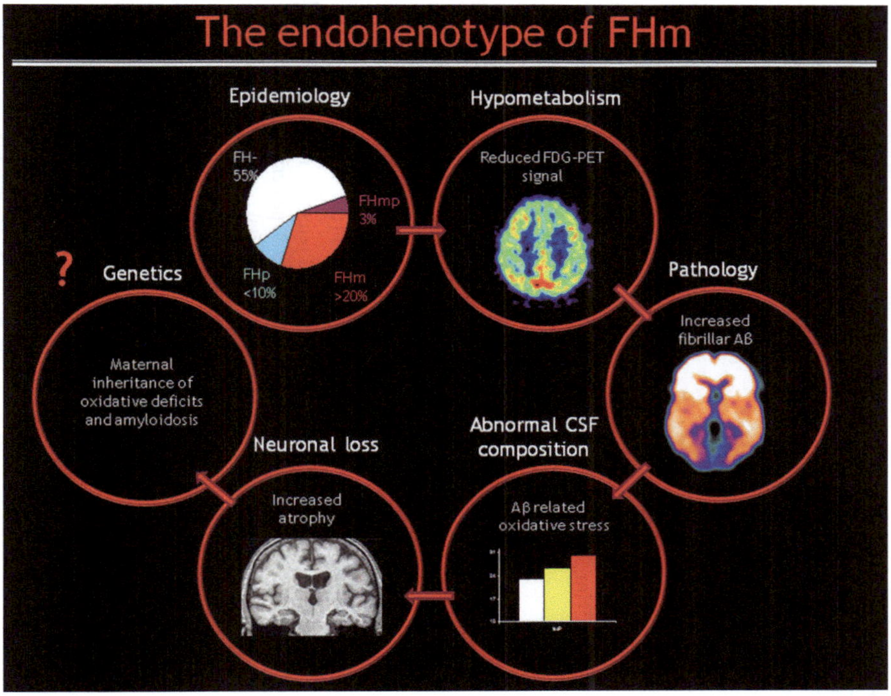


**Fig. 10.5** Summary of biomarkers studies in maternal transmission of AD, in chronological order. From *left* to *right*, *top* to *bottom*: in the 1990s, some epidemiology studies showed that approximately 30 % of all late-onset AD cases are associated with a maternal family history of AD (FHm) compared to <10 % cases with a paternal family history (FHp). About 3 % of cases have both parents affected (FHmp) while the majority have no parental family history of AD (FH−, i.e., sporadic AD). Starting in 2007, biomarker studies began to provide increasing evidence of biomarker abnormalities associated with FHm in cognitively normal elderly (NL). As compared to FHp and FH−, FHm shows progressive CMRglc reductions on FDG-PET, increased amyloid-beta load on PiB-PET, an abnormal CSF biomarker profile characterized by amyloid pathology and increased oxidative stress, and progressive brain atrophy on MRI. These findings suggest presence of genetic transmission not following Mendelian inheritance patterns

Approximately 40 % of all LOAD patients have at least one ApoE ε4 allele [57]. Despite its well-established association with AD, the ApoE ε4 genotype has no clear familial pattern of transmission and appears to act as a risk modifier by lowering the age at onset of clinical symptoms rather than as a casual determinant [58]. CMRglc reductions were observed in NL FHm independent of their ApoE genotype, indicating that other factors contribute to the etiology and phenotypic expression of disease in FHm. While LOAD does not follow classical Mendelian inheritance, the fact that FHm individuals have an increased risk of developing the disease suggests a genetic component. With all that is known about the genetic and molecular basis of glucose metabolism, FDG-PET findings of selectively reduced metabolic activity in NL with AD mothers suggest involvement of mitochondrial DNA (mtDNA). mtDNA is exclusively maternally inherited in humans and is transmitted equally to siblings, and mitochondrial genes are tightly involved with glucose metabolism [59].

**Table 10.1** Summary of biomarker findings in individuals with and without a parental history of late-onset AD

| Refs | Modality | Study type | Diagnosis | Age (years) | *N* | Main findings |
|---|---|---|---|---|---|---|
| 95 | FDG-PET | Cross-sectional | NL | 50–80 | $N=49$<br>16 FHm, 8 FHp, 25 FH– | Reduced CMRglc in AD-vulnerable regions (i.e., parietotemporal, posterior cingulate, and medial temporal cortices) in FHm compared with FH– and FHp<br>No CMRglc differences between FHp and FH– |
| 97 | FDG-PET | Cross-sectional | NL | 25–85 | $N=96$<br>36 FHm, 24 FHp, 36 FH– | Negative correlations between CMRglc in AD-vulnerable regions and maternal age at birth only in FHm<br>No associations between paternal age at birth and CMRglc in any group |
| 96 | FDG-PET | Longitudinal | NL | 50–82 | $N=75$<br>20 FHm, 9 FHp, 37 FH– | More severe CMRglc declines in AD-vulnerable regions in FHm compared to FH– and FHp<br>No longitudinal differences between FHp and FH– |
| 108 | PiB-PET | Cross-sectional | NL | 50–80 | $N=42$<br>14 FHm, 14 FHp, 14 FH– | Higher PiB retention in parietotemporal, frontal, posterior cingulate, and occipital cortices in FHm compared to FH– and FHp<br>Intermediate level of PiB retention in the PCC and frontal cortex in FHp compared with FHm and FH– |
| 101 | Plasma COX activity | Cross-sectional | NL | 27–71 | $N=36$<br>12 FHm, 12 FHp, 12 FH– | Reduced COX activity in platelet mitochondria in FHm compared to FHp and to FH–<br>No differences in COX activity between FHp and FH– |

| | | | | | | |
|---|---|---|---|---|---|---|
| 109 | CSF markers | Cross-sectional | NL | 40–80 | $N=59$<br>23 FHm, 14 FHp, 22 FH− | Higher IsoP and reduced Aβ42/40 CSF levels in FHm subjects compared with FH− and FHp<br>No differences between FH− and FHp<br>IsoP and Aβ42/40 levels are correlated only within the FHm group |
| 106 | MRI | Longitudinal | NL, MCI, AD | 55–90 | $N=484$<br>149 FHm, 50 FHp, 18 FHmp, 267 FH− | Smaller hippocampal volumes at baseline and at follow-up and greater 12-month atrophy rates in MCI FHm compared to FHp and FH−<br>No family history effects on hippocampal volume in NL |
| 103 | MRI | Cross-sectional | NL | 46–85 | $N=60$<br>20 FHm, 20 FHp, 20 FH− | Reduced gray matter volumes in frontal, parietal, and temporal cortices and precuneus in FHm compared with FH−<br>Reduced gray matter volumes in precuneus in FHm compared with FHp<br>No differences between FH− and FHp |
| 104 | MRI | Cross-sectional | NL | 60–85 | $N=67$<br>16 FHm, 8 FHp, 43 FH− | Reduced gray matter volumes in precuneus and frontal regions in FHm > FHp > FH− |
| 105 | MRI | Longitudinal | NL | 63–83 | $N=53$<br>11 FHm, 10 FHp, 32 FH− | Progressive atrophy in precuneus and parahippocampus/hippocampus regions in FHm compared to FH− and FHp<br>No longitudinal differences between FHp and FH− |

### 10.4.3 Maternal Family History of LOAD and Other Altered Biomarkers

One explanation for CMRglc reductions in NL FHm is that these deficits may result in part from mitochondrial dysfunction. Several studies have reported selective reductions cytochrome oxidase activity (COX, complex IV of the mitochondria ETC) in brain tissue, fibroblasts, and platelets in AD [14–16, 60–66]. Based on these reports, we examined COX activity in platelet mitochondria from NL with and without a parental family history of LOAD [67]. While groups were comparable for clinical and neuropsychological measures, COX activity was reduced by 30 % in FHm compared to FHp and to FH− , with and without correcting for citrate synthase activity (i.e., an index of mitochondrial mass) as a reference. There were no differences in COX activity between FHp and FH−. Although COX is encoded by both the mitochondrial and nuclear genomes, its three most catalytically important subunits are on mitochondrial DNA (mtDNA) [59]. This connection to maternal inheritance may help explain the increased risk for AD associated with FHm.

Amyloid PET and cerebrospinal fluid (CSF) studies have shown an association between maternal history of LOAD and increased Aβ levels. Among PET tracers for Aβ plaques, *N*-methyl-[$^{11}$C]2-(4′-methylaminophenyl)-6-hydroxybenzothiazole, aka Pittsburgh compound B (PiB) [68], has become the most widely utilized and best characterized in terms of tracer kinetics, modeling, and analytic methods. On $^{11}$C-PiB PET, NL FHm showed increased and more widespread PiB retention, reflecting higher Aβ burden, in brain regions typically affected in clinical AD patients as compared with NL FHp and FH− [69]. NL FHp subjects also displayed increased PiB retention in PCC and medial frontal cortex, which was intermediate between FHm and FH− [69]. Therefore, although both FHm and FHp groups showed PiB uptake retention in medial brain region (PCC and medial frontal gyrus) compared with controls, only the FHm group showed PiB retention in lateral neocortex. Braak's pathologic studies showed that deposition of Aβ plaques begins in the inferior frontal and inferior temporal cortex, spreads to PCC and medial frontal regions, and then to the lateral parietal, prefrontal, and temporal regions along with disease progression [2]. According to this postmortem staging, NL FHm subjects appear to be at a more advanced stage of brain amyloidosis than FHp. Longitudinal studies mapping the spreading of Aβ deposits are needed to determine whether FHm subjects develop amyloidosis at an earlier age than FHp subjects or show faster Aβ accumulation after middle age.

CSF studies showed increased $F_2$-isoprostanes (IsoP; a marker of lipid membrane peroxidation produced by oxidative stress) and reduced $A\beta_{42/40}$ CSF levels (reflecting increased Aβ sequestration in brain) in NL FHm vs. FHp and FH− [70]. In contrast, no differences were found between FHp and FH−. No group differences were found for markers of tau pathology. Lack of FHm effects on tau CSF markers is consistent with the view that Aβ-associated oxidative damage may precede NFT pathology and neuronal degeneration in the course of AD [8, 71], at least in individuals with a maternal history of LOAD.

Brain tissue loss (i.e., atrophy) on MRI is a well-established early imaging marker of AD [71]. Recent MRI studies found evidence of gray matter volume (GMV) reductions in middle-aged NL FHm when compared to FH− and FHp [72–74]. A longitudinal study examined hippocampal volume using radial distance mapping and conventional volumetric analyses in a large set of NL, MCI, and AD patients with and without a maternal history of dementia, including AD [75]. Results showed that MCI FHm had smaller hippocampal volumes at baseline and at follow-up and greater 12-month atrophy rates than FHp and FH− subjects [75].

Overall, these results indicate that NL FHm express a pathobiological phenotype characterized by Aβ-associated oxidative stress consistent with AD, which might reflect increased risk for developing the disease.

## 10.5 Conclusions

AD is characterized by progressive declines in memory, attention, and language. Often by the time patients express symptoms that warrant evaluation, too much irreversible brain damage has occurred for treatments to be effective. Therefore, it is of paramount importance to identify diagnostic markers as well as disease mechanisms for the preclinical stages of disease when symptoms are not yet apparent. Changes in brain structure and function have been estimated to occur as many as 20–30 years before symptoms onset in AD. Oxidative stress and increased ROS, which damages cell lipids and can eventually lead to neuronal death, are a natural by-product of oxidative phosphorylation, which the brain relies on for energy production. As metabolic changes appear to be an early event in AD pathophysiology, FDG-PET imaging is an invaluable tool in diagnosing AD thanks to the technique's capability to measure metabolic activity through regional glucose uptake and phosphorylation. Recent FDG-PET findings have shown that adult children of AD-affected mothers are more likely to express an AD endophenotype than adult children with a paternal or negative family history of AD. As glucose metabolism is highly regulated by mitochondria and mitochondrial DNA is maternally inherited, a possible upstream event in AD may be impaired bioenergetics in these maternal cases. Biomarker deficits in NL FHm and other at risk populations represent a unique opportunity for initiation of AD therapies and general preventive methods years, and possibly decades, prior to the onset of the clinical disease, presumably well before significant neuronal loss occurs. The combination of imaging and custom-tailored genetics may help clarify whether changes in the nuclear or mitochondrial genome and transcriptome account for pathogenic aging and AD.

**Acknowledgments** This study was supported by NIH/NIA grants AG035137, AG032554 and AG13616, NIH/NCRR grant M01-RR0096, and the Alzheimer's Association. The Authors are indebted to Prof. Alberto Pupi, University of Florence, Italy, for his many insightful comments.

## References


1. Kukull WA, Higdon R, Bowen JD, et al. Dementia and Alzheimer disease incidence: a prospective cohort study. Arch Neurol. 2002;59(11):1737–46.
2. Braak H, Braak E. Neuropathological stageing of Alzheimer-related changes. Acta Neuropathol. 1991;82(4):239–59.
3. Delacourte A, David JP, Sergeant N, et al. The biochemical pathway of neurofibrillary degeneration in aging and Alzheimer's disease. Neurology. 1999;52(6):1158–65.
4. Price JL, Ko AI, Wade MJ, Tsou SK, McKeel DW, Morris JC. Neuron number in the entorhinal cortex and CA1 in preclinical Alzheimer disease. Arch Neurol. 2001;58(9):1395–402.
5. Terry RD, Masliah E, Salmon DP, et al. Physical basis of cognitive alterations in Alzheimer's disease: synapse loss is the major correlate of cognitive impairment. Ann Neurol. 1991; 30(4):572–80.
6. Morrison JH, Hof PR. Life and death of neurons in the aging brain. Science. 1997; 278(5337):412–9.
7. Braak H, Braak E, Bohl J, Reintjes R. Age, neurofibrillary changes, A beta-amyloid and the onset of Alzheimer's disease. Neurosci Lett. 1996;210(2):87–90.
8. Haass C, Selkoe DJ. Soluble protein oligomers in neurodegeneration: lessons from the Alzheimer's amyloid beta-peptide. Nat Rev Mol Cell Biol. 2007;8(2):101–12.
9. Mattson MP, Magnus T. Ageing and neuronal vulnerability. Nat Rev Neurosci. 2006;7(4):278–94.
10. Sokoloff L. Relation between physiological function and energy metabolism in the central nervous system. J Neurochem. 1977;29(1):13–26.
11. Clarke DD, Sokoloff L. Circulation and energy metabolism of the brain. In: Siegel GJ, Agranoff BW, Albers RW, Molinoff PB, editors. Basic neurochemistry. New York: Raven; 1994. p. 645–80.
12. Beal MF. Aging, energy, and oxidative stress in neurodegenerative diseases. Ann Neurol. 1995;38(3):357–66.
13. Ames BN, Shigenaga MK, Hagen TM. Oxidants, antioxidants, and the degenerative diseases of aging. Proc Natl Acad Sci USA. 1993;90(17):7915–22.
14. Mecocci P, MacGarvey U, Beal MF. Oxidative damage to mitochondrial DNA is increased in Alzheimer's disease. Ann Neurol. 1994;36(5):747–51.
15. Subbarao KV, Richardson JS, Ang LC. Autopsy samples of Alzheimer's cortex show increased peroxidation in vitro. J Neurochem. 1990;55(1):342–5.
16. Kish SJ, Bergeron C, Rajput A, et al. Brain cytochrome oxidase in Alzheimer's disease. J Neurochem. 1992;59(2):776–9.
17. Cardoso SM, Proenca MT, Santos S, Santana I, Oliveira CR. Cytochrome c oxidase is decreased in Alzheimer's disease platelets. Neurobiol Aging. 2004;25(1):105–10.
18. Parker Jr WD, Mahr NJ, Filley CM, et al. Reduced platelet cytochrome c oxidase activity in Alzheimer's disease. Neurology. 1994;44(6):1086–90.
19. Valla J, Schneider L, Niedzielko T, et al. Impaired platelet mitochondrial activity in Alzheimer's disease and mild cognitive impairment. Mitochondrion. 2006;6(6):323–30.
20. Pratico D, Uryu K, Leight S, Trojanoswki JQ, Lee VM. Increased lipid peroxidation precedes amyloid plaque formation in an animal model of Alzheimer amyloidosis. J Neurosci. 2001; 21(12):4183–7.
21. Swerdlow RH. Mitochondria and cell bioenergetics: increasingly recognized components and a possible etiologic cause of Alzheimer's disease. Antioxid Redox Signal. 2012; 16(12):1434–55.
22. Reivich M, Kuhl D, Wolf A, et al. Measurement of local cerebral glucose metabolism in man with 18F-2-fluoro-2-deoxy-d-glucose. Acta Neurol Scand Suppl. 1977;64:190–1.
23. Sokoloff L, Reivich M, Kennedy C, et al. The [14C]deoxyglucose method for the measurement of local cerebral glucose utilization: theory, procedure, and normal values in the conscious and anesthetized albino rat. J Neurochem. 1977;28(5):897–916.
24. Rocher AB, Chapon F, Blaizot X, Baron JC, Chavoix C. Resting-state brain glucose utilization as measured by PET is directly related to regional synaptophysin levels: a study in baboons. Neuroimage. 2003;20(3):1894–8.

25. Attwell D, Iadecola C. The neural basis of functional brain imaging signals. Trends Neurosci. 2002;25(12):621–5.
26. Mazziotta JC, Phelps ME. Positron emission tomography studies of the brain. In: Phelps ME, Mazziotta JC, Schelbert H, editors. Positron emission tomography & autoradiography: principles & applications for the brain & heart. New York: Raven; 1986. p. 493–579.
27. Phelps ME, Huang SC, Hoffman EJ, Selin C, Sokoloff L, Kuhl DE. Tomographic measurement of local cerebral glucose metabolic rate in humans with (F-18)2-fluoro-2-deoxy-D-glucose: validation of method. Ann Neurol. 1979;6(5):371–88.
28. Sokoloff L. Modeling metabolic processes in the brain in vivo. Ann Neurol. 1984; 15(Suppl):S1–11.
29. Hawkins RA, Mans AM, Davis DW, Hibbard LS, Lu DM. Glucose availability to individual cerebral structures is correlated to glucose metabolism. J Neurochem. 1983;40(4):1013–8.
30. Iadecola C, Nakai M, Mraovitch S, Ruggiero DA, Tucker LW, Reis DJ. Global increase in cerebral metabolism and blood flow produced by focal electrical stimulation of dorsal medullary reticular formation in rat. Brain Res. 1983;272(1):101–14.
31. Fox PT, Raichle ME, Mintun MA, Dence C. Nonoxidative glucose consumption during focal physiologic neural activity. Science. 1988;241(4864):462–4.
32. Sandman CA, O'Halloran JP, Isenhart R. Is there an evoked vascular response? Science. 1984;224(4655):1355–7.
33. Petersen SE, Fox PT, Posner MI, Mintun M, Raichle ME. Positron emission tomographic studies of the cortical anatomy of single-word processing. Nature. 1988;331(6157):585–9.
34. Phelps ME, Mazziotta JC. Positron emission tomography: human brain function and biochemistry. Science. 1985;228(4701):799–809.
35. Schubert D. Glucose metabolism and Alzheimer's disease. Ageing Res Rev. 2005;4(2):240–57.
36. Magistretti PJ, Pellerin L. Astrocytes couple synaptic activity to glucose utilization in the brain. News Physiol Sci. 1999;14:177–82.
37. Mosconi L. Brain glucose metabolism in the early and specific diagnosis of Alzheimer's disease. FDG-PET studies in MCI and AD. Eur J Nucl Med Mol Imaging. 2005; 32(4):486–510.
38. Minoshima S, Giordani B, Berent S, Frey KA, Foster NL, Kuhl DE. Metabolic reduction in the posterior cingulate cortex in very early Alzheimer's disease. Ann Neurol. 1997;42(1):85–94.
39. Silverman DH, Small GW, Chang CY, et al. Positron emission tomography in evaluation of dementia: regional brain metabolism and long-term outcome. JAMA. 2001;286(17):2120–7.
40. Herholz K, Salmon E, Perani D, et al. Discrimination between Alzheimer dementia and controls by automated analysis of multicenter FDG PET. Neuroimage. 2002;17(1):302–16.
41. Herholz K. Cerebral glucose metabolism in preclinical and prodromal Alzheimer's disease. Expert Rev Neurother. 2010;10(11):1667–73.
42. Blass JP. Alzheimer's disease and Alzheimer's dementia: distinct but overlapping entities. Neurobiol Aging. 2002;23(6):1077–84.
43. Grady CL, Haxby JV, Schlageter NL, Berg G, Rapoport SI. Stability of metabolic and neuropsychological asymmetries in dementia of the Alzheimer type. Neurology. 1986; 36(10):1390–2.
44. Mosconi L, Tsui WH, Herholz K, et al. Multicenter standardized 18F-FDG PET diagnosis of mild cognitive impairment, Alzheimer's disease, and other dementias. J Nucl Med. 2008; 49(3):390–8.
45. Petersen RC, Roberts RO, Knopman DS, et al. Mild cognitive impairment: ten years later. Arch Neurol. 2009;66(12):1447–55.
46. Jagust W, Gitcho A, Sun F, Kuczynski B, Mungas D, Haan M. Brain imaging evidence of preclinical Alzheimer's disease in normal aging. Ann Neurol. 2006;59(4):673–81.
47. de Leon MJ, Convit A, Wolf OT, et al. Prediction of cognitive decline in normal elderly subjects with 2-[(18)F]fluoro-2-deoxy-D-glucose/positron-emission tomography (FDG/PET). Proc Natl Acad Sci USA. 2001;98(19):10966–71.
48. Mosconi L, De Santi S, Li J, et al. Hippocampal hypometabolism predicts cognitive decline from normal aging. Neurobiol Aging. 2008;29(5):676–92.
49. Mosconi L, Mistur R, Switalski R, et al. FDG-PET changes in brain glucose metabolism from normal cognition to pathologically verified Alzheimer's disease. Eur J Nucl Med Mol Imaging. 2009;36(5):811–22.

50. Mosconi L, Sorbi S, de Leon MJ, et al. Hypometabolism exceeds atrophy in presymptomatic early-onset familial Alzheimer's disease. J Nucl Med. 2006;47(11):1778–86.
51. Farrer LA, O'Sullivan DM, Cupples LA, Growdon JH, Myers RH. Assessment of genetic risk for Alzheimer's disease among first-degree relatives. Ann Neurol. 1989;25(5):485–93.
52. Green RC, Cupples LA, Go R, et al. Risk of dementia among white and African American relatives of patients with Alzheimer disease. JAMA. 2002;287(3):329–36.
53. Mosconi L, Berti V, Swerdlow RH, Pupi A, Duara R, de Leon M. Maternal transmission of Alzheimer's disease: prodromal metabolic phenotype and the search for genes. Hum Genomics. 2010;4(3):170–93.
54. Mosconi L, Brys M, Switalski R, et al. Maternal family history of Alzheimer's disease predisposes to reduced brain glucose metabolism. Proc Natl Acad Sci USA. 2007; 104(48):19067–72.
55. Mosconi L, Mistur R, Switalski R, et al. Declining brain glucose metabolism in normal individuals with a maternal history of Alzheimer disease. Neurology. 2009;72(6):513–20.
56. Mosconi L, Tsui W, Murray J, et al. Maternal age affects brain metabolism in adult children of mothers affected by Alzheimer's disease. Neurobiol Aging. 2012;33(3):624.e1–9.
57. Farrer LA, Cupples LA, Haines JL, et al. Effects of age, sex, and ethnicity on the association between apolipoprotein E genotype and Alzheimer disease. A meta-analysis. APOE and Alzheimer Disease Meta Analysis Consortium. JAMA. 1997;278(16):1349–56.
58. Tanzi RE, Bertram L. New frontiers in Alzheimer's disease genetics. Neuron. 2001; 32(2):181–4.
59. Lin MT, Beal MF. Mitochondrial dysfunction and oxidative stress in neurodegenerative diseases. Nature. 2006;443(7113):787–95.
60. Hirai K, Aliev G, Nunomura A, et al. Mitochondrial abnormalities in Alzheimer's disease. J Neurosci. 2001;21(9):3017–23.
61. Mutisya EM, Bowling AC, Beal MF. Cortical cytochrome oxidase activity is reduced in Alzheimer's disease. J Neurochem. 1994;63(6):2179–84.
62. Simonian NA, Hyman BT. Functional alterations in Alzheimer's disease: selective loss of mitochondrial-encoded cytochrome oxidase mRNA in the hippocampal formation. J Neuropathol Exp Neurol. 1994;53(5):508–12.
63. Valla J, Berndt JD, Gonzalez-Lima F. Energy hypometabolism in posterior cingulate cortex of Alzheimer's patients: superficial laminar cytochrome oxidase associated with disease duration. J Neurosci. 2001;21(13):4923–30.
64. Swerdlow RH, Parks JK, Cassarino DS, et al. Cybrids in Alzheimer's disease: a cellular model of the disease? Neurology. 1997;49(4):918–25.
65. Swerdlow RH, Kish SJ. Mitochondria in Alzheimer's disease. Int Rev Neurobiol. 2002; 53:341–85.
66. Curti D, Rognoni F, Gasparini L, et al. Oxidative metabolism in cultured fibroblasts derived from sporadic Alzheimer's disease (AD) patients. Neurosci Lett. 1997;236(1):13–6.
67. Mosconi L, de Leon M, Murray J, et al. Reduced mitochondria cytochrome oxidase activity in adult children of mothers with Alzheimer's disease. J Alzheimers Dis. 2011;27(3):483–90.
68. Klunk WE, Engler H, Nordberg A, et al. Imaging brain amyloid in Alzheimer's disease with Pittsburgh Compound-B. Ann Neurol. 2004;55(3):306–19.
69. Mosconi L, Rinne JO, Tsui WH, et al. Increased fibrillar amyloid-{beta} burden in normal individuals with a family history of late-onset Alzheimer's. Proc Natl Acad Sci USA. 2010;107(13):5949–54.
70. Mosconi L, Glodzik L, Mistur R, et al. Oxidative stress and amyloid-beta pathology in normal individuals with a maternal history of Alzheimer's. Biol Psychiatry. 2010;68(10):913–21.
71. Jack Jr CR, Knopman DS, Jagust WJ, et al. Hypothetical model of dynamic biomarkers of the Alzheimer's pathological cascade. Lancet Neurol. 2010;9(1):119–28.
72. Berti V, Mosconi L, Glodzik L, et al. Structural brain changes in normal individuals with a maternal history of Alzheimer's. Neurobiol Aging. 2011;32(12):2325.e17–26.

73. Honea RA, Swerdlow RH, Vidoni ED, Burns JM. Progressive regional atrophy in normal adults with a maternal history of Alzheimer disease. Neurology. 2011;76(9):822–9.
74. Honea RA, Swerdlow RH, Vidoni ED, Goodwin J, Burns JM. Reduced gray matter volume in normal adults with a maternal family history of Alzheimer disease. Neurology. 2010; 74(2):113–20.
75. Andrawis JP, Hwang KS, Green AE, et al. Effects of ApoE4 and maternal history of dementia on hippocampal atrophy. Neurobiol Aging. 2012;33(5):856–66.

# Chapter 11
# Stress and HPA Axis Dysfunction in Alzheimer's Disease

**Yash B. Joshi and Domenico Praticò**

**Abstract** Memory loss is the most prominent clinical aspect of Alzheimer's disease (AD) but, as recent clinical evidence has been revealed, intervening when memory difficulties are already apparent does little to alter the morbidity and mortality of the disease. Therefore, risk factors that accelerate the development of AD have recently received tremendous interest. Among those risk factors, interrogation of stress hormones/glucocorticoids have been particularly impactful because stress is an inherent aspect of life and unavoidable. Heightened indices of stress in mid-life predict greater risk for AD in late-life, stress hormone dysregulation in the aged increases AD vulnerability and higher levels of circulating glucocorticoid in AD patients correlates with faster cognitive decline. However, despite this evidence, the precise mechanism linking glucocorticoids and stress hormone to AD remain elusive.

In this chapter, we provide an overview of the hypothalamus–pituitary–adrenal (HPA) axis, and how stress, dysregulation of stress hormones and HPA axis dysfunction are currently thought to play a role in AD pathogenesis.

## 11.1 Introduction

The past several decades have dramatically improved understanding about the molecular pathogenesis of Alzheimer's disease (AD), especially its characteristic brain pathologies: plaques composed of amyloid beta (Aβ) peptides and neurofibrillary tangles composed of the microtubule-associated tau protein. In parallel, knowledge about the underlying genetic mutations found in patients with inherited, early-onset AD, have also lead to fruitful investigation of the Aβ precursor protein

Y.B. Joshi, B.Sc. • D. Praticò, M.D. (✉)
Department of Pharmacology, Center for Translational Medicine, Temple University School of Medicine, 947 Medical Education and Research Building, 3500 North Broad Street, Philadelphia, PA 19140, USA
e-mail: praticod@temple.edu

D. Praticò and P. Mecocci (eds.), *Studies on Alzheimer's Disease*, Oxidative Stress in Applied Basic Research and Clinical Practice, DOI 10.1007/978-1-62703-598-9_11,

(AβPP), and the proteases that cleave AβPP to form Aβ peptides, which include the β-secretase and the γ-secretase complex composed of the presenillin, presenillin enhancer-2, anterior pharynx-defective-1, and nicastrin proteins. However, well over 95 % of all AD cases are sporadic, without mutation in any of the above targets. Because of this evidence, environmental factors are thought to play a significant role in AD vulnerability, progression and severity. Of those environmental factors, recent attention has been placed on the role of stress and dysfunction of the hypothalamus–pituitary–adrenal (HPA) axis in AD. Because stress is an unavoidable aspect of life, understanding of how stress modulates AD vulnerability is useful both for the development of clinical preventative and therapeutic strategies as well as investigation of AD pathobiology.

## 11.2 Overview of the HPA Axis

In response to psychological or physiological stress, corticotrophin-releasing factor (CRF; also called corticotrophin-releasing hormone) is secreted from the paraventricular nucleus of the hypothalamus. CRF acts on the neuroendocrine cells of the anterior pituitary gland to release adrenocorticotropic hormone (ACTH; also called corticotropic hormone), which enters circulation induces secretion of stress hormones, such as glucocorticoids, from the adrenal glands (for a review on the HPA axis see [1]). The primary glucocorticoid in humans is cortisol which binds to mineralocorticoid (MRs) as well as glucocorticoid receptors (GRs), forming complexes that translocate to the nucleus to modulate patterns of gene expression. Emerging evidence also indicates that in the brain there exist membrane-associated receptors sensitive to corticosteroids which may explain rapid changes in cell physiology too fast for genomic modulation [2]. Glucocorticoids feedback at the level of paraventricular nucleus and anterior pituitary through GRs, with the GR-cortisol complex binding to CRF and ACTH, which suppress the HPA axis and is critical in maintaining a normal homeostasis. The vast majority of circulating endogenous glucocorticoids is bound by corticosteroid binding globulin (CBG) while a small portion is bound to serum albumin, with only the free steroids acting on tissues. Under basal physiological conditions, glucocorticoid levels follow patterns of discrete pulsatile release, approximately hourly, punctuated by low levels between release, with the highest circulating levels occurring in the early morning and the lowest levels occurring in the evening. However, this periodicity and circadian cycle of glucocorticoid release is plastic and can be greatly influenced by different physiological contexts, environmental factors or disease states.

## 11.3 The Glucocorticoid Hypothesis of Brain Aging

Decades of investigation have revealed that glucocorticoids and stress are critical modulators of memory, with significant attention being paid to the effects glucocorticoid on the hippocampus, a limbic system structure essential to memory.

Because aging is the strongest non-modifiable non-genomic risk factor for AD, Landfield and colleagues originally posited the hypothesis that glucocorticoids promote brain aging based on studies in rodents that suggested corticosteroids produce aging-associated neurodegenerative changes in hippocampus, a brain locus wherein GRs are richly expressed [3]. Since hippocampal neurons are important negative feedback regulators of the HPA axis, it was conjectured that extended stress or glucocorticoid exposure and aging acted cooperatively to produce cognitive decline. Animal and human studies initially provided support for this hypothesis, with elevations in circulating stress hormones correlating with cognitive decline, and longitudinal studies linking cortisol with loss of hippocampal volume. Extended activation of the HPA axis and elevation of stress hormones also structurally change the hippocampus, resulting in atrophy and altered metabolism. Positron emission tomography (PET) and functional magnetic resonance imaging have revealed, respectively, that stress/glucocorticoid administration reduces blood flow and decreases hippocampal activation during memory retrieval tasks [4–7]. However, these phenomena are influenced greatly by the magnitude and type of stressor as well as the length of exposure. Similarly, studies involving both humans as well as other model organisms show performance on memory tasks vary greatly and depend heavily on glucocorticoid dose and treatment duration. While glucocorticoids impair memory retrieval, they have been shown to enhance hippocampal-dependent memory consolidation [8]. In parallel with these behavioral findings, analysis of synapses in the hippocampus have revealed that acute exposure to low or moderate levels glucocorticoids strengthen and functionally enhance synaptic function, while chronic exposure reduces dendritic spine morphology [9]. Due to these observations, Landfield and colleagues have since reformulated this hypothesis to include molecular feedback between neuron and non-neuronal cells as well as context-dependent competing genomic actions of glucocorticoids. Regardless, to appropriately investigate the link between stress, aging, and hippocampal functioning, future human studies must use a combined approach that includes new imaging modalities, memory testing, and sensitive biomarkers of aging and stress. In particular, animal and in vitro studies must explore both genomic and non-genomic actions of glucocorticoids on not only neurons and other cell types in the hippocampus, but other brain areas known to be involved in learning and memory (e.g., cortex, other limbic structures such as the amygdala and fornix).

## 11.4 Stress and AD

Several human studies have been carried out suggesting that stress and stress hormones may be involved in AD pathogenesis. AD patients display higher basal salivary cortisol levels than controls and higher HPA activity, as measured by plasma cortisol, correlates with more severe disease progression in mild and moderate cases of AD [10–12]. Postmortem analyses of cerebrospinal fluid (CSF) cortisol levels also show a similar trend between AD and age-matched controls [13]. In elderly patients without detectable dementia, higher levels of chronic distress are associated with greater risk for development of mild cognitive impairment, which is considered by many in the field to be a prodrome for AD, as well as AD itself [14]. Higher urinary cortisol excretion is also associated with greater incidence of cognitive impairment [15]. In AD patients, hyperactivity of the HPA axis also correlates to hippocampal volume, with lower volume being associated with lower scores on neuropsychological batteries of episodic and visuospatial memory [16]. Administration of exogenous glucocorticoids, such as prednisone has also been reported to cause behavioral decline in AD patients [17].

Many of these observations have been recapitulated in several animal studies. In rodents, suppression of glucocorticoids from mid- to late-life increases neurogenesis in the hippocampi of aged animals while chronic long-term activation of the HPA axis results in cognitive dysfunction and reduction in neurogenesis [18, 19]. Work by multiple investigators has shown that behavioral stress, across a variety of paradigms (including restraint, isolation and/or immobilization stress), worsens AD-like pathology and exacerbates memory impairments in various rodent models of AD [20, 21]. Pharmacologic administration of synthetic glucocorticoids as well as endogenous corticosteroids also exacerbates the AD-like phenotype, while corticosteroid antagonists are protective [22–24]. In wild-type animals, behavioral stress also augments the detrimental cognitive effects of Aβ peptide infusions in the brain.

Despite these observations, the mechanism of how stress and glucocorticoids modulate the AD phenotype is elusive. In human trials, this issue is complicated because sensitive tests (i.e., PET labeling and CSF assays for Aβ and tau, and neuropsychiatric battery/examination) have not yet been widely adopted that would allow researchers to diagnose AD, and accurate biomarkers have not been developed that can track the trajectory and timeframe of cognitive decline from normal to mild cognitive impairment and full dementia. In animal models of the disease, there are also subtle idiosyncrasies in pathologic glucocorticoid-mediated Aβ and tau production. For example, in the 3×Tg animal model of AD developed by LaFerla and colleagues, which expresses the human AβPP Swedish mutation, presenilin-1, and tau, administration of glucocorticoids results in an elevation of AβPP and β-secretase expression [24]. This suggests that glucocorticoids exacerbate the symptomatology of AD through an elevation of starting substrate and its cleavage product, Aβ. While elevations in Aβ peptides are found in the Tg2576 model of AD developed by Hsiao and colleagues upon glucocorticoid administration (which expresses only human AβPP Swedish mutation), such changes in AβPP metabolism are not seen. This is an intriguing observation, likely attributable to the differences in promoters of the knock-in AβPP transgenes in different animal models (Thy1 with the 3×Tg, and hamster prion promoter with the Tg2576). However, given that there is a glucocorticoid response element in the promoter region of human AβPP a complete analysis of HPA axis dysfunction using these two animal models may not be possible. Similarly, differences in the phosphorylation of tau, a crucial step in the development of neurofibrillary tangle pathology, have also been reported that are not consistent and depend greatly on the stress paradigm used. For example,

Lee and colleagues have reported that restraint stress for 2 h/day for 16 days results in higher levels of ser199, thr231 and ser296 phosphorylated tau but not the ser202 tau phosphoepitope in the Tg2576 mice [25]. However, Jeong and colleagues have reported that transgenic mice expressing the AβPP London mutation displayed memory impairment and increased tau phosphorylation at the ser202/thr205 site after 8 months of immobilization and isolation stress, starting at 3 months, for 6 h/day for 4 day/week [20].

In addition to stress and glucocorticoids, a role for CRF has emerged in the pathogenesis of AD, despite earlier work that indicated CRF was protective in vitro. CRF acutely elevates brain Aβ levels and phosphorylated tau in transgenic AD animals which is prevented by using CRF antagonists [26]. Overexpression of CRF in an AD mouse also results in faster progression of the AD phenotype, while disruption of the CRF receptor results in normalization of pathology [27]. These early results seem to indicate that stress, in addition to elevating glucocorticoids, facilitates neurodegeneration through CRF. Other important aspects of the HPA axis, including the neuroactive properties of the mineralocorticoid receptor and ACTH, are less well studied and may also play a role in AD. Additionally, since the HPA axis can also be modulated by circulating catecholamines such as epinephrine and norepinephrine, further investigation of these in the context of stress may prove fruitful.

## 11.5 HPA Axis, Major Depressive Disorder, and Alzheimer's Disease

A recent body of work shows that potential links may exist between major depressive disorder (MDD) and AD [28]. MDD is characterized by a majority of the following symptoms for at least 2 weeks: depressed mood, anhedonia, weight loss or weight gain, insomnia or hypersomnia, psychomotor agitation or retardation, fatigue or loss of energy, feelings of worthlessness, diminished ability to concentrate, and suicidal ideation. HPA axis dysfunction, including an elevation of circulating glucocorticoids and CRF, is known to occur in patients with major depressive disorder. Interestingly, MDD that occurs early in life is correlated with the development of AD in later life, and the risk for developing symptoms of dementia increase by over 10 % per MDD-related hospitalization [29, 30]. Additionally, persons who develop MDD after the age of 50, termed late-life depression, many times also develop cognitive impairments [31]. In patients with mild cognitive impairment, co-incident MDD increases risk for development of AD, and MMD occurs in over 30 % of AD patients [32]. As with AD, in patients with MDD, there is volume loss in the hippocampus [33]. While at the present time it is unclear whether MDD is directly related to the development of AD, this relationship appears to be significant, and, given similar HPA axis dysfunction in both MDD and AD, further work must be done to understand how these diseases are related.

## 11.6 Conclusion

In summary, stress and HPA axis dysfunction appears to be a significant component of AD pathogenesis. Stress and stress hormones modulate important brain regions known to be crucial for learning and memory, such as the hippocampus. Stress increases the risk for the development of cognitive decline and many AD patient display dysregulation of the HPA axis. In AD animal models, stress leads to an increase in Aβ and tau pathology as well as cognitive decline. Finally, emerging data suggests that in other disease states where there is HPA axis dysfunction, including at least MDD, there is a greater risk for AD. While further work must be carried out to sufficiently dissect the pathological molecular mechanisms involved, current understanding of stress in the AD context suggests that behavioral or pharmacological management of stress should be a significant priority in researchers and clinicians who work not only with AD patients, but also with individuals bearing this risk to develop the disease.

## References

1. Aguilera G. HPA axis responsiveness to stress: implications for healthy aging. Exp Gerontol. 2011;46:90–5.
2. Dorey R, Pierard C, Shinkaruk S, et al. Membrnae mineralocorticoid but not glucocorticoid receptors of the dorsal hippocampus mediate the rapid effects of corticosterone on memory retrieval. Neuropsychopharmacology. 2011;36:2639–49.
3. Landfield PW, Blalock EM, Chen KC, et al. A new glucocorticoid hypothesis of brain aging: implications for Alzheimer's disease. Curr Alzheimer Res. 2007;4:205–12.
4. Symonds CS, McKie S, Elliott R, et al. Detection of the acute effects of hydrocortisone in the hippocampus using pharmacological fMRI. Eur Neuropsychopharmacol. 2012;22:867.
5. de Quervain DJ, Henke K, Aemi A, et al. Glucocorticoid-induced impairment of declarative memory retrieval is associated with reduced blood flow in the medial temporal lobe. Eur J Neurosci. 2003;12:1296–302.
6. Oei NY, Elzinga BM, Wolf OT, et al. Glucocorticoids decrease hippocampal and prefrontal activation during declarative memory retrieval in young men. Brain Imaging Behav. 2007;1:31–41.
7. Coluccia D, Wolf OT, Kollias S, et al. Glucocorticoid therapy-induced memory deficits: acute versus chronic effects. J Neurosci. 2008;28:3474–8.
8. Barsegyan A, Mackenzie SM, Kurose BD, et al. Glucocorticoids in the prefrontal cortex enhance memory consolidation and impair working memory by a common neural mechanism. Proc Natl Acad Sci U S A. 2010;107:16655–60.
9. Kassem MS, Lagopoulos J, Stait-Gardner T, et al. Stress-induced grey matter loss determined by MRI is primarily due to loss of dendrites and their synapses. Mol Neurobiol. 2013;47(2):645–61.
10. Arsenault-Lapierre G, Chertkow H, Lupien S. Seasonal effects on cortisol secretion in normal aging, mild cognitive impairment an Alzheimer's disease. Neurobiol Aging. 2010;31:1051–4.
11. Rasmuson S, Nasman B, Carlstrom K, et al. Increased levels of adrenocortical and gonadal hormones in mild to moderate Alzheimer's disease. Dement Geriatr Cogn Disord. 2002; 13:74–9.

12. Csernansky JG, Dong H, Fagan AM, et al. Plasma cortisol and progression of dementia in subjects with Alzheimer-type dementia. Am J Psychiatry. 2006;163:2164–9.
13. Hoogendiik WJ, Meynen G, Endert E, et al. Increased cerebrospinal fluid cortisol level in Alzheimer' disease is not related to depression. Neurobiol Aging. 2006;27:780.
14. Wilson RS, Schneider JA, Boyle PA, et al. Chronic distress and incidence of mild cognitive impairment. Neurology. 2007;68:2085–92.
15. Karlamangla AS, Singer BH, Chodosh J, et al. Increased cerebrospinal fluid cortisol level in Alzheimer' disease is not related to depression. Neurobiol Aging. 2005;Suppl 1:80–4.
16. Elgh E, Lindqvist A, Fagerlund M, et al. Cognitive dysfunction, hippocampal atrophy and glucocorticoid feedback in Alzheimer's disease. Biol Psychiatry. 2006;59:155–61.
17. Aisen PS, Davis KL, Berg JD, et al. A randomized controlled trial of prednisone in Alzheimer's disease. Alzheimer's Disease Cooperative Study. Neurology. 2000;54:588–9.
18. Hu P, Oomen C, can Dam AM, et al. A single-day treatment with mifepristone is sufficient to normalize chronic glucocorticoid induced suppression of hippocampal cell proliferation. PLoS One. 2012;7:e46224.
19. Montaron MF, Drapeau E, Dupret D, et al. Lifelong corticosterone level determines age-related decline in neurogenesis and memory. Neurobiol Aging. 2006;27:645–54.
20. Jeong YH, Park CH, Yoo J, et al. Chronic stress accelerates learning and memory impairments and increases amyloid deposition in APPv717I-CT100 transgenic mice, an Alzheimer's disease model. FASEB J. 2006;20:729–31.
21. Carroll JC, Iba M, Bangasser DA, et al. Chronic stress exacerbates tau pathology, neurodegeneration, and cognitive performance through a corticotrophin-releasing factor receptor-dependent mechanism in a transgenic mouse model of tauopathy. J Neurosci. 2011;31:14436–49.
22. Dong H, Yuede C, Yoo HS, et al. Corticosterone and related receptor expression are associated with increased beta-amyloid plaques in isolated Tg2576. Neuroscience. 2008;155:154–63.
23. Dong H, Goico B, Martin M, et al. Modulation of hippocampal cell proliferation, memory and amyloid plaque deposition in APPsw (Tg2576) mutant mice by isolation stress. Neuroscience. 2004;127(3):601–9.
24. Green KN, Billings LM, Roozendaal B, et al. Glucocorticoids increase amyloid- β and tau pathology in a mouse model of Alzheimer's disease. J Neurosci. 2006;265:9047–56.
25. Lee KW, Kim JB, Seo JS, et al. Behavioral stress accelerates plaque pathogenesis in the brain of Tg2576 mice via generation of metabolic oxidative stress. J Neurochem. 2009;108:165–75.
26. Kang JE, Cirrito JR, Dong H, et al. Acute stress increases interstitial fluid amyloid-beta via corticotrophin-releasing factor and neuronal activity. Proc Natl Acad Sci U S A. 2007; 104:10673.
27. Dong H, Murphy KM, Meng L, et al. Corticotrophin releasing factor accelerates neuropathology and cognitive decline in a mouse model of Alzheimer's disease. J Alzheimers Dis. 2012;28:579–92.
28. Ownby RL, Crocco E, Acevedo A, et al. Depression and risk for Alzheimer disease: systematic review, meta-analysis, and metaregression analysis. Arch Gen Psychiatry. 2006;63:530–8.
29. Robert PH, Shuck S, Dubois B, et al. Validation of the short cognitive battery (B2C). Value in screening for Alzheimer's disease and depressive disorders in psychiatric practice. Encéphale. 2003;29:266–72.
30. Kessing LV, Andersen PK. Does the risk of developing dementia increase with the number of episodes in patients with depressive disorder and in patients with bipolar disorder? J Neurol Neurosurg Psychiatry. 2004;75:1662–6.
31. Barnes DE, Yaffe K, Byers AL, et al. Midlife vs late-life depressive symptoms and risk of dementia: differential effects for Alzheimer disease and vascular dementia. Arch Gen Psychiatry. 2012;69:493–8.
32. Richard E, Reitz C, Honig LH, et al. Late-life depression, mild cognitive impairment, and dementia. Arch Neurol. 2012;31:1–7.
33. Sexton CE, Mackay CE, Ebmeier KP. A systematic review and meta-analysis of magnetic resonance imaging studies in late-life depression. Am J Geriatr Psychiatry. 2013;21:184–95.

# Part III
# Clinics

# Chapter 12
# Diabetes Mellitus and Its Impact on Sporadic Alzheimer's Disease

**Weili Xu**

**Abstract** The prevalence of type 2 diabetes mellitus (T2DM) is increasing, and it affects 338 million people worldwide. Approximately 36 million people worldwide have dementia. Alzheimer's disease (AD) is the most common form of dementia in elderly people. There is convincing evidence from epidemiology, neuropathology and neuroimaging showing an increased risk of AD in people with T2DM. The possible mechanisms linking T2DM and AD include vascular pathways, hyperglycemia, insulin resistance, inflammation, oxidative stress, and mitochondrial dysfunction as well as genetic factors. T2DM may lead to AD through a mixed pathology invoving both vascual and nonvascular changes in the brain. Among diabetic patients, diabetic treatment may play an imprtant role in the development of AD, and effective glyceamic control may attenuate the risk of AD, while hypoglycemia may increase the risk of dementia. Although T2DM and AD may share common pathological process, the two disorders may not be directly linked. A better understanding of the key mechanisms underlying diabetes and AD is needed for the design of preventive and therapeutic strategies.

---

W. Xu, M.D., Ph.D. (✉)
Aging Research Center, Karolinska Institutet-Stockholm University,
Gävlegatan 16, Stockholm 113 30, Sweden

Department of Epidemiology and Statistics, Tianjin Medical University,
Tianjin, People's Republic of China
e-mail: weili.xu@ki.se

D. Praticò and P. Mecocci (eds.), *Studies on Alzheimer's Disease*, Oxidative Stress in Applied Basic Research and Clinical Practice, DOI 10.1007/978-1-62703-598-9_12,

# 12.1 Introduction

## *12.1.1 Diabetes Mellitus*

Diabetes mellitus is a group of metabolic disorders with the common manifestation of hyperglycemia caused by defective insulin secretion, defective insulin action, or both. The World Health Organization (WHO) recognizes three main forms of diabetes: type 1, type 2, and gestational diabetes [1]. Type 1 diabetes is usually due to the autoimmune destruction of the pancreatic β-cells and accounts for 5–10 % of diabetes in the population. The prevalence of type 1 diabetes is higher in younger than in older age groups. Type 2 diabetes mellitus (T2DM) is characterized by insulin resistance in target tissues, and accounts for 90–95 % of all diabetes cases in the population.

T2DM is frequently undiagnosed for many years because the hyperglycemia is often not severe enough to provoke noticeable symptoms of diabetes. Approximately one-third of all people with diabetes may be undiagnosed [2]. Nevertheless, such patients are at increased risk of developing microvascular and macrovascular complications, and related disorders. In addition, T2DM is a slow onset disease, and preceded by impaired glucose regulation (impaired fasting glucose and impaired glucose tolerance), which refers to a metabolic state intermediate between normal glucose homeostasis and diabetes. Thus, two other categories of glucose intolerance, namely impaired glucose tolerance (IGT) and impaired fasting glucose (IFG) have been introduced. Impaired glucose regulation has been termed "prediabetes" or "borderline diabetes". Prediabetes often progresses to full-blown diabetes and increases the risk of cardiovascular disease [2]. Although many different diagnostic criteria for diabetes have previously been defined, they were rationalized in 1979/1980 in reports from the National Diabetes Data Group (NDDG) and the WHO (Table 12.1). An HbA1c of 6.5 % has been recently recommended as the cut

**Table 12.1** The World Health Organization criteria (1999) for the diagnosis of diabetes

| Diagnostic category | Plasma glucose (values/mmol) | | Whole-blood glucose (values/mmol) | |
|---|---|---|---|---|
| | Venous | Capillary | Venous | Capillary |
| *Diabetes* | | | | |
| Fasting or | ≥7.0 | ≥7.0 | ≥6.1 | ≥6.1 |
| 2-h post load | ≥11.1 | ≥12.2 | ≥10.1 | ≥11.1 |
| *IGT* | | | | |
| Fasting and | <7.0 | <7.0 | <6.1 | <6.1 |
| 2-h post load | 7.8–11.0 (incl.) | 8.9–12.1 (incl.) | 6.7–9.9 (incl.) | 7.8–11.0 (incl.) |
| *IFG* | | | | |
| Fasting and | 6.1–6.9 (incl.) | 6.1–6.9 (incl.) | 5.6–6.0 (incl.) | 5.6–6.0 (incl.) |
| 2-h post load | <7.8 | <8.9 | <6.7 | <7.8 |

Note: mmol/0.0555 = mg dl

*IGT* impaired glucose tolerance, *IFG* impaired fasting glucose

point for diagnosing diabetes from the WHO, but a value less than 6.5 % does not exclude diabetes diagnosed using glucose tests [3].

The number of people with diabetes is increasing due to population growth, aging, urbanization and the increasing prevalence of obesity and physical inactivity. The prevalence of diabetes for all age groups worldwide is estimated to be 2.8 % in 2000 and predicted to be 4.4 % in 2030 [4]. Diabetes occurs throughout the world, but is more common (especially type 2) in the more developed countries than in developing nations. The total number of people with diabetes is projected to rise from 171 million in 2000 to 366 million in 2030. Currently, T2DM affects 338 million people worldwide, with 6 million new cases each year. The incidence of diabetes increases with age. The annual incidence rates are 0.3–0.5 % in people aged 50–60 years, 0.5–1.0 % for 60–70 years, and 1 % above 70 years. In addition to diabetes, prediabetes also constitutes a major public health problem. In 2003, around 314 million people worldwide, or 8.2 %, in the age group 20–79 years had impaired glucose tolerance; by 2025 the number is projected to increase to 472 million, or 9.0 % in the adult population [4].

Diabetes is ranked among the leading causes of blindness, renal failure and lower limb amputation, and 70–80 % of people with diabetes die of cardiovascular disease. In fact, diabetes is one of the leading causes of death through its effects on the cardiovascular system. Chronic elevation of blood glucose, even when no symptoms are present to alert the individual to the presence of diabetes, will eventually lead to tissue damage, with consequent, and often serious disease [5]. Whilst evidence of tissue damage can be found in many organ systems, it is the kidneys, eyes, peripheral nerves and vascular tree that manifest the most significant, and sometimes fatal, diabetic complications due to microvascular and macrovascular damage [6, 7]. The mechanism by which diabetes leads to these complications is complex, and not yet fully understood, but involves the direct toxic effects of high glucose levels, along with the impact of elevated blood pressure, abnormal lipid levels and both functional and structural abnormalities of small blood vessels [6]. Diabetes and old age come together to increase the frequency and severity of vascular disease [8]. In the last decade, it has become increasingly evident that diabetes may also affect the central nervous system, a complication referred to as "diabetic encephalopathy" [9]. This complication is reflected in impaired cognitive functioning, and it is also associated with an increased risk of dementia.

### *12.1.2 Dementia and Sporadic Alzheimer's Disease*

Approximately 36 million people worldwide have dementia. Dementia is defined as a clinical syndrome, and characterized by the development of multiple cognitive deficits that are severe enough to interfere with daily functioning, including social and professional functioning. The cognitive deficits include memory impairment and at least one of the other cognitive domains, such as aphasia, apraxia, agnosia, or disturbances in executive functioning according to *Diagnostic and Statistical*

*Manual of Mental Disorders, Fourth Edition (DSM-IV)*. Alzheimer's disease (AD) is the most common cause of dementia in the elderly, accounting for 60–70 % of all demented cases [10]. AD is strictly a neuropathological diagnosis determined by the presence of neurofibrillary tangles and senile plaques in the brain of patients with dementia. The disease frequently starts with memory impairment, but is invariably followed by a progressive global cognitive impairment. Vascular dementia (VaD) is the second most common cause of dementia in the elderly after AD, and defined as loss of cognitive function resulting from ischemic, hypoperfusive, or haemorrhagic brain lesions due to cerebrovascular disease or cardiovascular pathology. Diagnosis of VaD requires cognitive impairment; vascular brain lesions, often predominantly subcortical, as demonstrated by brain imaging; a temporal link between stroke and dementia; and exclusion of other causes of dementia [11]. The combination of AD and VaD pathological changes in the brains of older people are extremely common, making mixed dementia probably the most common type of dementia [12].

The etiology of dementia and AD has been extensively studied trying to find efficacious prevention and treatment strategies. As described above, dementia is a multifactorial disorder caused by complex interaction between environmental and genetic factors. It has been estimated that 1–5 % of AD cases are due to genetic mutations, while the most part are ascribable to modifiable environmental factors and their interaction with genetic susceptibility [12]. Age is the most powerful determinant of dementia, suggesting that aging-related biological process may be involved in the pathogenesis of AD [13]. In actuality, the association between age and AD is mediated by the cumulative effect of other risk and protective factors over the lifespan. The major risk and protective factors for AD can be summarized basing on the different etiological hypotheses including genetic susceptibility hypothesis, vascular pathway hypothesis, psychosocial hypothesis, nutrition and dietary hypothesis, and others (e.g., toxic or inflammatory factors). While the role of genetic, vascular and psychosocial factors in the AD onset is supported by strong to moderate epidemiological, neuroimaging and neuropathological researches, the evidences for the other factors are controversial and insufficient. Following age, the presence of Apolipoprotein E ε4 allele (*APOE* ε4) is the most established genetic risk factor for developing late-onset AD [14].

## 12.2 Impact of Diabetes on Alzheimer's Disease

### *12.2.1 Epidemiological Evidence*

A potential link between diabetes and cognitive impairment was first reported more than 80 years ago. The association of diabetes with these cognitive changes is now well established [15–17]. Over the last decade, many population-based longitudinal studies have revealed a relationship between diabetes and an increased risk of

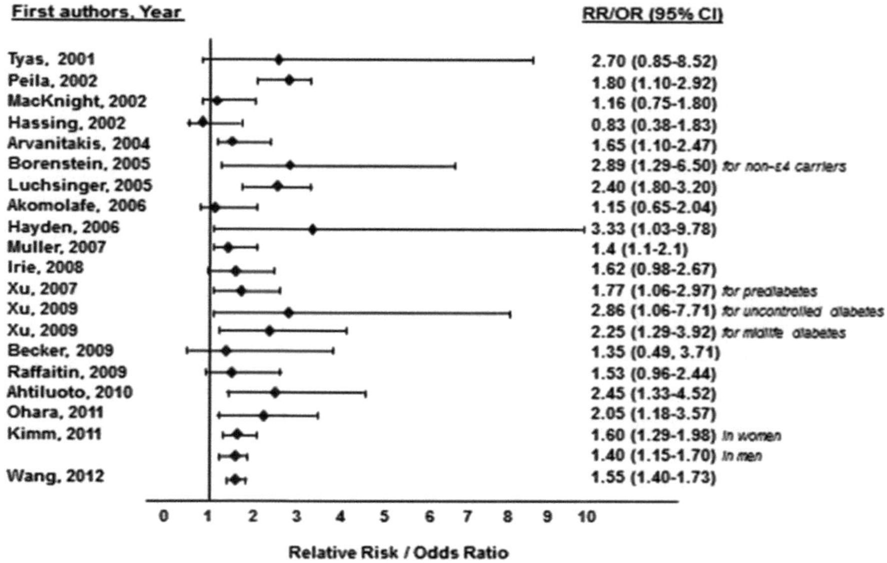


**Fig. 12.1** The impact of type 2 diabetes on Alzheimer's disease (findings from 20 population-based longitudinal studies)

dementia and VaD, although the results concerning the association of diabetes with the Alzheimer type of dementia are inconsistent. In fact, some prospective studies showed the association between diabetes and an increased risk of AD, or observed such an association only in specific subgroups, and others did not. Since 2000, 20 population-based studies have examined the longitudinal relation of diabetes and prediabetes to AD, and the incidence of AD was approximately 1.4–3.0-fold higher in people with diabetes or prediabetes than in those without the conditions in 13 studies (Fig. 12.1) [18–37]. In a recent meta-analysis the aggregate relative risk of AD for people with diabetes was 1.57 (95 % CI 1.41–1.75) based on 14 cohorts [38].

T2DM is a complex metabolic disorder that is closely associated with other established risk factors for dementia, such as hypertension and atherosclerotic vascular disease. These and other risk factors, such as demographic factors (e.g., age, education), diabetes-specific factors (e.g., diabetes duration, glycemic control), medication use, and genetic factors may be involved in the association between T2DM and dementia. The epidemiological studies on the association between T2DM and dementia took the potential confounding effects of age, gender and generally also ethnicity and education into account. However, the risk of dementia among those with T2DM is seldom reported separately for men and women. Only one study found a slightly increased risk in men, whereas others reported no difference [24]. There is no data on the influence of education and ethnicity on the risk of dementia specifically among those with T2DM, but comparisons of the relative risk of dementia associated with T2DM across different study populations, does not suggest major effects of ethnic background.

The association between diabetes and AD is largely independent of hypertension and hypercholesterolemia. The majority of studies assessed these vascular risk factors at old age and it is well known that the association between these factors and dementia is modulated by age, and that hypertension at midlife is a strong predictor for late life dementia while late-life hypertension is not. The limitation of clinical diagnostic criteria in the classification of dementia by pathological subtypes should be considered, especially in a complex disorder such as diabetes. Among these studies, detailed data on modulating or mediating effects of glycemic control, microvascular complications, and comorbidities were generally absent.

Genetic predisposition may contribute to the increased AD risk in people with diabetes. The *APOE* ε4 allele is the most widely examined genetic risk factor for dementia and is related to late-onset AD in the general population. Several studies have shown interactive effects between diabetes and the *APOE* ε4 allele on the risk of AD suggesting further increased risk of AD due to diabetes among ε4 carriers [18, 26, 27, 29, 32].

However, some points need to be considered, when drawing any conclusion from these studies addressing the association betweeen diabetes and AD. First, the characteristics of the study populations are different from study to study, especially regarding age and gender. Mortality risk is elevated in diabetes and diabetes-associated diseases. In a higher age range, selected survival might result in milder forms of diabetes being examined with respect to potential dementia. Therefore, it should be emphasized that these results from elderly populations cannot be generalized to younger persons. Second, different methods of assessment have been used in different studies to identify diabetic patients. A number of studies defined diabetes only based on the information from self-reports, medical records, or the use of antidiabetic medications, and they did not assess blood glucose concentration. As diabetes is often undiagnosed among elderly people, in these studies, a substantial proportion of people with diabetes might have been erroneously assigned to the nondiabetic group, which might have led to an underestimation of the disease risk attributable to diabetes. In addition, the diabetes-related cognitive deterioration may be attenuated by effective glycemic control, but studies that addressed the relation of diabetes to dementia generally did not take into account the effect of glycemic control and diabetes duration. Third, different criteria used in the diagnosis of dementia and its subtypes can have a large impact on the frequency of dementia. Within individual studies the reliability of these diagnostic criteria will be affected by the nature of the diagnostic work-up. The diagnosis of VaD is difficult in epidemiological studies, and the boundaries between AD and VaD remain controversial. Finally, an in-depth analysis of the modulating effect of comorbid disorders was generally not provided.

### *12.2.2 Neuropathological Evidence*

Although epidemiological studies also link diabetes to an increased risk of AD, this cannot be taken as definite proof of accelerated AD-type pathology in T2DM patients, because a clinical diagnosis of AD does not always match with the underlying neuropathology (i.e., AD- or vascular-type) [39]. Moreover, particularly in the oldest old, the neuropathology of dementia is most commonly a combination of AD-type and vascular lesions [40].

#### 12.2.2.1 Neuroimaging

A meta-analysis of studies on brain Magnetic Resonance Imaging (MRI) in patients with T2DM showed that there was a significant association between diabetes and lacunar infarcts (Odds Ratio 1.3, 95 % CI 1.1–1.6) [41]. However, most of these studies have been performed in non-demented populations. Case-control studies that used semiquantitative measures of white matter hyperintensities (WMH) observed a modest increase in WMH severity in T2DM patients compared to controls [42]. When it comes to effects of T2DM on the white matter, diffusion tensor imaging (DTI) is also of interest. This noninvasive technique provides quantitative measures of the microstructural integrity of the white matter. Recent studies with DTI report abnormalities in T2DM patients relative to controls [43]. In addition to vascular lesions, cross-sectional studies consistently report modest degrees of global atrophy in patients with T2DM [44]. Atrophy of the medial temporal lobe structures, including the hippocampus and amygdala, may also be relatively more pronounced [45, 46]. The pathological basis for this DM-associated global and regional atrophy needs to be resolved. It probably reflects microscopic neurodegeneration (loss of synapses, dendritic processes, and neurons) [47], which may occur as a result of AD-type pathologies, but it may also be secondary to vascular damage. Future studies, for example using amyloid Positron Emission Tomography (PET), or high resolution MRI of microvascular lesions, may provide more insight in these issues.

#### 12.2.2.2 Neuropathology

Several autopsy studies have addressed the relation between T2DM, cerebrovascular lesions and AD-type pathology. These studies found that T2DM patients were more likely to have small infarcts [21, 22, 48]. But a relation between the presence of T2DM and the severity of amyloid plaques and neurofibrillary tangles was found only in one study [18]. If anything, some autopsy studies reported that T2DM patients were even less likely to present these AD-pathologies [49].

Based on the current evidence, T2DM is associated with more pronounced vascular pathology in the brain. In some patients this vascular damage may lead to a clinical syndrome that matches clinical diagnostic criteria for VaD. However, in the majority of patients the clinical diagnosis will be AD, despite the fact that neuropathological studies thus far do not support a link between T2DM and accelerated AD-type pathology. A possible explanation is that these latter patients will generally have mixed pathologies, where DM-associated vascular damage lowers the threshold at which AD pathology becomes clinically manifest. This makes prevention of vascular damage the key target in the prevention of dementia in T2DM.

### 12.2.3 Diabetes Treatment and AD

The role of diabetes in increasing risk for cognitive decline and dementia leads to the exploration of the effect of antidiabetic medications on this association. To our knowledge, there are no published studies that have analyzed the association of the long-term effect of antidiabetic medications or long-term glucose control on cognitive function/dementia. Moreover, if antidiabetic medications are efficacious in dementia or AD prevention, it is unknown whether this effect is through diabetes control or other mechanisms. However, there is preclinical and clinical evidence demonstrating the potential beneficial effects of certain antidiabetic treatments in AD. In epidemiological studies, the role of antidiabetic medications in altering risk for dementia is controversial [50]. The apparent discrepancies among epidemiological studies are not necessarily contradictory from the mechanistic point of view. The type and number of prescribed antidiabetic medications may reflect difficulties in achieving glucose control, and therefore higher susceptibility to diabetes-related complications with cognitive compromise being an additional complication.

The importance of glucose control in dementia prevention was exemplified in studies showing that uncontrolled diabetes was associated with higher risk for AD, while better glucose control attenuated cognitive deterioration [27]. Moreover, even in subjects with HbA1c levels within the range considered to be normal, higher HbA1c is associated with higher risk for cognitive deterioration. On the other hand, certain antidiabetic medications, or combination of medications, may have a beneficial effect on cognition that is not necessarily related to glucose control.

The effect of antidiabetic medications on dementia has been assessed in several clinical trials of AD patients. When mild-to-moderate AD patients were randomized to rosiglitazone, an insulin sensitizer or placebo, rosiglitazone treatment was associated with improvement in cognition [51]. Intranasal insulin administration was associated with better performance in memory, attention and functional status compared with placebo in another trial [51]. In an observational follow-up study of subjects diagnosed at baseline as suffering from both AD and diabetes, participants were divided according to their baseline antidiabetic medication regimen (either oral hypoglycemic medications or a combination with insulin). At 12-month follow-up, the group treated with oral medications only performed significantly worse than the group treated with a combination of insulin and oral medication [51].

An additional consideration in the association of diabetes and antidiabetic treatment with dementia is hypoglycemia. Diabetic subjects, especially those treated with medications, are at increased risk for hypoglycemic episodes. Severe hypoglycemic episodes in subjects who were not demented at baseline were demonstrated by some, but not all to be associated with increased risk for dementia. Self-management of diabetes is complicated, probably requiring complex cognitive abilities. Therefore, it is not easy to distinguish whether severe hypoglycemia induces permanent brain damage, or, as demonstrated by some, dementia is a risk factor for hypoglycemic episodes [51]. There are several possible explanations for the discrepancies between different studies, particularly the different strategies used to ascertain hypoglycemic episodes. An additional complexity in research on the association between hypoglycemia and cognition is that the majority of hypoglycemic episodes is not considered to be severe and is difficult to record, thus, understanding of their role in cognition is limited.

## 12.3 Potential Mechanisms Liking Diabetes and AD

There are many pathophysiological mechanisms through which diabetes might affect the initiation and promotion of the many underlying pathologies associated with dementia.These mechanisms include those which are common to both AD and VaD, as well as aging itself. It is increasingly recognized that the brains of people with dementia, particularly in the very old, are likely to show a mixture of pathologies, particularly Alzheimer type and vascular changes. Figure 12.2 presents a simplified scheme, in which some of the endocrinologic, metabolic, and vascular abnormalities that are associated with diabetes and can lead to these different pathologies are indicated [16, 52]. T2DM and its comorbid conditions are associated with and increased risk of macrovascular and microvascular changes, leading to vascular pathology in the brain. Insulin resistance can cause complex changes in cerebral energy metabolism, induce inflammation and can substantially impact the vasculature. It can also directly interfere with amyloid and tau metabolism, giving rise to Alzheimer's type pathology (Fig. 12.2). Hyperglycemia can cause glucose toxicity

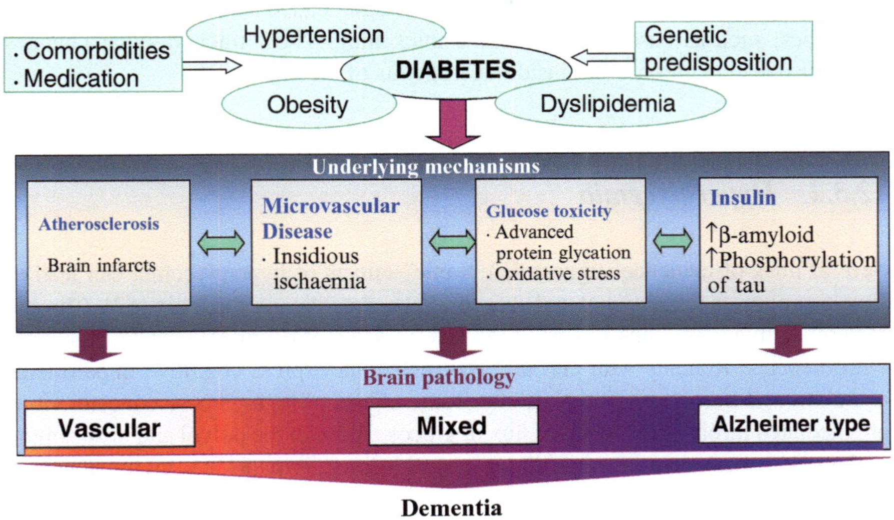


**Fig. 12.2** Mechanisms that may link diabetes and dementia [16, 52]. *Aβ* amyloid beta

leading to mitochondrial dysfunction and oxidative stress, giving rise to Alzheimer's type pathology.

These combined mechanisms can lead to mixed pathology. In some people with diabetes vascular damage will predominate, leading to a form of dementia that will be clinically classified as "pure vascular dementia". In other patients amyloid-related mechanisms may predominate leading to a clinical picture of "pure Alzheimer's disease". Most patients will present with intermediates between these two dementia syndromes.

### 12.3.1 Vascular Pathway

T2DM and associated vascular risk factors predispose to microvascular and macrovascular complications throughout the body and T2DM is an established risk factor for cerebral small vessel disease, as well as thromboembolic stroke. Stroke and vascular comorbidity are likely to be important determinants of the risk of dementia in individuals with diabetes. Diabetes is a known risk factor for stroke. This risk might not only be attributable to diabetes, but also to associated risk factors for vascular disease. T2DM may develop in the context of a cluster of these risk factors, including obesity, insulin resistance, atherogenic dyslipidaemia, hypertension, and prothrombotic and proinflammatory states. Together these factors constitute the metabolic syndrome, or insulin resistance syndrome. Several factors from the metabolic syndrome might be predictors of cerebrovascular disease, ischemic stroke, and accelerated cognitive decline and dementia. The combination of these risk factors in the metabolic syndrome and T2DM might reinforce these effects. Additionally, chronic exposure to hyperglycemia in diabetes might lead to abnormalities in cerebral capillaries, such as basement membrane thickening. These microvascular changes might also lead to chronic and insidious ischemia of the brain [52].

### 12.3.2 Hyperglycemia

Several lines of evidence suggest that "toxic" effects of hyperglycemia can lead to slowly progressive functional and structural abnormalities in the brain [52]. Chronic hyperglycemia could thus be one of the determinants of cognitive changes in people with diabetes. Rodents with chronic hyperglycemia express cognitive impairments and abnormalities in synaptic plasticity. Toxic effects of high glucose concentrations are mediated through an increased flux of glucose through the polyol and hexosamine pathways, disturbances of intracellular second messenger pathways, an imbalance in the generation and scavenging of reactive oxygen species, and by advanced glycation of important functional and structural proteins. These processes can affect brain tissue directly, but can also lead to microvascular changes. So, unlike causing circumscript vascular lesions, diabetes might evoke more generalized and widespread microvascular changes in the brain by this process, causing microinfarcts, probably leading to generalized atrophy and white-matter changes. The occurrence of microinfarcts and their relation to brain atrophy and cognitive decline in elderly people is an evolving concept in the area of dementia research.

These global-glucose mediated effects on cognition and brain structure might be referred to as "accelerated brain aging" [52]. Although this term is rather nonspecific, it might have importance from a conceptual point of view. Several of the above mechanisms that mediate the toxic effects of hyperglycemia, such as oxidative stress, the accumulation of advanced glycation end-products, and microvascular pathology, are also implicated in the aging process of the brain. In fact, the pattern of cognitive changes and brain atrophy in patients with diabetes without dementia mimics certain aspects of brain aging. Although this glucose-mediated "accelerated brain aging" in diabetes is unlikely to lead to frank dementia in itself, it could certainly reduce the threshold for dementia in combination with other pathological changes.

### 12.3.3 *Insulin Resistance*

Insulin resistance, at least in the early stages of T2DM, is associated with compensatory hyperinsulinemia. Several studies have identified hyperinsulinemia as a risk factor for accelerated cognitive decline and dementia. Part of this association is likely to be mediated through vascular disease because insulin has vasoactive effects. Several population-based studies in participants without diabetes have reported an increased risk of stroke in those with the highest insulin concentrations, an effect that persisted after adjustment for associated other vascular risk factors from the metabolic syndrome.

Additionally, insulin might have direct effects on the brain. Insulin is transported actively across the blood–brain barrier, and might even be produced locally in the brain. Insulin receptors are distributed throughout the brain, with particular abundance in the hippocampus and the cortex. Within the brain, insulin is a modulator of food intake and energy homoeostasis, and could also be associated with learning and memory. Aging is associated with changes in insulin and its receptor in the brain, and these changes might be even more pronounced in patients with AD. Observation that activation of the insulin receptor was impaired in brain autopsy samples of patients, has given rise to the notion that Alzheimer's disease could be qualified as "an insulin resistant brain state" [52].

### 12.3.4 *Oxidative Stress and Mitochondrial Dysfunction: Common Pathological Process*

There is strong epidemiological evidence that oxidative stress and mitochondrial dysfunction are involved in AD pathogenesis. Indeed, several key components of the tricarboxylic acid cycle, as well as enzymes of the mitochondrial electron transport chain, including cytochrome oxidase (COX), have been shown to be functionally impaired in AD brains. COX controls the final step of oxidative phosphorylation within the respiratory chain, where it catalyzes the reduction of molecular oxygen to water. Impaired COX function leads to an increase in the production of reactive oxygen species (ROS) and, subsequently, to oxidative stress.

ROS play a critical role in the defense against pathogens under physiological conditions; they may instead contribute to the severe neuronal damage in the AD setting by impairing mitochondrial function and energy metabolism [53]. A similar impairment of mitochondrial activity has been described in diabetic patients [54]. In fact, a reduction of mitochondrial function was found to be responsible for skeletal muscle insulin-resistance in diabetic subjects. Experiments with mouse models showed that diet-induced diabetes altered mitochondrial biogenesis, structure and function in muscle tissue. When lipids start to accumulate ectopically in muscle tissue, an impaired ability to oxidize fatty acids plays an important role in the development of diabetes, as reduced mitochondrial oxidative phosphorylation is directly associated with insulin resistance. In addition, a number of genes involved in oxidative phosphorylation, such as nuclear receptor co-activator PPARγ co-activator-1 (PGC-1) have been shown to be expressed to a lower extent in muscle tissue of T2DM patients, underlining the importance of mitochondrial dysfunction for T2DM pathophysiology [53]. Therefore, antioxidant treatment might, prove a promising tool for the treatment of T2DM and AD.

### *12.3.5 Inflammation*

Neurotoxic inflammatory processes seem to be linked to AD pathogenesis. In accordance, individuals who regularly consume nonsteroidal anti-inflammatory drugs (NSAIDs), as is the case for arthritis patients, have less severe Aβ pathology and consequently a reduced risk of developing AD [55]. Activated microglia and astrocytes in the AD brain emit several pro-inflammatory signals. Cytokines as well as components of the complement system can be detected whereby the interleukins seem to play a paramount role in the deleterious processes linked to AD. IL-1 plays a crucial part in AD-associated inflammation. Microglia and astrocytes also secrete IL-6, a cytokine whose effect is context-dependent, ranging from pro- to anti-inflammatory. IL-6 directly modulates tau phosphorylation status. Another pro-inflammatory cytokine connected to AD is tumor necrosis factor alpha (TNFα), again linking inflammation to oxidative stress. Pharmacological inhibition of TNFα leads to a reversal of formation of senile plaques after intra-cerebroventricular injection of Aβ. As in AD, T2DM has been characterized as a chronic, sub-acute inflammatory state. Inflammation factors such as C-reactive protein, IL-6 and TNFα are also elevated in diabetes. Taken together, there is substantial evidence that inflammation is more than only a peripheral factor in AD and it is not unlikely that inflammatory processes and alterations in the regulation and activity of pro-inflammatory molecules may play a crucial role in T2DM and AD pathogenesis [55].

### 12.3.6 Genetic Factors

It is well accepted that the *APOE* ε 4 allele is associated with an increased risk of AD. *APOE* ε4 may also modify the effect of insulin resistance in AD, but not all studies have supported this idea. Hyperinsulinemia might increase AD risk more in individuals with the ε 4 allele than those without the allele [56]. IDE (insulin degrading enzyme) is a metalloprotease that is ubiquitously expressed throughout the body and localizes primarily to the cytoplasm and peroxisomes of cells. IDE degrades insulin following internalization of insulin and its receptor, thus preventing over-accumulation of serum insulin levels [57]. Another major substrate of IDE is A$\beta$ peptide. IDE is secreted to the extracellular space by microglial cells in the brain, where it degrades monomeric but not oligomeric A$\beta$ peptide. This leads to reduced A$\beta$ peptide concentration in the brain, thus reducing aggregation and plaque formation. IDE levels have been reported to be decreased in the brains of AD patients, especially in the hippocampus. Genome-wide association studies have shown that the IDE gene links both AD and T2DM [57]. It has been hypothesized that increased serum concentrations of insulin during prediabetes effectively sequesters IDE, reducing A$\beta$ peptide degradation. This would increase levels of A$\beta$ peptide, and promote many of the pathological features of AD (including plaque formation). Alternatively, inactivation of IDE by or reduced expression could also lead to accumulation of A$\beta$ peptide and plaque formation.

### 12.3.7 Can Alzheimer's Disease Be a Form of Type 3 Diabetes?

Growing evidence that Alzheimer's is primarily a metabolic disease has led some researchers to propose reclassifying it as type 3 diabetes. As several pathological and histological features have been identified as common denominators of diabetic and Alzheimer's patients, such as insulin resistance, inflammation, and oxidative stress, there is a rapid growth in the literature pointing toword insulin deficiency and insulin resistance as mediators of AD-type nurodegeneration, but this view is riddled with conflicting andunsolved conccepts rergarding the potential contribution of T2DM to AD pathology. The term of "type 3 diabetes" accurately reflects the fact that AD represnets a form of diabetes that selectively involves the brain and has molecular and biochemical features that overlap with both type 1 and type 2 diabetes. T2DM and AD may not be directly linked, but may share common histological and pathological characteristics, which may at least partially explain the increased risk of AD in elderly diabetic patients and hold promiseas new therapeutic targets in the potentially combined treatment of T2DM and AD in the furture.

## 12.4 Conclusion and Implication

There is convincing evidence showing an increased risk of AD in people with T2DM. This relationship is so close that some authors have defined AD as type 3 diabetes [58]. Prevention and treatment of T2DM are important clinical and public health issues. AD is clearly a multifactorial disease that might be linked to several pathways, and increased understanding of the roles of T2DM, insulin, oxidative stress, inflammation, and IDE are important development. T2DM patients are at risk of AD through vascular and nonvascular mechanisms. Some evidence indicates that better control of T2DM might help to prevent cognitive decline. AD patients are not, however, routinely screened for T2DM or its correlates; nor are T2DM patients evaluated for cognitive deficits. Early detection of cognitive impairment in T2DM patients could help to prevent AD. Although treatment of AD patients with diabetic drugs cannot yet be supported by existing evidence, the benefits of T2DM treatment for prevention of AD in patients with diabetes are supportable in themselves. For a better understanding of risk factors for dementia in T2DM, a life-course perspective is needed for the design of preventive and therapeutic strategies.

**Acknowledgments** Weili Xu received grants from the Swedish Council for Working Life and Social Research (FAS2012-0022), the Loo and Hans Ostermans Foundation and the Foundation for Geriatric Diseases at Karolinska Institutet, the Gamla Tjänarinnor Foundation, Demensfonden and the Bertil Stohnes Foundation.

## References

1. American Diabetes Association. Diagnosis and classification of diabetes mellitus. Diabetes Care. 2006;29(Suppl 1):S43–48.
2. American Diabetes Association. Standards of medical care in diabetes – 2006. Diabetes Care. 2006;29(Suppl 1):S4–42.
3. World Health Organization. Use of glycated haemoglobin (HbA1c) in the diagnosis of diabetes mellitus. Geneva: WHO; 2011.
4. IDF Diabetes Atlas. International Diabetes Federation. 2009. http://www.idf.org/diabetesatlas/downloads/backgroundpapers. Accessed 10 Mar 2013.
5. Petrica L, Petrica M, Vlad A, et al. Cerebrovascular reactivity is impaired in patients with non-insulin-dependent diabetes mellitus and microangiopathy. Wien Klin Wochenschr. 2007;119:365–71.
6. Smith U, Laakso M, Eliasson B, et al. Pathogenesis and treatment of diabetic vascular disease – illustrated by two cases. J Intern Med. 2006;260:409–20.
7. Hwang JS, Wu TL, Chou SC, et al. Development of multiple complications in type 2 diabetes is associated with the increase of multiple markers of chronic inflammation. J Clin Lab Anal. 2008;22:6–13.
8. Chanudet X, Bonnevie L, Bauduceau B. Coronary heart disease and cardiovascular autonomic neuropathy in the elderly diabetic. Diabetes Metab. 2007;33 (Suppl 1):S19–31.
9. Gispen WH, Biessels GJ. Cognition and synaptic plasticity in diabetes mellitus. Trends Neurosci. 2000;23:542–9.
10. Fratiglioni L, Winblad B, von Strauss E. Prevention of Alzheimer's disease and dementia. Major findings from the Kungsholmen Project. Physiol Behav. 2007;92:98–104.

11. Roman GC. Vascular dementia: distinguishing characteristics, treatment, and prevention. J Am Geriatr Soc. 2003;51:S296–304.
12. Korczyn AD, Vakhapova V. The prevention of the dementia epidemic. J Neurol Sci. 2007;257:2–4.
13. Qiu C, Kivipelto M, von Strauss E. Epidemiology of Alzheimer's disease: occurrence, determinants, and strategies toward intervention. Dialogues Clin Neurosci. 2009;11:111–28.
14. Bennett DA, Wilson RS, Schneider JA, et al. Apolipoprotein E epsilon4 allele, AD pathology, and the clinical expression of Alzheimer's disease. Neurology. 2003;60:246–52.
15. Arvanitakis Z, Wilson RS, Bennett DA. Diabetes mellitus, dementia, and cognitive function in older persons. J Nutr Health Aging. 2006;10:287–91.
16. Exalto LG, Whitmer RA, Kappele LJ, et al. An update on type 2 diabetes, vascular dementia and Alzheimer's disease. Exp Gerontol. 2012;47:858–64.
17. Cheng G, Huang C, Deng H, et al. Diabetes as a risk factor for dementia and mild cognitive impairment: a meta-analysis of longitudinal studies. Intern Med J. 2012;42:484–91.
18. Peila R, Rodriguez BL, Launer LJ. Type 2 diabetes, APOE gene, and the risk for dementia and related pathologies: The Honolulu-Asia Aging Study. Diabetes. 2002;51:1256–62.
19. MacKnight C, Rockwood K, Awalt E, et al. Diabetes mellitus and the risk of dementia, Alzheimer's disease and vascular cognitive impairment in the Canadian Study of Health and Aging. Dement Geriatr Cogn Disord. 2002;14:77–83.
20. Luchsinger JA, Reitz C, Honig LS, et al. Aggregation of vascular risk factors and risk of incident Alzheimer disease. Neurology. 2005;65:545–51.
21. Arvanitakis Z, Schneider JA, Wilson RS, et al. Diabetes is related to cerebral infarction but not to AD pathology in older persons. Neurology. 2006;67:1960–5.
22. Ahtiluoto S, Polvikoski T, Peltonen M, et al. Diabetes, Alzheimer disease, and vascular dementia: a population-based neuropathologic study. Neurology. 2010;75:1195–202.
23. Ohara T, Doi Y, Ninomiya T, et al. Glucose tolerance status and risk of dementia in the community: the Hisayama study. Neurology. 2011;77:1126–34.
24. Kimm H, Lee PH, Shin YJ, et al. Mid-life and late-life vascular risk factors and dementia in Korean men and women. Arch Gerontol Geriatr. 2011;52:e117–22.
25. Hassing LB, Johansson B, Nilsson SE, et al. Diabetes mellitus is a risk factor for vascular dementia, but not for Alzheimer's disease: a population-based study of the oldest old. Int Psychogeriatr. 2002;14:239–48.
26. Akomolafe A, Beiser A, Meigs JB, et al. Diabetes mellitus and risk of developing Alzheimer disease: results from the Framingham Study. Arch Neurol. 2006;63:1551–5.
27. Xu WL, von Strauss E, Qiu CX, et al. Uncontrolled diabetes increases the risk of Alzheimer's disease: a population-based cohort study. Diabetologia. 2009;52:1031–9.
28. Raffaitin C, Gin H, Empana JP, et al. Metabolic syndrome and risk for incident Alzheimer's disease or vascular dementia: the Three-City Study. Diabetes Care. 2009;32:169–74.
29. Irie F, Fitzpatrick AL, Lopez OL, et al. Enhanced risk for Alzheimer disease in persons with type 2 diabetes and APOE epsilon4: the Cardiovascular Health Study Cognition Study. Arch Neurol. 2008;65:89–93.
30. Wang KC, Woung LC, Tsai MT, et al. Risk of Alzheimer's disease in relation to diabetes: a population-based cohort study. Neuroepidemiology. 2012;38:237–44.
31. Hayden KM, Zandi PP, Lyketsos CG, et al. Vascular risk factors for incident Alzheimer disease and vascular dementia: the Cache County study. Alzheimer Dis Assoc Disord. 2006;20:93–100.
32. Borenstein AR, Wu Y, Mortimer JA, et al. Developmental and vascular risk factors for Alzheimer's disease. Neurobiol Aging. 2005;26:325–34.
33. Xu W, Qiu C, Winblad B, et al. The effect of borderline diabetes on the risk of dementia and Alzheimer's disease. Diabetes. 2007;56:211–6.
34. Muller M, Tang MX, Schupf N, et al. Metabolic syndrome and dementia risk in a multiethnic elderly cohort. Dement Geriatr Cogn Disord. 2007;24:185–92.
35. Tyas SL, Manfreda J, Strain LA, et al. Risk factors for Alzheimer's disease: a population-based, longitudinal study in Manitoba, Canada. Int J Epidemiol. 2001;30:590–7.

36. Xu W, Qiu C, Gatz M, et al. Mid- and late-life diabetes in relation to the risk of dementia: a population-based twin study. Diabetes. 2009;58:71–7.
37. Becker JT, Chang YF, Lopez OL, et al. Depressed mood is not a risk factor for incident dementia in a community-based cohort. Am J Geriatr Psychiatry. 2009;17:653–63.
38. Vagelatos NT, Eslick GD. Type 2 diabetes as a risk factor for Alzheimer's disease: the confounders, interactions, and neuropathology associated with this relationship. Epidemiol Rev. 2013; [Epub ahead of print].
39. Echavarri C, Burgmans S, Caballero MC, et al. Co-occurrence of different pathologies in dementia: implications for dementia diagnosis. J Alzheimers Dis. 2012;30:909–17.
40. Schneider JA, Arvanitakis Z, Bang W, et al. Mixed brain pathologies account for most dementia cases in community-dwelling older persons. Neurology. 2007;69:2197–204.
41. van Harten B, de Leeuw FE, Weinstein HC, et al. Brain imaging in patients with diabetes: a systematic review. Diabetes Care. 2006;29:2539–48.
42. van Harten B, Oosterman JM, Potter van Loon BJ, et al. Brain lesions on MRI in elderly patients with type 2 diabetes mellitus. Eur Neurol. 2007;57:70–4.
43. Hsu JL, Chen YL, Leu JG, et al. Microstructural white matter abnormalities in type 2 diabetes mellitus: a diffusion tensor imaging study. Neuroimage. 2012;59:1098–105.
44. Jongen C, Biessels GJ. Structural brain imaging in diabetes: a methodological perspective. Eur J Pharmacol. 2008;585:208–18.
45. Korf ES, White LR, Scheltens P, et al. Brain aging in very old men with type 2 diabetes: the honolulu-Asia aging study. Diabetes Care. 2006;29:2268–74.
46. Brundel M, van den Berg E, Reijmer YD, et al. Cerebral haemodynamics, cognition and brain volumes in patients with type 2 diabetes. J Diabetes Complications. 2012;26:205–9.
47. Jack Jr CR. Alzheimer disease: new concepts on its neurobiology and the clinical role imaging will play. Radiology. 2012;263:344–61.
48. Nelson PT, Smith CD, Abner EA, et al. Human cerebral neuropathology of Type 2 diabetes mellitus. Biochim Biophys Acta. 2009;1792:454–69.
49. Beeri MS, Silverman JM, Davis KL, et al. Type 2 diabetes is negatively associated with Alzheimer's disease neuropathology. J Gerontol A Biol Sci Med Sci. 2005;60:471–5.
50. Xu WL, Qiu CX, Wahlin A, et al. Diabetes mellitus and risk of dementia in the Kungsholmen project: a 6-year follow-up study. Neurology. 2004;63:1181–6.
51. Ravona-Springer R, Schnaider-Beeri M. The association of diabetes and dementia and possible implications for nondiabetic populations. Expert Rev Neurother. 2011;11:1609–17.
52. Biessels GJ, Staekenborg S, Brunner E, et al. Risk of dementia in diabetes mellitus: a systematic review. Lancet Neurol. 2006;5:64–74.
53. Moreira PI, Custodio JB, Nunes E, et al. Mitochondria from distinct tissues are differently affected by 17beta-estradiol and tamoxifen. J Steroid Biochem Mol Biol. 2011;123:8–16.
54. Reijmer YD, Brundel M, de Bresser J, et al. Microstructural white matter abnormalities and cognitive functioning in type 2 diabetes: a diffusion tensor imaging study. Diabetes Care. 2013;36:137–44.
55. Jones A, Kulozik P, Ostertag A, et al. Common pathological processes and transcriptional pathways in Alzheimer's disease and type 2 diabetes. J Alzheimers Dis. 2009;16:787–808.
56. Cole AR, Astell A, Green C, et al. Molecular connexions between dementia and diabetes. Neurosci Biobehav Rev. 2007;31:1046–63.
57. Qiu WQ, Folstein MF. Insulin, insulin-degrading enzyme and amyloid-beta peptide in Alzheimer's disease: review and hypothesis. Neurobiol Aging. 2006;27:190–8.
58. Accardi G, Caruso C, Colonna-Romano G, et al. Can Alzheimer disease be a form of type 3 diabetes? Rejuvenation Res. 2012;15:217–21.

# Chapter 13
# Peripheral Biomarkers of Oxidative Stress in Alzheimer's Disease

**Fabio Di Domenico and Marzia Perluigi**

**Abstract** Alzheimer's disease is a multifactorial dementing disorder with a complex etiology affecting millions of people worldwide. One of the major challenges of both researchers and clinicians is to identify putative peripheral biomarkers which may facilitate disease diagnosis at early stages, monitor disease progression, and assesses the response to treatments by the time that disease-modifying treatments become available in clinical practice. Among proposed candidates, particular attention has been devoted to peripheral markers of oxidative stress (OS). Indeed, several studies have demonstrated the accumulation of consistent oxidative damage in specific brain regions of both MCI and AD patients. These findings support the view that OS is an early event in the pathogenesis of AD and thus assessment of the presence of peripheral oxidative damage that correlates with the occurrence of the disease is intriguing. Biological fluids, such as CSF and plasma/serum, represent the best potential source for biomarker discovery. However, the complexity of their composition and wide variability among population has led to limited results. This chapter summarizes the current knowledge on peripheral biomarkers of oxidative stress in biological fluids and their relevance for Alzheimer research. Further efforts are needed to improve and widen such knowledge, which is fundamental for a better understanding, diagnosis and treatment of AD.

F. Di Domenico, Ph.D. • M. Perluigi, Ph.D. (✉)
Department of Biochemical Sciences, Sapienza University of Rome, Piazzale Aldo Moro 5, 00185 Rome, Italy
e-mail: marzia.perluigi@uniroma1.it

D. Praticò and P. Mecocci (eds.), *Studies on Alzheimer's Disease*, Oxidative Stress in Applied Basic Research and Clinical Practice, DOI 10.1007/978-1-62703-598-9_13,

## 13.1 From the Brain to the Periphery: Do Body Fluids Reflect CNS Pathology?

Though a definite diagnosis of AD is obtained by postmortem neuropathological examination, brain tissue cannot be used for early diagnosis of cognitive decline. Thus, in recent years growing studies have been focused in establishing a direct link between tissue specific damage and systemic alteration, as well as in identifying biochemical markers of brain dysfunction that can be measured in body fluids such as cerebrospinal fluid (CSF), plasma, and urine (Fig. 13.1). Biomarkers detected in body fluids could contribute to increase the accuracy of diagnosis and prediction of progression to AD (Table 13.1).

*CSF* among other extracellular fluids represents the closest approximation to CNS for its direct contact with brain parenchyma. The CSF is a clear and colorless ultrafiltrate of plasma produced from arterial blood by the choroid plexuses (CP). Ependymal cells also produce a small amount of CSF. Albumin is the most abundant protein in the CSF (50–70 %), and normally γ-globulin levels are very low (5–12 %). Among multiple functions, CSF protects the brain during blood pressure fluctuations associated with posture, respiration, and exertion, regulates the chemical environment of the CNS, playing an important role in the homeostasis and metabolism, and assures the excretion of the potentially toxic by-products, and it is a vehicle for intracerebral transport. The brain's ability to turn over or renew CSF is crucially important in maintaining metabolic balance in the CNS.

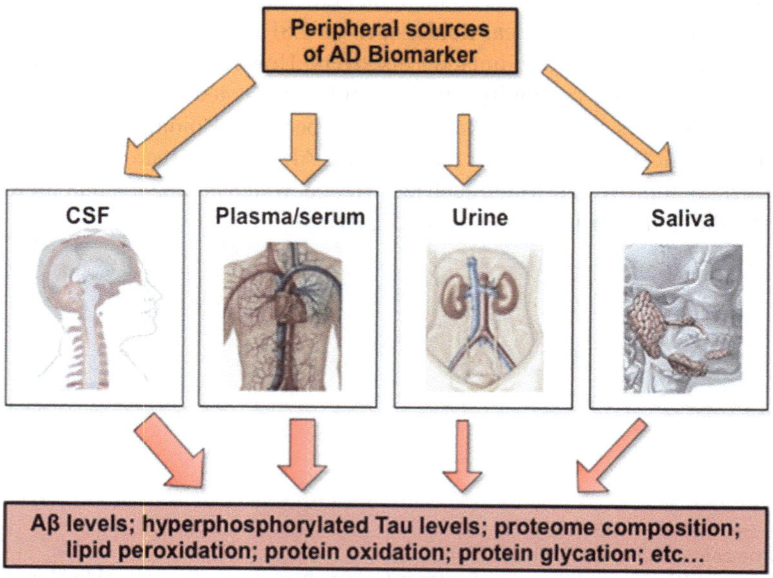


**Fig. 13.1** Peripheral sources and main types of AD biomarkers. The *arrow thickness* indicates the importance of the biofluid as a diagnostic marker

**Table 13.1** OS markers measured in different fluids from AD patients

| | | | | |
|---|---|---|---|---|
| Levels of total carbonylated proteins | ↑ [12]≈[69, 31] | ≈[31, 56] | ↑ [1]≈[31] | / |
| Levels of total nitrated proteins | / | ↑ [56] | ↑ [1,65] | / |
| Lipid peroxidation | ≈ F-2, F-4 isoprostanes [26, 44] | ↑ MDA [46] | ↑ F-2 isoprostanes [50, 40–42, 21] | ↑ F-2 isoprostanes [49, 66])≈F-2, F-4 isoprostanes [43] |
| Oxidazed LDL | ↓ [4] | ↑ [56] | / | / |
| Redox proteomics | Fibrinogen $\gamma$ chain precursor and alpha 1 antitrypsin [10] Hemopexin, transferrin [68] | / | $\lambda$ Chain precursor [32] | / |
| DNA oxidation | ↑ Lymphocyte 8-OHdG | | ↑ 8-OHdG [19,37] | ↓8-OHdG [35] |

↑ = increased oxidation; ↓ = decreased oxidation; ≈ = no difference

Accurate analysis of CSF provides a wide range of information about the neurological health of the patient. Moreover its relative availability make possible to conduct longitudinal molecular analyses of changes in CSF during the course of diseases. Historically, CSF compositional analyses have been used widely to monitor distortions in brain metabolism, evaluate disruptions of barrier transport and permeability functions, obtain pharmacokinetic parameters for drugs targeting the brain parenchyma, and identify biomarkers for aiding the diagnosis and prognosis of CNS diseases [27]. Thus, the CSF contains valuable biochemical and cellular information that can be advantageous for more effective clinical management of brain complications. Aging- or disease-induced alterations in CSF turnover rate adversely impact neuronal performance with deleterious consequences for the brain. The BBB separates circulating blood from the interstitial space of the CNS and affects in a critical way the traffic of molecules in and out of the brain. When the BBB is damaged, protein leaks from the blood into the CSF. Toxic catabolites in CSF, generated especially by peroxidative phenomena, increase with age and neurodegeneration. Oxidized and glycated proteins in CSF are elevated in AD. A vicious cycle ensues as toxic neuronal products (e.g., $A\beta$) enter CSF from brain and inflict further harm to the CP already debilitated by aging. CSF biomarkers reflecting the amyloid cascade ($A\beta40$, $A\beta42$) and cytoskeletal degeneration (total tau, and phosphorylated-tau) have been studied extensively and these biochemical markers have been established as reliable biomarkers for AD pathology [15].

*Blood-related body fluids* analysis being less invasive that CSF, represent an ideal tool in prognosis and in monitoring therapeutic interventions, especially for large scale studies and for repeated measures.

Blood is a convoluted body fluid containing proteins, peptides, lipids and metabolites that reflect physiological activity and pathology in numerous body organs, including the central nervous system (CNS). In humans about 500 mL of CSF is absorbed into blood daily, making blood an appropriate source of neurodegenerative disease biomarkers [58]. Plasma is the straw-colored liquid component of blood in which the blood cells in whole blood are normally suspended. It makes up about 55 % of the total blood volume. It is the intravascular fluid part of extracellular fluid (all body fluid outside of cells). It is mostly water (93 % by volume) and contains dissolved proteins, glucose, clotting factors, mineral ions, hormones and carbon dioxide (plasma being the main medium for excretory product transportation). Plasma also serves as the protein reserve of the human body. It plays a vital role in intravascular osmotic effect that keeps electrolyte in balance form and protects the body from infection and other blood disorders. Serum is made up of non-clotting proteins, glucose, nutrients, electrolytes, hormones, antigens, antibodies and other particles. The components of plasma are same as that of serum, except for fibrinogens and clotting factors.

A variety of potential blood-related biomarkers for AD have been identified, however, the application of these candidate biomarkers have yet to achieve the diagnostic power, sensitivity, and reproducibility necessary for widespread use in a clinical setting. Among other biological fluid of human body, the analysis of *urine and saliva* captioned some interest in the search of AD biomarker. Urine is an aqueous solution of greater than 95 % water, with the remaining constituents, in order of decreasing concentration urea 9.3 g/L, chloride 1.87 g/L, sodium 1.17 g/L, potassium 0.750 g/L, creatinine 0.670 g/L and other dissolved ions, inorganic and organic compound. Urine, produced by the kidneys, contains the byproducts of metabolism—salts, toxins, and water—that end up in the blood. The nephrons in the kidneys filter and eliminate waste substances from blood avoiding their buildup to dangerous levels. Biochemical testing of urine composition for AD prediction is currently an ongoing issue and several biomarker of the pathology has been proposed; however, the real suitability of urine for AD diagnosis is still under debate. Indeed the accuracy and relevance of a urinary analyte about brain status, given that it need to pass the BBB, enter the circulation, reach the kidney, be filtered through the glomeruli, and be stored in the bladder at body temperature for unknown time periods, is questionable.

Saliva is produced from salivary glands and mucous membranes. Salivary fluid is an exocrine secretion consisting of approximately 99 % water, containing a variety of electrolytes (sodium, potassium, calcium, chloride, magnesium, bicarbonate, phosphate) and proteins, represented by enzymes, immunoglobulins, and other antimicrobial factors, mucosal glycoproteins, traces of albumin and some polypeptides and oligopeptides of importance to oral health. Saliva is not considered an ultrafiltrate of plasma. Salivary proteins concentration is dependent by several factors like stress, inflammation or infection [24]. In humans, in addition to the classical accumulation in the brain, Aβ protein deposits are found in peripheral regions, including lingual glands. Salivary alterations may reflect changes in CSF and recent studies showed association of activity and levels of salivary acetylcholinesterase with AD [5].

## 13.2 Peripheral Biomarkers: Limits and Prospective

Identification of biomarkers for AD is challenging and complicated by diagnostic imprecision, long asymptomatic prodromal stages, variability in clinical features and rates of progression, and complex disease genetics.

At present, diagnostic criteria for AD and MCI are based on clinical features allowing a "probabilistic" diagnosis and exclusion of other types of dementia. This low specificity also is a major limit for the therapeutical management of these subjects to test the efficacy of disease-modifying drugs [14].

Body fluids from living subjects are without any doubt the best potential source of information on healthy and diseased subjects. So far, CSF biomarkers have been focused on the amyloid cascade hypothesis and cytoskeletal degeneration, measuring the levels of total tau and phosphorylated-tau. Results obtained were found to be promising and reliable biomarkers of AD pathology. Indeed, increase in total tau and p-tau, and decrease in Aβ42 level and Aβ42/Aβ40 ratio have been documented in CSF from AD patients and from MCI subjects [6,20,23]. However, the sensitivity of these markers is still unsatisfactory and they are not useful for monitoring the progression of the disease.

The lumbar punctures used to obtain CSF, still remains an invasive procedure followed by several side effects. Further, patient consent is very low and this aspect remains the major limit for clinical routine analysis. Thus, the majority of research aimed to identify biochemical changes in blood, which is relatively easy to collect and may reflect those occurring in diseased brain regions. Nevertheless, only few studies were well correlated with the disease state and the development of methods for measuring biomarkers in blood is complicated by the potential influence of the blood brain barrier which affects the dynamics of release of these proteins into blood.

A major difference between CSF and plasma is that protein concentration in CSF is about 200-fold less than that in plasma. This could be due to the exposure to some plasma proteins that might modulate properties of both neuronal and non-neuronal components of the CNS, or following injuries that result in increased permeability of the blood–brain barrier. For example, ischemic stroke is associated with pathophysiological changes affecting both glial and neuronal brain tissue. These changes are mirrored in the release of specific proteins into peripheral blood. Neuron-specific enolase, protein S100B and glial fibrillary acidic protein are those proteins investigated most often as peripheral surrogate markers of brain damage after stroke in humans [8,17]. Other limitations are the influence of other pathologies at systemic level, the standardization of the methods and the reproducibility of the results.

Studies evaluating the diagnostic performance of a biomarker for AD should set up the molecule's sensitivity (the ability of a biomarker to identify patients who have disease), specificity (the ability of a biomarker to identify patients who do not have disease) and reproducibility (variation in biomarker levels between different centers probably resulting from variations in clinical procedures—such as the protocols for sample processing and other laboratory practices—as well as batch-to-batch variation in the biomarker assays). These types of variations are well known in clinical chemistry, and are routinely controlled by external control programs. Standardized protocols should minimize variation caused by differences in pre-analytical and laboratory procedures and, thus, allow direct comparisons of biomarker levels between laboratories and between publications [7]. Increased markers of protein, lipid and nucleic acid oxidation and reduced activities of antioxidant enzymes in AD brain support a role of OS in neurodegeneration in AD. OS is present in AD since the early stages [38,49] of the disease and the search for altered levels of free radical by-products or oxidatively modified molecules might facilitate the selection of biomarkers of early AD diagnosis. Indeed, redox modifications of proteins can significantly alter their biological activity even in cases where the relative abundance may not, necessarily change. Several studies have shown that free radical products are present in the CSF, serum, and plasma of AD patients [43,48,61].

Recently NIH has supported a coordinated study, the Biomarker of Oxidative Stress Study (BOSS), to assess which biochemical products of oxidative damage represent a valuable in vivo biomarker in an experimental model of oxidative injury to rat liver [28]. Of these, MDA, F2-IsoPs, and 8-OHdG were acceptable quantitative in vivo biomarkers of oxidative damage.

In the past years the analysis of body fluid composition by proteomics and redox proteomics has been performed by different groups [2,10,32,63]. However, as previously stated, the discovery and validation of protein biomarkers from body fluids is complicated by the enormous difference in quantity between the high and the low abundant proteins with a dynamic range of about 12 orders of magnitude. Some proteins are present at high concentration, while others at very low concentration, thus making their detection highly challenging.

Serum albumin, the most abundant circulating protein, represents about two-thirds of the entire protein content of plasma, and with IgGs and few other high-abundance proteins constitute greater than 90 % of total protein mass, interfering with the detection of lower abundance proteins that might represent the biologically interesting population [59]. To increase the number of proteins that can be identified by MS, depletion of abundant plasma proteins is a common strategy, which however can affect also detection of the low abundant fraction that might interacts/binds the high abundant one [3]. This strategy takes advantage of different biochemical and biophysical features of proteins, such as molecular weight, hydrophobicity and isoelectric point. Among different techniques commonly utilized, the most common rely on antibody-based retention of a chosen set of the most abundant proteins. Commercial removal kits are available, and depletion of high-abundant proteins—ranging from 2 to 20—allow the removal of about 85–95 % of the total protein content. Several studies have been carried out to validate this strategy by also comparing the various depletion approaches. Overall, improved quality of proteomics separation and quantitation is achieved after removal of high-abundant proteins [47].

## 13.3 Markers of Oxidative Damage in Body Fluids

In an effort to give insight into the mechanism of oxidative stress-induced neuronal loss growing studies are focused on the search for circulating biomarkers of oxidative stress in patients with a clinical diagnosis of AD. Indeed, postmortem investigations cannot easily address the question of whether oxidative stress is an early event in the pathogenesis of the disease, or a common final step of the neurodegenerative process. Lipid peroxidation products are the most broadly analyzed molecules in the search of oxidative stress markers in biofluids and their presence was often correlated with the progression of the pathology. Protein oxidation, indexed as protein carbonyls, was extensively studied in CSF while few reports are present in plasma and none in urine. The investigation of peripheral oxidative stress biomarker, even if potentially precious in AD diagnosis, meets several conceptual and experimental limitations. The paucity of data on oxidized proteins in body fluids is perhaps due to the difficulty of finding oxidized proteins in complex matrices, the multiplicity of possible oxidative modifications, and low abundance [33].

### *13.3.1 CSF*

As previously stated, CSF is the most proximal biofluids to CNS and therefore the logical source to find any biomarker directly related to AD pathology.

The most reliable CSF markers in AD are Aβ42 and tau. Low CSF Aβ42 is associated with amyloid pathology in the brain and high Tau is linked with neurofibrillary pathology [18]. Most subjects with decreased CSF Aβ42 and high tau develop AD during the follow-up [22]. Therefore, these CSF markers may reflect brain pathology and identify preclinical AD.

Since the involvement of OS was largely investigated in the brain much effort was done to parallel brain OS with CSF in order to find a molecular markers able to reveal the ongoing oxidation in the brain.

The first report on increased presence of oxidative stress markers in CSF concern the products of lipid peroxidation, indeed increased levels of HNE and elevated levels of F2-isoprostane (F2-IsoP), formed by free radical-mediated peroxidation of polyunsaturated fatty acid has been observed in CSF of AD and MCI subjects compared to CTR [36,40–42,50]. In a longitudinal study, levels of CSF F2-IsoPs in AD patients were significantly increased during the follow-up period, and also significantly declined in patients accepting anti-oxidant treatment [53]. Following studies on protein oxidation by Tohgi et al. [65] demonstrated that 3-nitrotyrosine moderately but significantly increased with advancing aging CSF showing a remarkable increase in patients with AD. The increased 3-nitrotyrosine was likely due to an increase in nitrated tyrosines in proteins or increased degradation of 3-nitrotyrosine-containing proteins, which are highly vulnerable to degradation.

Ahmed et al. [1] measured in CSF the levels of protein glycation, oxidation, and nitration. Finding that these were all increased in subjects with AD, a combination of nitration and glycation adduct estimates of CSF may conceivably provide an indicator for the diagnosis of AD.

Previous studies on CSF nitrite and nitrate levels in patients with AD have provided contradictory results, with some showing decreased nitrate levels [34], others showing unaltered nitrite/nitrate levels [25], and still others increased nitrate levels [64]. However, it was reported that nitrite/nitrate levels in AD were stage-dependent, being elevated only in the early phase of AD and decreasing to control levels with disease progression [64]. This finding was interpreted to reflect progressive reduction of neurons. In contrast, free 3-nitrotyrosine levels increased significantly in parallel with the severity of AD, suggesting that protein degradation increases with disease progression, resulting in increased release of free 3-nitrotyrosine from tyrosine residues that have been nitrated. 3-nitrotyrosine and the 3-nitrotyrosine/tyrosine ratios in the CSF, both of which are believed to reflect degradation of nitrated tyrosine-containing proteins, increased significantly with age and were remarkably higher in patients with AD than in controls.

Interestingly, the levels of CSF Aβ42 showed a tendency to correlate positively with serum oxidative markers in the whole study population and with plasma nitrotyrosines in AD patients. The correlation between CSF AD markers and blood oxidative markers may suggest that oxidative metabolism is changed in AD. This hypothesis is further supported by the finding of decreased CSF protein carbonylation in APOE ε4 carriers, which is considered an important risk factor for developing AD [54] and correlates with redox proteomics studies that identified metabolic proteins as oxidatively modified and dysfunctional [11].

A few previous studies measuring total levels of protein carbonyls or nitrated proteins in CSF and plasma have yielded contradictory results [1,9,69]. Korolainen et al. [31,43,44] found that protein carbonyl levels did not differ in CSF between AD patients and controls. However, by using 2-D oxyblot, an increased degree of carbonylation for one single protein, lambda-chain, in CSF of AD patients as compared with controls [32] was showed. Further studies are needed to replicate this finding and also to add new information in this topic. Yet the analysis of CSF oxidized proteome present a number of challenges and it is likely that additional changes in protein carbonylation will be revealed by using more sensitive methods.

Studies by Lovell et al. demonstrated an elevation of 8-OHdG in DNA and a decrease in free 8-OHdG in ventricular CSF in AD that correlate with disease severity [37]. These studies support the concept, already expressed by other studies, about increased oxidative stress in AD brain, proposing DNA damage as a more reliable biomarker than protein oxidation, and suggest for the first time that AD brain might present deficiencies in essential DNA repair mechanisms.

A similar result was obtained by Gackowski et al. that demonstrated increased DNA damage in CSF of mixed dementia subjects, which comprises, among others, also AD patients [19].

### *13.3.2 Plasma/Serum*

Increasing number of studies demonstrate that oxidative damage in AD patients also occurs outside the brain and can be detected in different body fluids. Conrad et al. [12] were the first to measure the levels of oxidized proteins in plasma from AD patients compared with controls. The levels of total oxidized proteins were measured by HPLC, while specific protein oxidation was assayed by western blot using anti-DNP antibody. The results showed a statistically significant increase of total carbonyl groups in plasma proteins while a specific increased oxidation of a 78 KDa protein appeared to be differentially oxidized in subjects with AD.

Yu et al. [68] focused attention on plasma glycoproteins to evaluate both native and oxidized fraction in plasma from AD patients. The glycoprotein fraction was firstly isolated by affinity chromatography on heparin-agarose and concanavalin A-agarose columns, followed by separation on one/two-dimensional polyacrylamide gels. In-strip derivatization with DNPH and anti-DNP immunoblotting was performed to detected carbonylated glycoproteins. The results demonstrated increased levels of carbonylation for glycosylated hemopexin and transferrin in the AD subjects as compared to controls. This finding suggested that systemic perturbation in the heme/iron/redox homeostasis and activation of the acute phase response occurred in sporadic AD [68].

Choi et al. [10] firstly applied 2DE and immunochemical detection to identify carbonylated proteins in AD plasma and non-AD controls. They found that the increased oxidation was not a generalized phenomenon. Though more than 300 spots were detected, less than 20 spots were positive by immunostaining with anti-DNP antibody and among the most intensively oxidized proteins, their relative levels of oxidation differed. The most significant increase of carbonyl levels in AD compared with controls was restricted to fibrinogen gamma chain precursor and alpha-1-antitrypsin precursor. Though fibrinogen comprised six different chains (a2; b2; c2), the alteration of only one of these as a result of its oxidative modification may result in increased activation of plasminogen thereby contributing to fibrinolysis and proteolysis in areas of inflammation [60]. Similarly, alpha-1-antitrypsin oxidation may be particularly relevant to the disease. Alpha-1-antitrypsin is one of the major serine proteinase inhibitors (serpins) in human plasma, functions to inhibit overexpressed proteinases during inflammation [29]. These inhibitors tightly regulate proteinase activity, under normal physiological conditions. However, under some pathological conditions, proteinase activity may exceed the capacity of such proteinase inhibitors as alpha-1-antitrypsin. This could be caused by oxidation inactivation of the inhibitor [67]. Recently, plasma carbonyl content has been found to be unchanged in AD patients as compared with age-matched controls [69].

In the same study, total antioxidant activity in plasma and the activity of endogenous antioxidant enzymes such as glutathione peroxidase, glutathione reductase and superoxide dismutase were also evaluated. The total antioxidant plasmatic status of the patients with AD both in light-moderate phase and in advanced phase was lower than in the controls. No significant differences of protein oxidation levels were observed between the different stages of the disease. Peroxidation was higher in patients in the advanced stage of the disease than in the control group. However, no significant differences were observed between different stages.

The most recent study is the one published by Korolainein et al. in 2009 that reported total levels of protein oxidation (carbonyl and 3-NT) in CSF, plasma, and serum in AD. Protein carbonyls were found not differing in CSF between AD patients and controls and corroborate previous findings from the same group [31]. Accordingly, no differences were found in plasma while the levels of serum protein carbonyls were decreased in AD subjects when compared with controls. The reliability of these results is limited by the fact that most values measured in blood samples were under the detection limit. Nevertheless, these data agree with the study by Zafrilla and collaborators [69] of similar plasma carbonyl concentrations among AD patients and controls. Overall, it seems that levels of protein carbonyls may vary in different compartments; or alternatively, the separation of serum and plasma from whole blood may result in differences in measurable carbonyls. Reasons for the tendency of decreased levels of serum protein carbonyls remain unclear. Decreased serum carbonyls may thus be an epiphenomenon and not linked to AD per se. On the other hand, also previous studies have reported decreased levels of oxidative stress markers such as nitric oxide and oxidized high-density lipoprotein in blood from AD patients [4,13]. Bergt et al. [4] found significantly lower oxHDL levels in plasma of AD patients compared to age-matched, cognitive healthy individuals. The decrease was statistically significant in males (32 % reduction) and it followed the same tendency in females (34 % reduction), however, without reaching statistical significance. However, they did not observe a marked decrease in Apo A-I plasma levels as reported by others [30]. These results are in contrast with the findings of Sinem et al. [56] and Dildar et al. [16] that demonstrated that serum NO and oxLDL levels in patients with AD were significantly higher than in both controls and patients with vascular dementia (VaD). However, no significant differences in plasma NO and ox-LDL levels were found between VaD and controls. On the other hand, they did not find any significant difference in serum 3-NT values of both AD and VaD patients when compared with controls. Although it has been suggested that nitrotyrosine levels were increased in the brain of AD patients, several factors might be responsible for these results, for example 3-NT is not stable in serum or protein nitration occurs at undetectable levels. Some previous studies [1,65] reported evidence of increased CSF nitrotyrosine concentrations in AD patients as compared with controls, whereas another study revealed no differences [55]. Overall, these results point to the fact that the concentrations of nitrotyrosines in individuals may vary greatly. However, further studies with more sensitive methods are needed to assess nitrotyrosine levels in AD plasma or serum.

As well as in the CSF, F-2 and F-4 isoprostane levels were also investigated in plasma of patients with AD and subjects with MCI. The results showed were somehow controversial, however, in most of the cases no differences in F-2 and F-4 isoprostane, between AD and control subjects, were showed [26,44,45,52].

Studies from Padurario et al. demonstrated a progressive increase of the peripheral levels of MDA in patients with MCI and AD [46]. The end products of lipid peroxidation, so called lipofuscin-like pigments (LFP), were also found increased in erythrocytes of AD patients compared to controls. The specific fraction of LFP in AD patients, which was isolated, might represent a disease-specific product in the blood [57].

Recently, markers of oxidative damage have been found to be elevated in mitochondria isolated from lymphocytes from AD patients compared with their age-matched controls [62]. Indeed, mitochondrial dysfunction has been widely implicated in the etiology of AD. This is the first report to show mitochondrial alteration in peripheral lymphocytes, thus suggesting that the oxidative stress indices could potentially be used as putative biomarkers for AD. Very few studies investigated the extent of oxidative damage to DNA at peripheral levels. Mecocci et al. showed that levels of DNA 8-OHdG content in lymphocytes were significantly higher and plasma levels of antioxidants (with the exception of lutein) were significantly lower in patients with AD compared with controls [39]. In patients with AD, a significant inverse relationship between lymphocyte DNA 8-OHdG content and plasma levels of lycopene, lutein, alpha-carotene, and beta-carotene, respectively, was observed. Previous work from the same group work and others showed increased amounts of 8-OHdG in mitochondrial and nuclear DNA from AD-affected brains compared with controls [70]. In the same study, an inverse trend between plasma antioxidants and the content of 8-OHdG was observed. These findings suggest that lymphocyte DNA 8-OHdG content in patients with AD reflects a condition of increased oxidative stress related to a poor antioxidant status.

### *13.3.3 Urine and Saliva*

Their easy way of collection, made urine samples an interesting source of AD biomarker and several studies on Aβ, tau protein, NBT protein as well as OS markers has been performed. Increased oxidative damage in the brain is thought to drive the presence of increased OS markers at peripheral level, and the search of these has been the main focus of several AD studies. However, the production of oxidized molecule in all other tissues contributes to peripheral OS masking the brain production. Data about lipid peroxidation products, F2-IsoPs, are indeed quite controversial in urine. Praticò et al. showed that F2-IsoP levels were increased in AD patients urine compared to controls [51]. In addition, Tuppo et al. demonstrate a significant increase of isoprostanes in urine of patients assessed to have mild to moderate dementia as compared to non-demented patients, suggesting that patients with mild to moderate dementia associated with probable AD are experiencing significant oxidative stress even at peripheral level [66]. In contrast, Montine et al. showed no difference of lipid peroxidation marker in urine of AD patients and age-matched controls or definite AD patients and probable AD patients supporting the model of oxidative damage restricted to brain only [44]. A further step, in the study of urine as a source of oxidative stress biomarker, was given by analyses on DNA oxidative damage. A study of free 8-OHG in urine indicated that excretion decreases with age whereas in the brain the levels of 8-OHG in intact DNA increases with age raising additional doubts about the use of urine as in the search of oxidative stress biomarkers [35].

None of the OS biomarkers in peripheral fluids has been analyzed in saliva, which has been only considered, until now, only a good source for Aβ levels and acetylcholinesterase levels investigations that apparently reflect changes in the CSF [5].

## 13.4 Concluding Remarks

Most of the above reported results demonstrated the presence of peripheral oxidative damage that correlates with the occurrence of the disease. However, some these findings are controversial and do not fully elucidate the complex cascade of reactions involved. On the one hand, it is reasonable to hypothesize that oxidative stress first develops in the periphery as a result of different causes, and then it contributes to perturb neuronal homeostasis, either by increasing the production of ROS or by depleting antioxidant defense. On the other hand, it is also possible that oxidative stress starts in the CNS and several different metabolic end products are released into the blood stream. Understanding the timing of appearance of such products, or their derivatives, at systemic level could allow correlating with the onset of AD.

The development of more sensitive methods to detect a biochemical marker in AD are highly desirable to increase diagnostic accuracy, to identify MCI subjects who will progress to clinical AD, to monitor pharmacological and biological effects of drugs. So far important steps have been accomplished, through the evolution of modern techniques and advanced proteomics platforms, but there is still a lot of effort to be directed towards the discovery, testing, and validation of a panel of assays that could serve all the requirements for an ideal biomarker.

Nevertheless, the major limitation is represented is the wide variability among different studies that led to contrasting results. Thus, there is an urgent need to standardize protocols for replicate experiments on large population, which may allow to better understand the effect of systemic oxidative damage in the pathogenesis and progression of AD. The essential goal in biomarker discovery studies is the identification of "preclinical marker", which facilitates disease diagnosis at early stages, monitors disease progression and assesses the response to treatments by the time that disease-modifying treatments become available in clinical practice.

## References


1. Ahmed N, Ahmed U, et al. Protein glycation, oxidation and nitration adduct residues and free adducts of cerebrospinal fluid in Alzheimer's disease and link to cognitive impairment. J Neurochem. 2005;92(2):255–63.
2. Aluise CD, Sowell RA, et al. Peptides and proteins in plasma and cerebrospinal fluid as biomarkers for the prediction, diagnosis, and monitoring of therapeutic efficacy of Alzheimer's disease. Biochim Biophys Acta. 2008;1782(10):549–58.
3. Bandow JE. Comparison of protein enrichment strategies for proteome analysis of plasma. Proteomics. 2010;10(7):1416–25.
4. Bergt C, Nakano T, et al. Oxidized plasma high-density lipoprotein is decreased in Alzheimer's disease. Free Radic Biol Med. 2006;41(10):1542–7.
5. Bermejo-Pareja F, Antequera D, et al. Saliva levels of Abeta1-42 as potential biomarker of Alzheimer's disease: a pilot study. BMC Neurol. 2010;10:108.
6. Blennow K, De Meyer G, et al. Evolution of Abeta42 and Abeta40 levels and Abeta42/Abeta40 ratio in plasma during progression of Alzheimer's disease: a multicenter assessment. J Nutr Health Aging. 2009;13(3):205–8.
7. Blennow K, Hampel H, et al. Cerebrospinal fluid and plasma biomarkers in Alzheimer disease. Nat Rev Neurol. 2010;6(3):131–44.
8. Brouns R, De Vil B, et al. Neurobiochemical markers of brain damage in cerebrospinal fluid of acute ischemic stroke patients. Clin Chem. 2010;56(3):451–8.
9. Calabrese V, Sultana R, et al. Nitrosative stress, cellular stress response, and thiol homeostasis in patients with Alzheimer's disease. Antioxid Redox Signal. 2006;8(11–12):1975–86.
10. Choi J, Malakowsky CA, et al. Identification of oxidized plasma proteins in Alzheimer's disease. Biochem Biophys Res Commun. 2002;293(5):1566–70.
11. Choi J, Forster MJ, et al. Proteomic identification of specific oxidized proteins in ApoE-knockout mice: relevance to Alzheimer's disease. Free Radic Biol Med. 2004;36(9):1155–62.
12. Conrad CC, Marshall PL, et al. Oxidized proteins in Alzheimer's plasma. Biochem Biophys Res Commun. 2000;275(2):678–81.
13. Corzo L, Zas R, et al. Decreased levels of serum nitric oxide in different forms of dementia. Neurosci Lett. 2007;420(3):263–7.
14. Di Domenico F, Coccia R, et al. Circulating biomarkers of protein oxidation for Alzheimer disease: expectations within limits. Biochim Biophys Acta. 2011;1814(12):1785–95.
15. Di Terlizzi R, Platt SR. The function, composition and analysis of cerebrospinal fluid in companion animals: part II – analysis. Vet J. 2009;180(1):15–32.
16. Dildar K, Sinem F, et al. Serum nitrosative stress levels are increased in Alzheimer disease but not in vascular dementia. Alzheimer Dis Assoc Disord. 2010;24(2):194–7.
17. Ehrenreich H, Kastner A, et al. Circulating damage marker profiles support a neuroprotective effect of erythropoietin in ischemic stroke patients. Mol Med. 2011;17(11–12):1306–10.
18. Frey HJ, Mattila KM, et al. Problems associated with biological markers of Alzheimer's disease. Neurochem Res. 2005;30(12):1501–10.
19. Gackowski D, Rozalski R, et al. Oxidative stress and oxidative DNA damage is characteristic for mixed Alzheimer disease/vascular dementia. J Neurol Sci. 2008;266(1–2):57–62.
20. Ganzer S, Arlt S, et al. CSF-tau, CSF-Abeta1-42, ApoE-genotype and clinical parameters in the diagnosis of Alzheimer's disease: combination of CSF-tau and MMSE yields highest sensitivity and specificity. J Neural Transm. 2003;110(10):1149–60.
21. Grossman M, Farmer J, Leight S, Work M, Moore P, Van Deerlin V, Pratico D, Clark CM, Coslett HB, Chatterjee A, Gee J, Trojanowski JQ, Lee VM Cerebrospinal fluid profile in frontotemporal dementia and Alzheimer's disease. 2002;359(9309):881–4.
22. Herukka SK, Helisalmi S, et al. CSF Abeta42, Tau and phosphorylated Tau, APOE epsilon4 allele and MCI type in progressive MCI. Neurobiol Aging. 2007;28(4):507–14.
23. Hulstaert F, Blennow K, et al. Improved discrimination of AD patients using beta-amyloid(1–42) and tau levels in CSF. Neurology. 1999;52(8):1555–62.

24. Humphrey SP, Williamson RT. A review of saliva: normal composition, flow, and function. J Prosthet Dent. 2001;85(2):162–9.
25. Ikeda M, Sato I, et al. Nitrite, nitrate and cGMP in the cerebrospinal fluid in degenerative neurologic diseases. J Neural Transm Gen Sect. 1995;100(3):263–7.
26. Irizarry MC, Yao Y, et al. Plasma F2A isoprostane levels in Alzheimer's and Parkinson's disease. Neurodegener Dis. 2007;4(6):403–5.
27. Johanson CE, Duncan 3rd JA, et al. Multiplicity of cerebrospinal fluid functions: new challenges in health and disease. Cerebrospinal Fluid Res. 2008;5:10.
28. Kadiiska MB, Gladen BC, et al. Biomarkers of oxidative stress study II: are oxidation products of lipids, proteins, and DNA markers of CCl4 poisoning? Free Radic Biol Med. 2005;38(6):698–710.
29. Kalsheker NA. Molecular pathology of alpha 1-antitrypsin deficiency and its significance to clinical medicine. QJM. 1994;87(11):653–8.
30. Kawano M, Kawakami M, et al. Marked decrease of plasma apolipoprotein AI and AII in Japanese patients with late-onset non-familial Alzheimer's disease. Clin Chim Acta. 1995;239(2):209–11.
31. Korolainen MA, Pirttila T. Cerebrospinal fluid, serum and plasma protein oxidation in Alzheimer's disease. Acta Neurol Scand. 2009;119(1):32–8.
32. Korolainen MA, Nyman TA, et al. Multiplexed proteomic analysis of oxidation and concentrations of cerebrospinal fluid proteins in Alzheimer disease. Clin Chem. 2007;53(4):657–65.
33. Korolainen MA, Nyman TA, et al. An update on clinical proteomics in Alzheimer's research. J Neurochem. 2010;112(6):1386–414.
34. Kuiper MA, Visser JJ, et al. Decreased cerebrospinal fluid nitrate levels in Parkinson's disease, Alzheimer's disease and multiple system atrophy patients. J Neurol Sci. 1994;121(1):46–9.
35. Lee SH, Kim I, et al. Increased urinary level of oxidized nucleosides in patients with mild-to-moderate Alzheimer's disease. Clin Biochem. 2007;40(13–14):936–8.
36. Lovell MA, Ehmann WD, et al. Elevated 4-hydroxynonenal in ventricular fluid in Alzheimer's disease. Neurobiol Aging. 1997;18(5):457–61.
37. Lovell MA, Gabbita SP, et al. Increased DNA oxidation and decreased levels of repair products in Alzheimer's disease ventricular CSF. J Neurochem. 1999;72(2):771–6.
38. Markesbery WR. Oxidative stress hypothesis in Alzheimer's disease. Free Radic Biol Med. 1997;23(1):134–47.
39. Mecocci P, Polidori MC, et al. Lymphocyte oxidative DNA damage and plasma antioxidants in Alzheimer disease. Arch Neurol. 2002;59(5):794–8.
40. Montine TJ, Markesbery WR, et al. Cerebrospinal fluid F2-isoprostane levels are increased in Alzheimer's disease. Ann Neurol. 1998;44(3):410–3.
41. Montine TJ, Beal MF, et al. Increased CSF F2-isoprostane concentration in probable AD. Neurology. 1999;52(3):562–5.
42. Montine TJ, Kaye JA, et al. Cerebrospinal fluid abeta42, tau, and f2-isoprostane concentrations in patients with Alzheimer disease, other dementias, and in age-matched controls. Arch Pathol Lab Med. 2001;125(4):510–2.
43. Montine TJ, Neely MD, et al. Lipid peroxidation in aging brain and Alzheimer's disease. Free Radic Biol Med. 2002;33(5):620–6.
44. Montine TJ, Quinn JF, et al. Peripheral F2-isoprostanes and F4-neuroprostanes are not increased in Alzheimer's disease. Ann Neurol. 2002;52(2):175–9.
45. Mufson EJ, Leurgans S. Inability of plasma and urine F2A-isoprostane levels to differentiate mild cognitive impairment from Alzheimer's disease. Neurodegener Dis. 2010;7(1–3): 139–42.
46. Padurariu M, Ciobica A, et al. Changes of some oxidative stress markers in the serum of patients with mild cognitive impairment and Alzheimer's disease. Neurosci Lett. 2010;469(1): 6–10.
47. Polaskova V, Kapur A, et al. High-abundance protein depletion: comparison of methods for human plasma biomarker discovery. Electrophoresis. 2010;31(3):471–82.

48. Pratico D. Evidence of oxidative stress in Alzheimer's disease brain and antioxidant therapy: lights and shadows. Ann N Y Acad Sci. 2008;1147:70–8.
49. Pratico D, Delanty N. Oxidative injury in diseases of the central nervous system: focus on Alzheimer's disease. Am J Med. 2000;109(7):577–85.
50. Pratico D, L. V MY, et al. Increased F2-isoprostanes in Alzheimer's disease: evidence for enhanced lipid peroxidation in vivo. FASEB J. 1998;12(15):1777–83.
51. Pratico D, Clark CM, et al. Increased 8,12-iso-iPF2alpha-VI in Alzheimer's disease: correlation of a noninvasive index of lipid peroxidation with disease severity. Ann Neurol. 2000;48(5):809–12.
52. Pratico D, Clark CM, et al. Increase of brain oxidative stress in mild cognitive impairment: a possible predictor of Alzheimer disease. Arch Neurol. 2002;59(6):972–6.
53. Quinn JF, Montine KS, et al. Suppression of longitudinal increase in CSF F2-isoprostanes in Alzheimer's disease. J Alzheimers Dis. 2004;6(1):93–7.
54. Raber J, Huang Y, et al. ApoE genotype accounts for the vast majority of AD risk and AD pathology. Neurobiol Aging. 2004;25(5):641–50.
55. Ryberg H, Soderling AS, et al. Cerebrospinal fluid levels of free 3-nitrotyrosine are not elevated in the majority of patients with amyotrophic lateral sclerosis or Alzheimer's disease. Neurochem Int. 2004;45(1):57–62.
56. Sinem F, Dildar K, et al. The serum protein and lipid oxidation marker levels in Alzheimer's disease and effects of cholinesterase inhibitors and antipsychotic drugs therapy. Curr Alzheimer Res. 2010;7(5):463–9.
57. Skoumalova A, Ivica J, et al. The lipid peroxidation products as possible markers of Alzheimer's disease in blood. Exp Gerontol. 2011;46(1):38–42.
58. Song F, Poljak A, et al. Plasma biomarkers for mild cognitive impairment and Alzheimer's disease. Brain Res Rev. 2009;61(2):69–80.
59. Stempfer R, Kubicek M, et al. Quantitative assessment of human serum high-abundance protein depletion. Electrophoresis. 2008;29(21):4316–23.
60. Stief TW, Marx R, et al. Oxidized fibrin(ogen) derivatives enhance the activity of tissue type plasminogen activator. Thromb Res. 1989;56(2):221–8.
61. Straface E, Matarrese P, et al. Oxidative imbalance and cathepsin D changes as peripheral blood biomarkers of Alzheimer disease: a pilot study. FEBS Lett. 2005;579(13):2759–66.
62. Sultana R, Mecocci P, et al. Increased protein and lipid oxidative damage in mitochondria isolated from lymphocytes from patients with Alzheimer's disease: insights into the role of oxidative stress in Alzheimer's disease and initial investigations into a potential biomarker for this dementing disorder. J Alzheimers Dis. 2011;24(1):77–84.
63. Thambisetty M, Hye A, et al. Proteome-based identification of plasma proteins associated with hippocampal metabolism in early Alzheimer's disease. J Neurol. 2008;255(11):1712–20.
64. Tohgi H, Abe T, et al. The cerebrospinal fluid oxidized NO metabolites, nitrite and nitrate, in Alzheimer's disease and vascular dementia of Binswanger type and multiple small infarct type. J Neural Transm. 1998;105(10–12):1283–91.
65. Tohgi H, Abe T, et al. Alterations of 3-nitrotyrosine concentration in the cerebrospinal fluid during aging and in patients with Alzheimer's disease. Neurosci Lett. 1999;269(1):52–4.
66. Tuppo EE, Forman LJ, et al. Sign of lipid peroxidation as measured in the urine of patients with probable Alzheimer's disease. Brain Res Bull. 2001;54(5):565–8.
67. Twining SS, Fukuchi T, et al. Corneal synthesis of alpha 1-proteinase inhibitor (alpha 1-antitrypsin). Invest Ophthalmol Vis Sci. 1994;35(2):458–62.
68. Yu HL, Chertkow HM, et al. Aberrant profiles of native and oxidized glycoproteins in Alzheimer plasma. Proteomics. 2003;3(11):2240–8.
69. Zafrilla P, Mulero J, et al. Oxidative stress in Alzheimer patients in different stages of the disease. Curr Med Chem. 2006;13(9):1075–83.
70. Zhu X, Smith MA, et al. Mitochondrial failures in Alzheimer's disease. Am J Alzheimers Dis Other Demen. 2004;19(6):345–52.

# Chapter 14
# Nutrition, Lifestyle and Oxidative Stress: Prevention of Alzheimer's Disease

**M. Cristina Polidori and Ludger Pientka**

**Abstract** The debate emerging from the frustrating results obtained while pursuing the search of a single curative compound, a single preventive drug or lifestyle strategy against dementia onset and progression has highlighted the importance of a comprehensive preventive strategy implementing vascular risk control and healthy lifestyle. In this respect, nutritional interventions aimed at keeping a condition of oxidative stress as low as possible in the aging organism play a particularly important role. The consumption of higher amounts of fruits and vegetables and the adherence to Mediterranean diet are strongly encouraged to maintain cognitive health.

## 14.1 Introduction

As the Western population ages and so-called "civilization" diseases including cognitive impairment become more and more prevalent in the developing countries [16], the use of a homogeneous term to describe and diagnose severe cognitive impairment and the search for a homogeneous preventive and therapeutic approach against it remain a public health priority. Nevertheless, dementia and especially its most common cause, Alzheimer's disease (AD), have been proven to be mostly heterogeneous and age-related conditions against which no individual preventive strategy and no single curative drug have been found to date. The identification of a single antidementia strategy is hindered by a problematic combination of several

M.C. Polidori, Ph.D. (✉)
Institute of Geriatrics, University of Cologne, Cologne, Germany
e-mail: polidori@uni-duesseldorf.de

L. Pientka
Department of Geriatrics, Marienhospital Herne, Ruhr University, Bochum, Germany

D. Praticò and P. Mecocci (eds.), *Studies on Alzheimer's Disease*, Oxidative Stress in Applied Basic Research and Clinical Practice, DOI 10.1007/978-1-62703-598-9_14,

factors, mostly related to the difficulty of applying an in vitro well-working chemical substance targeted toward a single system to patients with a multi-systemic disease who are usually of advanced age and have several other diseases, each treated with various medications. Despite the amount of progress on knowledge of AD pathophysiology achieved in the past decades, a commentary stating that "no agent used thus far has been able to reduce, reverse or abolish the progressive pathology of this dementia" and that "medicines [...] have largely addressed a fully developed stage of AD when mild to severe cognitive dysfunction is unravelling the mind and when cerebral tissue is already irreparably damaged." is still appropriate [7].

In the present work, the traditional terminology will be used for the sake of concept accessibility. Therefore, the acronym "AD" will be used to describe a progressive neurodegenerative alteration that gradually reduces cognitive and functional abilities of the affected subject. In this context, loss of synapses and neurons as well as accumulation of extracellular β-amyloid plaques and intraneuronal neurofibrillary tangles will be further on considered as the main hallmarks of AD brain and aging the major risk factor for AD. As demographics show 35.6 million people living with dementia worldwide in 2010, a projection to 65.7 million by 2030 and 115.4 million by 2050 (http://www.alz.co.uk/research/statistics) and as AD is also a highly disabling disorder [16], the remarkable progress in understanding AD pathogenesis and the troublesome search for a cure against the disease leave slowly but inexorably an unsubstitutable place to multifaceted preventive strategies. Fortunately, there are factors with a large potential to be modified before the onset or during the course of the disease. These include vascular and lifestyle factors, which of course are strictly connected to each other. Unfortunately, clinical efforts to improve lifestyle and vascular function at any age with the scope of maintaining cognitive health are largely insufficient to produce evident results, and the complete avoidance of about 23 million of the globally expected AD cases by 2050 if it had been possible to delay AD onset by 2 years starting in 2010 [37] remains a goal.

The aim of this work is to present the state of knowledge of the preventive effects of selected lifestyle habits—cognitive, social, and leisure activities as well as physical activity and nutrition—with particular attention to those aspects of nutrition and diet related to the potential of lowering oxidative stress status. Interestingly, most of the protective substances mentioned in this chapter constitute a substantial part of the Mediterranean diet, recently inscribed on the Representative List of the Intangible Cultural Heritage of Humanity (http://www.unesco.org/culture/ich/en/RL/00394). They are integrative part of a healthy lifestyle, which in fact was called "diaeta" by the Romans referring to the word "διά ήτα" with which the ancient Greeks intended the "way of living". Ultimately, we hope to provide the reader with some hints of the great potential of immediate preventive strategies against cognitive impairment as well as of specific needs for designing future experiments and trials.

## 14.2 Modifiable Risk Factors for AD

### *14.2.1 Considerations on Lifestyle, Vascular Risk Factors, and Oxidative Stress*

Although most of the risk factors for AD are not modifiable, modifiable ones include type 2 diabetes, atherosclerosis, hyperlipidemia, hypertension, hypercholesterolemia, cardiac disorders, cerebrovascular pathology, body mass index, and several unhealthy lifestyle conditions ([8], third issue of the 20th volume of the Journal of Alzheimer's disease; [26]). It is intended to mention them here as each one of them has been shown to be related not only to dementia risk but also, probably primarily and mainly—in light of the primacy of free radical-related damage in AD development [35]—, to oxidative stress [7, 20–23, 27, 33]. While social, cognitive and leisure activities appear to act beneficially against cognitive deterioration mainly through improvement of cognitive reserve [14], other lifestyle-related healthy habits are likely to positively affect dementia risk mainly through their effect on minimizing oxidative damage, improving vascular tone and endothelial function as well as decreasing vascular pathology [2, 6, 28, 38].

Lifestyle components such as cognitive and leisure activities have been associated with the concept of "cognitive reserve" [36] and have been shown to strongly influence the risk of cognitive impairment with and without dementia [8, 9, 36]. Engagement in leisure activities of intellectual and social nature is indeed associated with slower cognitive decline in the healthy elderly and may reduce the risk of incident dementia, being the evidence from functional imaging studies that subjects engaging in such leisure activities can clinically tolerate more AD pathology [8]. Intellectual challenge and life experience through leisure may result in functionally more efficient cognitive networks and therefore provide a cognitive reserve that delays the onset of clinical manifestations of dementia. The interpretation of the results of studies on cognitive and leisure activities in dementia prevention is, unfortunately, strongly hindered by the poor comparability of studies using different activity measures as well as by the frequent inclusion of physical activity among leisure activities and by difficulties in measuring cognitive reserve. Despite this, and similarly to the preventive aspects of physical exercise against dementia, there are still a number of reasons that should motivate physicians to encourage continuous cognitive stimulation and moderate and constant physical exercise in healthy and cognitively impaired subjects [10], above all the "use it or lose it" concept. As far as physical exercise in dementia prevention is concerned, the self-addressed nature of cognitive performance and performed physical activity as well as to the populations studied, compliance, time-window of applied physical intervention, and dyshomogeneity of the training frequency applied hinders clear and definitive guidelines. Although the comparability of studies on physical activity and dementia prevention (either primary or secondary) in the elderly is poor—and guidelines for the primary and secondary prevention of dementia cannot be drafted—, there is a wealth of relevant articles on the best-designed and performed recent studies confirming longitudinal and short-term RCT evidence that physical activity improves cognitive function in older subjects (reviewed in [10]). Therefore, the current recommendations of regular physical activity as a key component of successful aging can be given to both healthy adults and elderly subjects with and without cognitive impairment. While awaiting for the results of large ongoing randomized controlled trials in the coming decade, it is to recommend that persons with cognitive impairment with and without dementia pursue a moderate but regular, variable exercise program consisting of at least 30 min three times weekly of walking alternating with aerobically challenging exercise and group sports [4].

### *14.2.2 Nutrition: An Example of Oxidative Stress-Related Modifiable Risk Factor for AD*

There is strong epidemiologic evidence that a poor diet is one of the leading causes of death for Americans [18] and Europeans (http://ec.europa.eu/health/reports/european). Decreased food intakes, eating behavior disturbances, and loss of body weight are particularly significant problems among AD patients [34]. While nutrition is a wide and complex field including the concepts of caloric restriction, fat and alcohol consumption as well as fiber and cereal intake and weight maintenance, the discussion of these topics goes beyond the scope of this chapter which focuses on antioxidant strategies related to natural nutrition.

Several epidemiological studies confirmed a reduced risk of AD in antioxidant vitamin supplement users, such as the Rotterdam study [12] and the Cache County Study [39], although data on supplemental intake and antioxidant plasma levels seemed to become contradictory at some point [13, 17]. The prospective evaluation of 36 community-dwelling patients in early stages of AD and 58 age-matched cognitively intact community-based controls over a 18-months period led to the observation that nutrient intakes from diet and supplements were higher in control subjects, with significant differences in energy, the micronutrients calcium, iron, zinc, vitamin K, vitamin A, and dietary fiber as well as n-3 and n-6 fatty acids [34]. The authors suggested that suboptimal diet is early in the onset of the disease and that AD patients would benefit from systematic dietary assessment and intervention to prevent further deterioration in food consumption and increased nutritional risk [34]. Nutritional problems in the elderly in general and in aged subjects with dementia in particular, together with the knowledge of the critical role played by oxidative stress in the onset and progression of neurodegeneration have prompted the design of several studies using mostly individual substances with direct or indirect antioxidant effect in the last years.

The main base for the rationale of nutritional interventions for the prevention of cognitive impairment resides in the modulation of oxidative stress. Interestingly, if on one side it is rather obvious to connect oxidative stress with the concept of antioxidant defense, it is known to date that several substances that cannot be defined as antioxidants also may influence redox reactions. They are therefore able to diminish a condition of oxidative stress which, in the presence of unfavorable circumstances such as preclinical pathology, may trigger the onset and/or worsening of cognitive impairment, thereby delaying it.

As mentioned above, single molecules with strong antioxidant activity have proven to be markedly consumed in the presence of oxidative stress and in the complex neurodegeneration-redox imbalance in several experimental models and human studies (reviewed in the whole 38th issue of BioFactors, [29]), so that most of the RCT and prospective cohort studies have focussed on dietary restriction, single antioxidants, fish oil, omega-3 fatty acid and other individual supplements. As probably in general expected, due to the fact that antioxidant nutrients alone cannot account for the preventive effects of whole healthy nutrition and fruit and vegetable intake against dementia, most of the RCTs on antioxidants in dementia prevention and treatment led to inconsistent results [29].

#### 14.2.2.1 Fruits and Vegetables, Mediterranean Diet, and Antioxidants

Fruits and vegetables are thought to represent the best source of antioxidant micronutrients due to synergisms of their components, because they may allow a better bioavailability of protective compounds than single vitamins and due to their low content in saturated fats, but it is only in the last two decades that fruits and vegetables are considered the best polypharmacy against diseases and have been related to oxidative stress for their effects (reviewed in [11]). Interestingly and accordingly, the strongest evidence for antioxidant protection against AD from a single compound remains in favor of vitamin E, whose food sources are vegetable oils, nuts, seeds, fruits (avocado, melon, apple) and vegetables such as collard greens. One key point which favors the discrepancy between (1) epidemiological studies on fruit/vegetable intake and resistance against disease, (2) observational studies on antioxidant levels in demented subjects and (3) RCTs of antioxidant supplements against dementia is the largely unexplored relationship between intake of fruits and vegetables, antioxidant micronutrient status, a condition of oxidative stress and cognitive performance in healthy subjects. There are indeed some hints of biological interactions between these components after evaluation of independent measurements in healthy subjects [1, 25], and it is likely that when the clinical symptoms of AD appear a large proportion of neuronal cells is already destroyed and the intervention with antioxidants, particularly that with a single compound, comes too late.

In an observational case–control study, 39 community-dwelling healthy subjects aged 65 years and older consuming high intakes of fruits and vegetables as over 400 g daily and 48 healthy age-matched subjects consuming low intakes of fruit and vegetables daily (less than 100 g) were enrolled and plasma levels of retinol, tocopherols, carotenoids and malondialdehyde (MDA) as well as content of protein carbonyls in Ig G were measured [1]. Plasma levels of retinol, tocopherols and carotenoids were found to be significantly higher in healthy subjects with a high intake of fruits and vegetables compared to individuals consuming less than one portion per day [1]. Being the results independent of age and gender [1], it is possible to hypothesize that the antioxidant profiles observed in several diseases [22] and in centenarians [24] reflect the role of fruit and vegetable intake as one major determinant of antioxidant micronutrient status. This study strongly highlighted the modification of nutritional habits as a tool to lower prevalence of oxidative stress-related disease in later life and prompted the performance of an interventional study aimed to modify nutritional habits in healthy subjects. Indeed, dietary counseling influences fruit and vegetable intake and improves levels of circulating antioxidants and biomarkers of oxidative stress, as shown in a sample of 129 employees of the Heinrich-Heine University hospital [25]. Subjects followed a diet consisting of at least five portions of fruits and vegetables per day over 3 months. During this time, fruit and vegetable intake was monitored by means of a self-administered, 2-week based food-frequency questionnaire, counseling sessions were offered every 4 weeks and blood samples were obtained on the day of the counseling. A significant increase in several plasma antioxidant micronutrients was observed at the end of the 3-month period in comparison to baseline even in this population of mostly females, relatively young and well educated [25]. However, there were no changes of biomarkers of oxidative stress, supporting the knowledge that optimal antioxidant intake may vary substantially from population to population, and that health-conscious subjects may present oxidative stress at levels difficult to further lower. Overall, these results suggested that a nutritional counseling program can lead to improvement in plasma antioxidant status even in a health conscious population in which a relevant decrease in biomarkers of oxidative stress is not to be expected. Another study published in 2009 aimed at considering the evaluation of cognitive performance as a third parameter in addition to fruit and vegetable intake and oxidant-antioxidant balance of the organism in healthy subjects [24]. In this study including 193 healthy subjects aged 45–102 years, individuals of any age assuming 400 g of fruit and vegetables per day and over had significantly higher cognitive test scores, higher levels of carotenoids, $\alpha$- and $\gamma$-tocopherol as well as lower levels of F2$\alpha$ isoprostanes than individuals with an intake of fruits and vegetables less than one daily portion [24]. Cognitive scores were directly correlated with blood levels of $\alpha$-tocopherol and lycopene and negatively correlated with F2$\alpha$ isoprostanes and protein carbonyls independent of age, gender, body mass index, education, total cholesterol, LDL- and HDL-cholesterol, triglycerides, and albumin [24]. A targeted nutritional intervention aimed at increasing plasma antioxidant levels and at decreasing the ongoing condition of oxidative stress might prove beneficial in addition to standard therapeutic options in patients with AD showing increased circulating levels of biomarkers of oxidative stress. In a prospective cohort study of over 3,700 older participants of the Chicago Health and Age Project, high vegetable consumption was associated with a slower rate of cognitive decline over 6 years after adjusting for age, gender, race, education, cardiovascular-related conditions and risk factors [19]. In this study the consumption of green leafy vegetables, rich in antioxidant micronutrients like carotenoids, showed the strongest inverse linear association with the rate of cognitive decline. The specific protection shown by vegetables and particularly by the green leafy ones appears to be in disagreement with the concept that fruit and vegetable consumption might be beneficial in the frame of a generally healthy lifestyle, as health-conscious individuals tend to consume both fruits and vegetables. In the Three-City Cohort Study fruit and vegetable consumption frequently but also consumption of fish and omega-3 fat was associated with a lowered risk of dementia and of AD especially among ApoE ε4 noncarriers [3].

Healthy diet in general and the Mediterranean (a diet characterized by high intake of fish, vegetables, legumes, fruits, cereals, and unsaturated fatty acids mostly in the form of olive oil, low intake of dairy products, meat, and saturated fatty acids and a regular but moderate intake of alcohol) regimen in particular have been recently shown to affect risk for and mortality from AD ([3, 30–32]). The existing evidence does not support the recommendation of specific supplements, foods, or diets for the prevention of AD. However, a review of 34 studies in the areas of dietary restriction, antioxidants and Mediterranean diet provided evidence that nutritional interventions against dementia and AD have a great potential of influencing dementia development [5].

## 14.3 Conclusive Remarks

In their recent article on the evolving classification of dementia, George, Whitehouse and Ballenger [15] state that "if the history of AD [Alzheimer's Disease] tells us anything, it is that the quest for scientific truth is an ongoing process that is surely subject to the vicissitudes of historical and cultural circumstance. With the DSM-V due for publication in 2 years, the culturally salient question remains: To what extent ought Western culture continue medicalizing the continuum of brain aging and creating specific thresholds of disease given both the heterogeneity and the attendant social meanings of conditions such as AD and other forms of cognitive and functional loss? The publication of the DSM-V in 2013 will be revealing not only of the evolution of our psychiatric classifications, but also of our values as a culture. Once again, we will bear witness to how the Western biomedical system reflects Western society at large" [15]. Indeed, the need for preventing cognitive impairment with and without dementia is clearly overcoming that of waiting for a cure. Together with the absence of a cure, the longevity-related prevalence increase of dementia is its major feature currently; far from the intention of neglecting the enormous advance permitted by Dr. Alzheimer in the field of neurodegeneration, it is time that health practitioners in the field awake from the comfort of giving obsolete diagnoses and therapies to get actively and convincingly engaged in the battle against cognitive impairment. The control of specific bioclinical conditions whose presence is associated to increased risk for cognitive impairment is the first step in this process and will save an enormous amount of time and money to every society. The debate emerging from the frustrating results obtained while pursuing the search of a single curative compound, a single preventive drug or lifestyle strategy against dementia onset and progression has highlighted the importance of a comprehensive preventive strategy implementing vascular risk control and healthy lifestyle. In this respect, nutritional interventions aimed at keeping a condition of oxidative stress as low as possible in the aging organism play a particularly important role. The consumption of higher amounts of fruits and vegetables and the adherence to Mediterranean diet are strongly encouraged to maintain cognitive health.

## References

1. Anlasik T, Sies H, Griffiths HR, Stahl W, Polidori MC. Dietary habits are major determinants of the plasma antioxidant status in healthy elderly subjects. Br J Nutr. 2005;94:639–42.
2. Aoi W, Sakuma K. Oxidative stress and skeletal muscle dysfunction with aging. Curr Aging Sci. 2011;4:101–9.
3. Barberger-Gateau P, Raffaitin C, Letenneur L, Berr C, Tzourio C, Dartigues JF, Alperovitch A. Dietary patterns and risk of dementia. Neurology. 2007;69:1921–30.
4. Blankevoort CG, van Heuvelen MJG, Boersma F, Luning H, de Jong J, Scherder EJA. Review oft he effects of physical activity on strength, balance, mobility and ADL performance in elderly subjects with dementia. Dement Geriatr Cogn Disord. 2010;30:392–402.
5. Burgener SC, Buettner L, Coen Buckwalter K, Beattie E, Bossen AL, Fick DM, Fitzsimmons S, Kolanowski A, Richeson NE, Rose K, Schreiner A, Pringle Specht JK, Testad I, Yu F, McKenzie S. Evidence supporting nutritional interventions for persons in early stage Alzheimer's disease. J Nutr Health Aging. 2008;12:18–21.
6. Crimi E, Ignarro LJ, Napoli C. Microcirculation and oxidative stress. Free Radic Res. 2007;41:1364–75.
7. De la Torre JC. Alzheimer's disease: how does it start? J Alzheimers Dis. 2002;4:497–512.
8. De la Torre JC, et al. Basics of dementia prevention. J Alzheimers Dis. 2010;20(3):687–8.
9. De la Torre JC. Vascular risk factor detection and control may prevent Alzheimer's disease. Ageing Res Rev. 2010;9:218–25.
10. Denkinger MD, Nikolaus T, Denkinger C, Lukas A. Physical activity for the prevention of cognitive decline: current evidence from observational and controlled studies. Z Gerontol Geriatr. 2012;45:11–6.
11. Eastwood MA. Interaction of dietary antioxidants in vivo: how fruit and vegetables prevent disease? QJM. 1999;92:527–30.
12. Engelhart MJ, Geerlings MI, Ruitenberg A, van Swieten JC, Hofman A, Witteman JC, Breteler MM. Dietary intake of antioxidants and risk of Alzheimer's disease. JAMA. 2002;287:3223–9.
13. Engelhart MJ, Ruitenberg A, Meijer J, Kiliaan A, van Swieten JC, Hofman A, Witteman JC, Breteler MM. Plasma levels of antioxidants are not associated with Alzheimer's disease or cognitive decline. Dement Geriatr Cogn Disord. 2005;19:134–9.
14. Fratiglioni L, Wang HX. Brain reserve hypothesis in dementia. J Alzheimers Dis. 2007;12:11–22.
15. George DR, Whitehouse PJ, Ballenger J. The evolving classification of dementia: placing the DSM-V in a meaningful historical and cultural context and pondering the future of "Alzheimer's". Cult Med Psychiatry. 2011;35:417–35.
16. Kalaria RN, Maestre GE, Arizaga R, Friedland RP, Galasko D, Hall K, Luchsinger JA, Ogunniyi A, Perry EK. Alzheimer's disease and vascular dementia in developing countries: prevalence, management, and risk factors. Lancet Neurol. 2008;7:812–26.
17. Luchsinger JA, Tank MX, Shea S, Mayeux R. Antioxidant vitamin intake and risk of Alzheimer's disease. Arch Neurol. 2003;60:203–8.
18. Mokdad AH, Marks JS, Stroup DF, Gerberding JL. Actual causes of death in the United States 2000. JAMA. 2004;291:1238–45.

19. Morris MC, Evans DA, Tangney CC, Bienias JL, Wilson RS. Associations of vegetable and fruit consumption with age-related cognitive change. Neurology. 2006;67:1370–6.
20. Octavia Y, Brunner-La Rocca HP, Moens AL. NADPH oxidase-dependent oxidative stress in the failing heart: from pathogenic roles to therapeutic approach. Free Radic Biol Med. 2012;52:291–7.
21. Polidori MC, Marvardi M, Cherubini A, Senin U, Mecocci P. Heart disease and vascular risk factors in the cognitively impaired elderly: implications for Alzheimer's dementia. Aging Clin Exp Res. 2001;13:1–9.
22. Polidori MC, Stahl W, Eichler O, Niestroj I, Sies H. Profiles of antioxidants in human plasma. Free Radic Biol Med. 2001;30(5):456–62.
23. Polidori MC, Mariani E, Mecocci P, Nelles G. Congestive heart failure and Alzheimer's disease. Neurol Res. 2006;28:588–94.
24. Polidori MC, Praticó D, Mangialasche F, Mariani E, Aust O, Anlasik T, Mang N, Pientka L, Stahl W, Sies H, Mecocci P, Nelles G. High fruit and vegetable intake is positively correlated with antioxidant status and cognitive performance in healthy subjects. J Alzheimers Dis. 2009;17:921–7.
25. Polidori MC, Carrillo JC, Verde PE, Stahl W, Sies H, Siegrist J. Plasma micronutrient status is improved after a 3-month dietary intervention with 5 daily portions of fruits and vegetables: implications for optimal antioxidant levels. Nutr J. 2009;8:10–1.
26. Polidori MC, Pientka L, Mecocci P. A review of the major vascular risk factors related to Alzheimer's disease. J Alzheimers Dis. 2012;32(3):521–30.
27. Polidori MC, Pientka L. Bridging the pathophysiology of Alzheimer's disease with vascular pathology: the feedback, the feed forward and oxidative stress. J Alzheimers Dis. 2012;28:1–9.
28. Polidori MC, Stahl W, De Spirt S, Pientka L. Influence of vascular comorbities on the antioxidant defense system in Alzheimer's disease. Dtsch Med Wochenschr. 2012;7:305–8.
29. Polidori MC. Preface to the special issue of biofactors "Biofactors and cognitive impairment". Biofactors. 2012;38:77.
30. Scarmeas N, Stern Y, Mayeaux R, Luchsinger GA. Mediterranean diet, Alzheimer disease, and vascular mediation. Arch Neurol. 2006;63:1709–17.
31. Scarmeas N, Stern Y, Tang MX, Mayeux R, Luchsinger JA. Mediterranean diet and the risk for Alzheimer's disease. Ann Neurol. 2006;59:912–21.
32. Scarmeas N, Luchsinger JA, Mayeux R, Stern Y. Mediterranean diet and Alzheimer's disease mortality. Neurology. 2007;69:1084–93.
33. Schaffer SW, Jong CJ, Mozaffari M. Role of oxidative stress in diabetes-mediated vascular dysfunction: unifying hypothesis of diabetes revisited. Vascul Pharmacol. 2012;57(5–6):139–49.
34. Shatenstein B, Kergoat MJ, Reid I, Chicoine ME. Dietary intervention in older adults with early-stage Alzheimer dementia: early lessons learned. J Nutr Health Aging. 2008;12:461–9.
35. Smith MA, Nunomura A, Lee HG, Zhu X, Moreira PI, Avila J, Perry G. Chronological primacy of oxidative stress in Alzheimer's disease. Neurobiol Aging. 2005;26:579–80.
36. Tucker AM, Stern Y. Cognitive reserve in aging. Curr Alzheimer Res. 2011;8:354–60.
37. Viswanathan A, Rocca WA, Tzourio C. Vascular risk factors and dementia. How to move forward? Neurology. 2009;72:368–74.
38. Vita JA. Endothelial function. Circulation. 2011;124:e906–12.
39. Zandi PP, Anthony JC, Khachaturian AS, Stone SV, Gustafson D, Tschanz JT, Norton MC, Welsh-Bohmer KA, Breitner JC, Cache County Study Group. Reduced risk of Alzheimer's disease in users of antioxidant vitamin supplements: the Cache County Study. Arch Neurol. 2004;61:82–8.

# Chapter 15
# Are Antioxidant Food and Nutrients Useful in Preventing Cognitive Decline?

**Luc Letenneur, Catherine Feart, and Pascale Barberger-Gateau**

**Abstract** With the aging of the population, cognitive impairment is increasingly common, and dementia, the most common disorder that affects the brain of elderly adults, is increasing worldwide. At present, no cure is available against dementia and prevention strategies to delay cognitive decline are sought. Epidemiological studies have reported several risk factors and a consensus emerged that low educational level, vascular factors and dietary habits may be important factors. Among the latter, intake of dietary antioxidants could contribute to limit oxidative damage associated with brain aging and neurodegenerative disease. The aim of this chapter is to critically review evidence from observational and intervention studies regarding the link between intake of antioxidants and cognitive function in older adults.

## 15.1 Background

The free radical theory of aging states that organisms age because cells accumulate free radical damage over time [1]. Free radicals such as reactive oxygen species (ROS) are produced during normal metabolism: a certain amount of ROS production is, in fact, necessary for good health. For instance, it helps the body's immune system to kill microorganisms. ROS are mainly produced in mitochondria [2] and are oxidants, i.e., molecules or atoms which can oxidize a substrate and are reduced

L. Letenneur, Ph.D. (✉) • C. Feart, Ph.D. • P. Barberger-Gateau, Ph.D.
ISPED, Centre INSERM U897 – Epidemiologie-Biostatistique, University of Bordeaux, Bordeaux 33000, France

INSERM, ISPED, Centre INSERM U897–Epidemiologie-Biostatistique, Bordeaux 33000, France
e-mail: luc.letenneur@inserm.fr

D. Praticò and P. Mecocci (eds.), *Studies on Alzheimer's Disease*, Oxidative Stress in Applied Basic Research and Clinical Practice, DOI 10.1007/978-1-62703-598-9_15,

in this reaction. They are able to damage several key cellular components like membrane lipids, nucleic acids, carbohydrates and proteins, thereby severely disturbing major cellular and organic physiologic functions. This type of damage occurs when the host defenses against oxidants are quantitatively and/or qualitatively unable to counteract the production and effects of oxidants themselves. This state is called oxidative stress.

The antioxidant defense system provides protection against oxidative reactions. A network of endogenous antioxidant enzymes is available to scavenge ROS once they are generated. Superoxide dismutase, catalase, glutathione and glutathione-dependent enzymes as well as other sulfur- or selenium-containing proteins and low molecular weight compounds are synthesized by the organism for defense.

Natural dietary antioxidants are exogenous molecules and include vitamins A, C, and E, carotenoids, flavonoids and other polyphenols and selenium. Vitamin C is rapidly distributed to all tissues, whereas vitamin E is incorporated into lipoproteins in the liver, and is then secreted together with them into plasma [3]. Vitamin C can scavenge many reactive species and may stabilize catecholamine from forming ROS. Vitamin E is a family of several fat soluble compounds including tocopherols and tocotrienols. Vitamin E is a powerful antioxidant that inhibits lipid peroxidation [4]. Carotenoids can scavenge singlet oxygen and a range of other ROS in vitro, but there is still little evidence that they contribute significantly to the antioxidant defense system in the central nervous system [3]. Flavonoids belong to a group of natural substances with variable phenolic structures and are found in fruits, vegetables, grains, flowers, tea, and wine [5]. Selenium is a cofactor associated with the antioxidant enzyme activity of glutathione peroxidase. These dietary antioxidants are an essential component of the antioxidant defense network.

The brain is particularly sensitive to oxidative stress [6]. Weighing about 2 % of the body mass, the brain utilizes 20 % of the total oxygen consumption. It is enriched with readily peroxidizable polyunsaturated fatty acids. In addition, the brain is not particularly rich in antioxidant defenses: it has a very low level of catalase activity and only moderate amounts of the endogenous antioxidant enzymes, superoxide dismutase and glutathione peroxidase. Additionally, the brain has high levels of iron and vitamin C, which are the key catalysts for lipid peroxidation.

## 15.2 Data from Observational Epidemiological Studies

### *15.2.1 Association Between Biomarkers of Antioxidant Status and Cognition*

The association between plasmatic concentration of antioxidants and cognitive performance was assessed in normal subjects. Goodwin et al. found a correlation between memory test scores and plasma levels of vitamin C in 260 healthy individuals aged 60 years and older [7]. In a population based sample of 885 individuals aged

74–79 years from the SENECA study, higher cognitive performance measured by the Mini Mental State Examination (MMSE) was positively correlated with higher plasmatic concentrations of lycopene, alpha-carotene, beta-carotene, total carotene, beta-cryptoxanthin, and alpha-tocopherol [8]. In 1,389 elderly volunteers from the EVA study, low levels of carotenoids were associated with poor cognitive performance in tests assessing visual attention (Odds ratio (OR) = 1.34, $p = 0.055$) or logical reasoning (OR = 1.38, $p = 0.04$) but low levels of other antioxidants were not related to poor cognitive functioning [9]. In the third National Health and Nutrition Examination Survey, Perkins et al. measured serum antioxidant concentrations and administered two tests of verbal memory to 4,809 multiethnic men and women sampled from the US general population [10]. They found that the risk of a low score on their two tests was significantly lower among subjects with the highest vitamin E concentrations than among those with the lowest concentrations (Relative risk (RR) = 0.48; 95 % Confidence Interval (CI) [0.34:0.98]) but that vitamin C had no effect.

When analyzing a pathological state such as dementia, cross-sectional analyses, based mainly on case–control studies, have shown conflicting results. Rivière et al. reported that patients with Alzheimer's disease (AD) had lower plasma vitamin C concentrations (despite similar intakes) as compared to control subjects [11]. Indeed, in Alzheimer subjects, vitamin C plasma levels decreased in proportion to the severity of the cognitive impairment despite similar vitamin C intakes. In the group of hospitalized Alzheimer, patients had normal vitamin C intakes, but their plasma vitamin C was lower than that of controls. Institutionalized Alzheimer patients had normal vitamin C level and vitamin C intakes compared with community-dwelling subjects of similar degree of cognitive impairment. Interestingly, in these patients, vitamin E levels did not correlate with the degree of cognitive impairment.

Sinclair et al. reported that plasma concentrations of vitamin C and beta-carotene were not different in AD vs. controls. On the opposite, plasma vitamin E was lower in AD versus controls [12]. Plasma lipid peroxides and total antioxidant capacity were not different across groups.

However, Rinaldi et al. reported that the plasma concentrations of several antioxidant micronutrients, including vitamins A, C, E, and carotenoids, were lower in AD patients and in individuals affected by Mild cognitive impairment (MCI) as compared to control subjects, independently of the apolipoprotein E (ApoE) genotype [13].

Few studies have investigated the relationship between selenium (Se) status and cognitive function. In a population-based sample of 1,389 volunteers aged 59–71 years included in the EVA study, levels of selenium, carotenoids, and thiobarbituric reactive substances in plasma, and of vitamin E, glutathione peroxidase and Cu–Zn superoxide dismutase in red blood cells were measured [14]. Cognitive functioning was evaluated through several psychometric tests. Low levels of total carotenoids were associated with poor performance in two tests (visual attention and executive functioning). After controlling for demographic factors, alcohol and tobacco intake and history of cardiovascular diseases, low levels of selenium are no longer associated with poor performance in visual attention.

Cardoso et al. measured selenium levels in plasma, erythrocytes, and nail samples of 28 AD subjects and 29 healthy elderly controls. Se levels in plasma (50.99 μg/l) and erythrocytes (76.19 μg/l) were significantly higher in the control group than in AD subjects (32.59 and 43.74 μg/l in plasma and erythrocytes respectively) [15]. In nails, higher values in the control group were also observed when compared with the AD group (0.400 vs. 0.302 μg/l).

A cross-sectional study performed in a large population-based sample of elderly Chinese also used selenium levels measured in nail samples as a biomarker of selenium status [16]. Lower selenium levels were significantly associated with lower cognitive score ($p<0.0087$ for all tests).

The main difficulty in interpreting cross-sectional study results lies in the fact that it is impossible to assess whether the lower plasma concentrations of antioxidant, micronutrients, including vitamins A, C, E, and carotenoids in AD vs. controls are due to their lower intake or to increased demand.

Several studies aimed at investigating the occurrence of dementia according to antioxidant biomarkers. In a nested case–control study, Helmer et al. matched 46 incident cases of dementia with 136 controls and measured plasmatic vitamin A and E levels as well as Malondialdehyde level (MDA, a lipoperoxidation product and therefore a marker of the oxidative stress level) [17]. The risk of dementia was significantly increased in the lowest vitamin E tertile (<21.0 mmol/l) (OR=3.12, $p=0.033$) compared to the highest one (>25.5 mmol/l). The risk of Alzheimer's disease was also increased, with borderline significance (OR=3.06, $p=0.053$). Similarly, there was a trend to an increased risk of dementia in the highest tertile of MDA (OR=2.44, $p=0.13$).

In a dementia free sample of 232 subjects aged 80+years and followed for 6 years, higher level of plasma vitamin E were associated with a reduced risk of developing AD (HR=0.55, 95 % CI [0.32:0.94]) [18]. According to the several isoforms of vitamin E, the risk of developing AD was reduced only in association with high plasma levels of beta-tocopherol (HR=0.62, 95 % CI [0.39:0.99]).

In the EVA cohort, a sample of 1,389 subjects followed for 9 years, cognitive decline was correlated with plasma selenium change [19]. Cognitive decline was associated with decrease of plasma selenium over time. Among subjects who had a decrease in their plasma selenium levels, the greater the decrease in plasma selenium, the higher the probability of cognitive decline. Among subjects who had an increase in their plasma selenium levels, cognitive decline was greater in subjects with the smallest selenium increase. However, there was no association between short-term (2-year) selenium change and cognitive changes.

### *15.2.2 Cross-sectional Association Between Dietary Antioxidants and Cognition*

Dietary intake of antioxidants may better reflect the actual exposure since some vitamins are highly regulated in the body. In a sample of 5,182 subjects aged 55–95 years living in Rotterdam (Netherlands), Jama et al. showed that lower dietary intake of beta-carotene was associated with a lower cognitive performance, but no association was found with dietary intake of vitamin C and vitamin E [20]. Cross-sectional studies have also shown an association between fruit and vegetable consumption and cognitive impairment. In a sample of 260 non institutionalized elderly aged 65–90 years, Ortega et al. showed that better cognitive functioning (characterized by good performance to the MMSE) was associated with higher intake of fruits [21]. Subjects with poorer performance tended to have poorer fruit intake (388 g/d (sd = 194 g/d) in men and 318 g/d (sd = 188 g/d) in women) than did those who performed better on the MMSE (398 g/d (sd = 214 g/d) in men and 331 g/d (sd = 225 g/d) in women). As expected, subject with better cognitive performance had higher intake of fiber and vitamin C. However, no association was found with beta-carotene or vitamin E.

Dietary intake of selenium was assessed in AD 28 subjects compared to 29 elderly controls, both aged between 60 and 89 years [15]. Se intake was evaluated by using a 3 day dietary food record and deficient Se intake was largely observed in the AD group.

The major problem in cross-sectional studies is that dietary intake and cognition are recorded at the same time. Therefore, people who have poorer cognitive functioning may have changed their dietary intake because of cognitive impairment. In such studies, it is not possible to assess whether dietary intake caused poor cognitive functioning or cognitive impairment caused bad dietary habits. Longitudinal studies are more powerful in this respect as dietary intake is recorded some time prior to the occurrence of a cognitive deficit.

### 15.2.3 Longitudinal Association Between Dietary Antioxidants and Cognition

In Zutphen (Netherlands), 342 men were followed for 3 years [22]. The decline of more than two points to the MMSE was not associated with dietary intake of vitamin C ($p < 0.9$), vitamin E ($p < 0.7$), beta-carotene ($p < 0.6$) or flavonoids ($p < 0.06$). However, the follow-up was probably too short to capture a big decline of cognitive performance.

In South–West of France, a sample of 1,642 subjects aged 65 years and older was followed for 10 years [23]. At the initial visit, individuals in the higher quartile of flavonoid intake had better MMSE score than those in the lower quartiles. In addition, after 10 years of follow-up, subjects of the highest quartile tended to have lower decline than the others. Moreover, a gradient in cognitive decline was observed according to flavonoid intake since MMSE decline increased as dietary flavonoid intake decreased.

Fruits and vegetables are major providers of combinations of antioxidant nutrients, including vitamin C, carotenoids and polyphenols, and to a lesser extent, vitamin E. In the US, 13,388 nurses were followed for 2 years and cognitive functioning was assessed using psychometric tests [24]. The authors showed that vegetable intake was associated with the decline in cognitive functioning. In a dose-dependent manner, women who consumed more green leafy vegetables experienced a lower decline. Apparent benefits generally increased linearly with each level of intake. For cruciferous vegetables, significantly less memory decline was found in those at the highest quintile of intake. No linear dose–response relations were observed; instead a threshold effect at the fourth quartile was seen.

In Manhattan, 980 subjects were followed for 4 years and 242 incident cases of dementia were diagnosed [25]. Luchsinger et al. found no association between developing dementia and dietary intake of beta-carotene, vitamin C, or vitamin E. In Rotterdam, 5,395 subjects were followed for 6 years and 197 incident cases of dementia were identified. A lower risk of dementia was observed with higher intake of vitamin C (HR=0.82, $p<0.05$) and vitamin E (Hazard Ratio (HR)=0.82, $p<0.04$), but no association was found with beta-carotene (HR=0.87) or flavonoids (HR=0.99). When the analyses were stratified for smoking habits, the risk of AD associated with higher intake of vitamin C and vitamin E was lower in current smokers than in former or nonsmokers (HR=0.65 vs. 0.91 and 0.83 respectively). High intake of flavonoids and beta-carotene was also associated with reduced risk of AD in current smokers (HR=0.54 and HR=0.49 respectively).

In Chicago, 815 elderly aged 65 years and older were followed for 4 years. Although 131 incident cases were identified [26], total vitamin E intake (from foods and supplements) did not predict the incidence of the disorder. Vitamin E intake from foods had a statistically significant dose–response protective effect in the age-adjusted model ($p=0.04$). The risk for persons in the top fifth of intake was lower by 67 % compared with that of persons in the lowest fifth of intake. Among persons who were ApoE ε4 negative, vitamin E from foods showed a strong linear protective association with AD. Vitamin C intake from foods appeared to have an inverse relationship with AD but was statistically significant in the fourth quintile only, and no dose–response relationship was seen. Therefore, intake of vitamin E from food was inversely associated with incident AD. There was no association with the use of vitamin E as a supplement. Vitamin C and beta-carotene also had no statistically significant association with AD. The linear protective association of vitamin E was found only among persons who were ApoE ε4 negative.

Occurrence of dementia has also been associated with fruit and vegetable intake. In King County (US), 1,589 Japanese American were followed for 6 years [27]. During this period, 81 new cases of dementia were diagnosed. Compared to subjects with a low fruit and vegetable juice intake (less than one per week), subjects with high intake (three or more juices per week) had a lower risk of developing dementia (HR=0.24, 95 % CI [0.09:0.61]). Subjects with moderate intake (1 or 2 times per week) had a nonsignificant reduced risk (HR=0.84, 95 % CI [0.31:2.29]). The inverse association between fruit and vegetable juices and AD appeared in all strata of education, smoking status, tea drinking, regular physical activity, ApoE genotype, and total fat intake. However, the association tended to be stronger among those who were former or current smokers, drank tea less often, were positive for the ApoE ε4 allele, and were less physically active.

In the Three City study, 8,085 elderly were followed for 4 years, and 281 incident cases of dementia were diagnosed [28]. A lower risk of developing dementia was observed (HR = 0.72, $p < 0.02$) in subjects with frequent (every day) fruit and vegetable intake. The strength of the association remained almost unchanged after controlling for ApoE genotype, body mass index and diabetes.

## 15.3 Results from Clinical Trials

Few clinical trials have studied the effect of antioxidant components on cognition [29]. We shall first review the evolution of cognition in random clinical trials designed initially for investigating the effect of vitamin supplementation on prevention of major health conditions. A cognitive sub-study was performed in the Women's Health study. This study was a randomized double-blind placebo-controlled supplementation with vitamin E (600 IU) and low dose aspirin (100 mg) for the prevention of cardiovascular disease and cancer [30]. A total of 39,876 women were enrolled in 1992–1995. In 1998, 5.6 years after randomization, 6,377 women performed cognitive tests and were followed on average 4 years with a contact every 2 years. Whatever the type of measure considered (mean performance, mean cognitive change over time, risk of substantial decline), there were no difference in global performance between vitamin E and placebo groups. However, the vitamin E group experienced less adverse cognitive change compared with the placebo group among women with dietary intake below the median of 6.1 mg/day. In contrast, among women with high intakes of dietary vitamin E, the two groups were similar in their change.

In the Physicians' Health Study II (an extension of the Physicians' Health Study (PHS) which began in 1982 when 22,071 physicians were randomized in a trial for prevention of cardiovascular disease and cancer), 4,052 subjects from PHS agreed in 1997 to participate to PHS II and 4,052 new physicians were included from 1998 to 2001 [31]. Half were included in the treatment group (50 mg beta-carotene, alternate days) and the others in the placebo group. Mean duration of treatment ranged from 2 months to 20 years. Cognitive functioning was evaluated using psychometric tests administered on telephone at baseline in 1998 and at follow-up in 2002. In newly recruited participants, there was no evidence of cognitive benefits with short-term beta-carotene supplementation. In participants with long-term treatment, those assigned to beta carotene performed significantly better on the global score compared with the placebo group (mean difference in z scores, 0.047 standard units; $p = 0.03$). Although beta-carotene supplementation was associated with higher mortality [32], long term beta-carotene supplementation may provide cognitive benefits.

An ancillary study of the SU.VI.MAX trial was designed to assess the long term effects of antioxidant nutrient supplementation on the cognitive performance of the participants, 6 years after the end of the trial [33]. The SU.VI.MAX study is a double-blind, placebo-controlled, randomized trial that enrolled 5,583 subjects aged 45–60 years. From 1994 to 2003, participants received daily vitamin C (120 mg), beta-carotene (6 mg), vitamin E (45 IU), selenium (100 μg), and zinc (20 mg) in combination or a placebo. In 2007–2009, the cognitive performance of 4,447 participants was assessed with 4 neuropsychological tests. Subjects receiving active antioxidant supplementation showed better episodic memory scores. Verbal memory scores were better only in supplemented subjects who were nonsmokers or who had low serum vitamin C concentrations at baseline. Although this study showed that supplemented subjects have better performance on specific cognitive domains, the study did not assessed cognition at baseline and initial comparability of the groups was not assured.

Other trials have investigated the effect of vitamin supplement in a secondary prevention perspective. Petersen et al. treated 769 subjects with MCI which is considered to be a prodromal state of AD [34]. Three arms of the study were constituted: Donepezil, vitamin E (2,000 IU), or placebo and the main outcome was to delay the conversion to dementia. After 3 years of follow-up, no difference was observed between treatments and placebo. However, in a more recent follow-up publication, the same authors reported that brain imaging showed that changes in the volumes of some areas of the brain (hippocampus, enthorinal cortex) were lower in the group that received vitamin E rather than placebo [35].

Finally, trials were also designed to evaluate the effect of supplements in diseased subjects. Sano et al. conducted a double-blind placebo controlled trial in patients with AD [36]. A total of 341 AD patients were divided into four arms: Vitamin E (2,000 IU a day), Selegiline (a selective monoamine oxidase inhibitor), both, and no treatment. The primary outcome was the time to the occurrence of any of the following: death, institutionalization, loss of the ability to perform basic activities of daily living, or severe dementia. After a follow-up of 2 years, high doses of vitamin E resulted in a longer time to institutionalization and a delayed time to deterioration of activities of daily living although cognitive function did not appear to be improved.

More recently, Lloret et al. treated 57 AD patients with 800 IU of vitamin E and assessed responsiveness to the antioxidant treatment [37]. The authors checked the blood oxidative status of the patients using the blood total glutathione levels and oxidized glutathione (GSSG). They found that not all patients respond equally to antioxidant treatment and identified "responders" (vitamin E treatment resulted in a reduction of the GSSG levels) and "nonresponders." In responders to vitamin E, cognitive performance measured by the MMSE was maintained, whereas in nonresponders, cognition decreased sharply, to levels even lower than those of patients taking placebo. Although the number of subjects is low, this result highlights that giving vitamin E to AD patients may be harmful, especially if the oxidative stress status of the patients is not carefully monitored.

This aspect of differential response to supplementation is probably a key factor that is underestimated. Indeed, another clinical trial showed cognitive difference according to the response to a formulation of six vitamins and nutraceuticals (30 IU of vitamin E, 6 μg of vitamin B12, 400 μg of folic acid, 400 mg of *S*-adenosyl methionine, 600 mg of *N*-acetyl cysteine, 500 mg of Acetyl L-carnitine) [38]. Adults (age 18–86 years) of both genders without dementia received the treatment or a placebo.

After 3 months of follow-up, participants who received the treatment had improved to the Trail making test that evaluated executive function ($p<0.03$), while those receiving placebo did not improve. However, unlike younger participants, participants ≥74 years of age receiving treatment did not on an average demonstrate improvement versus placebo. The percentage of responders (i.e., subjects that showed an improvement to the test after 3 months of treatment) tended to decrease with age. Nonresponders within all age groups up to 74 years of age displayed similar performance but lower than responders, while those ≥74 years of age displayed substantially poorer scores at 3 months than did all other age groups. Therefore, elder non responders showed poorer performance than younger non responders. This age-related decline may be due to decreased absorption of nutrients, and/or decreased basal vitamin levels due to suboptimal nutrition. This result may explain why clinical trials performed on elderly individuals did not show a beneficial effect of antioxidant vitamins since it may be too late and/or the supplement may not be adapted to the physiology of the older adult.

In view of these results, one may wonder if antioxidants reduce oxidative stress in the brain and have any influence on the pathological pathway of AD. Galasko et al. enrolled 78 AD patients and looked for change in potential CSF biomarkers (Aβ42, tau, and phospho-tau) and oxidative stress (F2-isoprostane) as well as cognition (Mini-Mental State Examination) and daily function (ADCS Activities of Daily Living Scale) at the end of a 16-week treatment period [39]. One-third of the participants were randomized into a placebo group. The others received either a daily supplement of 800 IU vitamin E, 500 mg vitamin C, and 900 mg α-lipoic acid (E/C/ALA group) or 400 mg coenzyme Q (CoQ group). None of the AD biomarkers changed whatever the treatment, leading to the conclusion that supplements did not alter the pathological pathway. CSF levels of F2-isoprostane fell about 19 % in the E/C/ALA group, suggesting that this antioxidant combination lowered oxidative stress in the brain. In contrast, CoQ did not change CSF F2-isoprostane levels. From the cognitive point of view, the E/C/ALA group seemed to decline faster with a 2.8 point change on MMSE scores from baseline, compared to 0.9–1.0 point change in the placebo and CoQ groups.

## 15.4 Discussion

Clinical trials have found that antioxidant supplementation does not delay or avoid cognitive decline. It however reduces oxidative stress as measured by GSSG or F2 isoprostanes. The increased decline in cognitive performance observed in some trials suggests that higher dose of vitamin E may be harmful in subjects that are non-responders. Even observational studies have produced inconsistent results with anti-oxidant micronutrients. In contrast, fruit and vegetable intake has tended to be more consistently associated with a lower risk of developing dementia, although the number of available studies is small. This indicates that specific antioxidant nutrients such as vitamin E are not sufficient to protect against cognitive deficit.

This is also consistent with studies that have shown that people consuming a Mediterranean diet have a lower risk of developing cognitive impairment. The traditional Mediterranean diet is characterized by high consumption of plant foods (vegetables, fruits, legumes, and cereals), high intake of olive oil as the principal source of monounsaturated fat but low intake of saturated fat, moderate intake of fish, low to moderate intake of dairy products, low consumption of meat and poultry, and wine consumed in low to moderate amounts, normally with meals. In a cohort study of a large community-based population without dementia in New York, higher Mediterranean diet adherence was associated with a reduced risk for mild cognitive impairment and AD [40]. A cohort study in France also showed less cognitive decline in subjects who adhered to a Mediterranean diet [41]. The biological basis for the apparent health benefits of a Mediterranean diet involves a decrease in oxidative stress, inflammation, and vascular disease, which also participate in the pathophysiology of neurodegenerative diseases. The Mediterranean diet pattern probably does not fully explain the better health of persons who adhere to it, but it may contribute directly. A Mediterranean diet also may indirectly constitute an indicator of a complex set of favorable social and lifestyle factors that contribute to better health.

Do these observations mean that the oxidative-stress hypothesis of AD is not valid anymore? First, considering the complexity of the redox system in vivo, we may probably need better antioxidant drugs and, in certain cases, a combinatory approach would be preferable to a single antioxidant. Second, before starting any antioxidant therapy trial, it will be extremely important to have clear information on the endogenous antioxidant levels of the participating subjects. This aspect is important for patient selection in order to identify potential responders versus nonresponders to a drug with antioxidant properties.

In conclusion, fruits and vegetables are associated with a better cognitive evolution. It has not been possible to demonstrate that specific antioxidant nutrients such as vitamin E were at the origin of the protection. Fruit and vegetable intake is more complex than the addition of specific nutrients. Dietary behavior may also be involved, and Mediterranean diet is an illustration of a combination of several foods that may be beneficial. In a preventive point of view, rather than recommending antioxidant nutrients intake in the form of vitamins, it would be wiser to promote fruits and vegetables consumption that may bring benefit not only on cognitive aging but also on other pathologies like cardiovascular disease and cancer.

## References

1. Harman D. Aging: a theory based on free radical and radiation chemistry. J Gerontol. 1956;11:298–300.
2. Sastre J, Pallardó F, Viña J. The role of mitochondrial oxidative stress in aging. Free Radic Biol Med. 2003;35:1–8.
3. Halliwell B. Role of free radicals in the neurodegenerative diseases: therapeutic implications for antioxidant treatment. Drugs Aging. 2001;18:685–716.

4. McCay P. Vitamin E: interactions with free radicals and ascorbate. Annu Rev Nutr. 1985;5:322–40.
5. Middleton E. Effect of plant flavonoids on immune and inflammatory cell function. Adv Exp Med Biol. 1998;439:175–82.
6. Floyd RA, Hensley K. Oxidative stress in brain aging. Implications for therapeutics of neurodegenerative diseases. Neurobiol Aging. 2002;23:795–807.
7. Goodwin JS, Goodwin JM, Garry PJ. Association between nutritional status and cognitive functioning in a healthy elderly population. J Am Med Assoc. 1983;249:2917–21.
8. Haller J, Weggemans RM, Ferry M, Guigoz Y. Mental health: mini mental state examination and geriatric depression score of elderly Europeans in the SENECA study of 1993. Eur J Clin Nutr. 1996;50 suppl 2:s112–6.
9. Berr C, Richard M, Gourlet V, Garrel C, Favier A. Enzymatic antioxydant balance and cognitive decline in aging. The EVA study. Neuroepidemiology. 2004;19:133–8.
10. Perkins A, Hendrie H, Callahan C, Gao S, Unverzagt F, et al. Association of antioxidants with memory in a multiethnic elderly sample using the Third National Health and Nutrition Examination Survey. Am J Epidemiol. 1999;150:37–44.
11. Riviere S, Birlouez-Aragon I, Nourhashemi F, Vellas B. Low plasma vitamin C in Alzheimer patients despite an adequate diet. Int J Geriatr Psychiatry. 1998;13:749–54.
12. Sinclair AJ, Bayer AJ, Johnston J, Warner C, Maxwell SR. Altered plasma antioxidant status in subjects with Alzheimer's disease and vascular dementia. Int J Geriatr Psychiatry. 1998; 13:840–5.
13. Rinaldi P, Polidori MC, Metastasio A, Mariani E, Mattioli A, et al. Plasma antioxidants are similarly depleted in mild cognitive impairment and Alzheimer's disease. Neurobiol Aging. 2003;24:915–9.
14. Berr C, Richard M, Roussel A, Bonithon-Kopp C. Systemic oxidative stress and cognitive performance in the population-based EVA study. Free Radic Biol Med. 1998;24:1202–8.
15. Cardoso B, Prates Ong T, Jacob-Filho W, Jallul O, D'Avila-Freitas I, et al. Nutritional status of selenium in Alzheimer's disease patients. Br J Nutr. 2009;103:803–6.
16. Gao S, Jin Y, Hall K, Liang C, Unverzagt F, et al. Selenium level and cognitive fonction in rural elderly Chinese. Am J Epidemiol. 2007;165:955–65.
17. Helmer C, Peuchant E, Letenneur L, Bourdel-Marchasson I, Larrieu S, et al. Assocaition between antioxidant nutritional indicators and the incidence of dementia: results from the PAQUID prospective cohort study. Eur J Clin Nutr. 2003;57:1555–61.
18. Mangialasche F, Kivipelto M, Mecocci P, Rizzuto D, Palmer K, et al. High plasma level of vitamin E forms and reduced Alzheimer's disease risk in advanced age. J Alzheimers Dis. 2010;20:1029–37.
19. Akbaraly T, Hininger-Favier I, Carrière I, Arnaud J, Gourlet V, et al. Plasma selenium over time and cognitive decline in the elderly. Epidemiology. 2007;18:52–8.
20. Jama JW, Launer LJ, Witteman JC, den Breeijen JH, Breteler M, et al. Dietary antioxidants and cognitive function in a population-based sample of older persons. The Rotterdam Study. Am J Epidemiol. 1996;144:275–80.
21. Ortega R, Requejo A, Andrès P, Ldpez-Sobaler A, Elena Quintas M, et al. Dietary intake and cognitive function in a group of elderly people. Am J Clin Nutr. 1997;66:803–9.
22. Kalmijn S, Feskens E, Launer LJ, Kromhout D. Polyunsaturated fatty acids, antioxidants, and cognitive function in very old men. Am J Epidemiol. 1997;145:33–41.
23. Letenneur L, Proust-Lima C, Le Gouge A, Dartigues JF, Barberger-Gateau P. Flavonoid intake and cognitive decline over a 10 year period. Am J Epidemiol. 2007;165:1364–71.
24. Kang JH, Ascherio A, Grodstein F. Fruit and vegetable consumption and cognitive decline in aging women. Ann Neurol. 2005;57:713–20.
25. Luchsinger JA, Tang M, Shea S, Mayeux R. Antioxidant vitamin intake and risk of Alzheimer disease. Arch Neurol. 2003;60:203–8.
26. Morris M, Evans D, Bienias J, Tangney C, Bennet D, et al. Dietary intake of antioxidant nutrients and the risk of incident Alzheimer disease in a biracial community. JAMA. 2002;287: 3230–7.

27. Dai Q, Borenstein A, Wu Y, Jackson J, Larson E. Fruit and vegetable juices and Alzheimer's Disease: The Kame Project. Am J Med. 2006;119:751–9.
28. Barberger Gateau P, Raffaitin C, Letenneur L, Berr C, Tzourio C, et al. Dietary patterns and risk of dementia: The Three-City cohort study. Neurology. 2007;69:1921–30.
29. Barberger Gateau P, Féart C, Samieri C, Letenneur L. Dietary patterns and dementia. In: Yaffe K, editor. Chronic medical disease and cognitive aging: toward a healthy body and brain. Oxford: Oxford University Press; 2013.
30. Kang JH, Cook N, Manson J, Buring J, Grodstein F. A randomized trial of vitamin E supplementation and cognitive function in women. Arch Intern Med. 2006;166:2462–8.
31. Grodstein F, Kang JH, Glynn R, Cook N, Gaziano M. A randomized trial of beta carotene suplementation and cognitive function in men. Arch Intern Med. 2007;167:2184–90.
32. Virtamo J, Pietinen P, Huttunen J, Korhonen P, Malila N, et al. Incidence of cancer and mortality following alpha-tocopherol and beta-carotene supplementation: a postintervention follow-up. JAMA. 2003;290:476–85.
33. Kesse-Guyot E, Fezeu L, Jeandel C, Ferry M, Andreeva V, et al. French adults' cognitive performance after daily supplementation with antioxidant vitamins and minerals at nutritional doses: a post hoc analysis of the Supplementation in Vitamins and Mineral Antioxidants (SU.VI.MAX) trial. Am J Clin Nutr. 2011;94:892–9.
34. Petersen R, Thomas R, Grundman M, Bennet D, Doody R, et al. Vitamin E and donepezil for the treatment of mild cognitive impairment. N Engl J Med. 2005;352:2379–88.
35. Jack C, Petersen R, Grundman M, Jin S, Gamst A, et al. Longitudinal MRI findings from the vitamin E and donepezil treatment study for MCI. Neurobiol Aging. 2008;29:1285–95.
36. Sano M, Ernesto C, Thomas R, Klauber M, Schafer K, et al. A controlled trial of selegiline, alpha-tocopherol, or both as treatment for Alzheimer's disease. N Engl J Med. 1997;336:1216–22.
37. Lloret A, Badia MC, Mora N, Pallardo F, Alonso MD, et al. Vitamin E paradox in Alzheimer's disease: it does not prevent loss of cognition and may even be detrimental. J Alzheimers Dis. 2009;17:143–9.
38. Chan A, Remington R, Kotyla E, Lepore A, Zemianek J, et al. A vitamin/nutriceutical formulation improves memory and cognitive performance in community-dwelling adults without dementia. J Nutr Health Aging. 2010;14:224–30.
39. Galasko D, Peskind E, Clark C, Quinn J, Ringman J, et al. Antioxidants for Alzheimer disease: a randomized clinical trial with cerebrospinal fluid biomarker measures. Arch Neurol. 2012. doi:10.1001/archneurol.2012.85.
40. Scarmeas N, Stern Y, Tang M, Mayeux R, Luchsinger JA. Mediterranean diet and risk for Alzheimer's disease. Ann Neurol. 2006;59:912–21.
41. Féart C, Samieri C, Rondeau V, Amieva H, Portet F, et al. Adherence to a Mediterranean diet, cognitive decline, and risk of dementia. JAMA. 2009;302:638–48.

# Chapter 16
# Antioxidant Clinical Trials in Mild Cognitive Impairment and Alzheimer's Disease

**Patrizia Mecocci, M. Cristina Polidori, and Domenico Praticó**

**Abstract** Many studies have been published so far showing the significant role of oxidative stress in the pathogenesis of Alzheimer's disease (AD), and the role of antioxidants as protective factors have been also shown in epidemiological survey. On this basis, clinical trials have been conducted in AD as well as in mild cognitive impairment (MCI) with different antioxidants, alone or in combination. In this chapter we review results or protocols of completed or ongoing studies with different molecules. Briefly, most of these studies are characterized by extremely poor comparability and the absence of a substantial clinical benefit but many aspects in their design and lack of fundamental prerequisites, such as pretreatment antioxidant status, cause several biases and do not offer a possibility to really understand if an antioxidant therapy can be effective both in preventing MCI conversion to dementia or in treating AD. Since oxidative stress is undoubtedly involved in AD, new RCTs with better defined rationale, quality of design, and clear outcomes are needed.

Since the time at which Alois Alzheimer described what for many years has been a "simple" case report, several steps have been taken in the past few decades to understand the pathophysiological mechanisms of the disease but, despite remarkable progresses, the cure against the disease is troublesome and frustrating, especially in

P. Mecocci, M.D. Ph.D. (✉)
Department of Clinical and Experimental Medicine, Institute of Gerontology and Geriatrics, Ospedale S. Maria della Misericordia Perugia, University of Perugia, Block C Floor 4, S.Andrea delle Fratte, 06156 Perugia, Italy
e-mail: mecocci@unipg.it

M.C. Polidori, Ph.D.
Institute of Geriatrics, University of Cologne, Cologne, Germany

D. Praticó, M.D.
Department of Pharmacology, Center for Translational Medicine, Temple University School of Medicine, 947 Medical Education and Research Building, 3500 North Broad Street, Philadelphia, PA 19140, USA

D. Praticò and P. Mecocci (eds.), *Studies on Alzheimer's Disease*, Oxidative Stress in Applied Basic Research and Clinical Practice, DOI 10.1007/978-1-62703-598-9_16,

light of its epidemiologic growing. Most of the research on therapy for AD transfers one chemical substance targeted on a single biological system to subjects with age-modified biology, suffering from multimorbidity and under polypharmacy: a single answer to a multifaceted puzzling problem, but the "one alteration, one disease, one drug" strategy does not seem to work well in AD, a typical example of a complex multifactorial disease. This is clearly shown by the approved acetylcholinesterase inhibitors (donepezil, rivastigmine, galantamine) as anti-AD therapy. The rationale of their use is supported by the decrease of cholinergic transmission in the basal forebrain cholinergic neurons, but they have mainly, if not only, symptomatic effects, with large interindividual responses. Symptomatic is also the effect of the approved NMDA receptor antagonist memantine. Several anti-AD strategies such as anti-amyloid and anti-tau therapies, anti-mitochondrial dysfunction drugs and neurotrophins, are under evaluation in RCTs of safety and efficacy [27].

The free radical and oxidative stress theory of aging ([16, 40]; Sies, this book) suggests that oxidative damage is a major player in neuronal degeneration and several studies have demonstrated that oxidative stress is an early occurring condition in AD (reviewed in [26, 29]). On these bases, a number of trials have been performed in the last few years aimed at exploring the efficacy of antioxidants in AD. Since the confirmation of the chronological primacy of oxidative stress in AD development [41] through its demonstration in mild cognitive impairment (MCI) [36, 37], antioxidant trials have been also prompted in MCI. Based on data derived from several observational and epidemiological studies, compounds with antioxidant activity have been proposed for prevention of cognitive decline and treatment of MCI or AD. Rationale for their use in these conditions as well the main results of relative clinical trials are presented hereafter.

Flavonoids, the most common group of polyphenolic compounds in the human diet, are found ubiquitously in plants and seeds like cocoa beans and grape seeds. These compounds may be divided into several subclasses, of which catechins are found in high concentration in the tea plant leaves. The major tea catechins include (−)-epicatechin (EC), (−)-epigallocatechin (EGC), (−)-EC gallate (ECG), and (−)-EGC gallate (EGCG). Their antioxidant activity is multifactorial but is mainly represented by the scavenging effect of superoxide anions and hydroxyl radicals by the C-ring gallate group and the A ring [53]. Plasma concentration of polyphenols is very low compared to other antioxidants, but they effectively act as modulators of cell-signaling pathways and of inflammatory responses rather than to a direct antioxidant activity [23]. In addition, since high blood levels of homocysteine may be increased in AD and hyperhomocysteinemia may contribute to disease pathophysiology by vascular and direct neurotoxic mechanisms [17], the effect of an antioxidant beverage rich in polyphenols on the plasmatic levels of homocysteine in patients with AD has been recently evaluated [31]. The antioxidant drink versus placebo drink attenuated homocysteine increase in the control group and AD, especially in the moderate phase, but no other effects where achieved [31].

Prospective cohort studies on flavonoids intake and risk of developing dementia led to inconsistent results [6, 7, 10, 20, 22, 49]. There is an ongoing phase II/III RCT on patients with early AD already under donepezil treated with EGCG versus placebo for 18 months recruitment phase. The primary endpoint of the study is the ADAS-cog score at study end [clinicaltrials.gov NCT00951834].

Resveratrol is a polyphenolic compound of the family of stilbenes found in grapes and red wine. Due to the observation of antioxidant and anti-inflammatory properties in vitro as well as cardioprotective and longevity effects in animal models [8], it has been tested in a double-blind, placebo-controlled, crossover investigation on young healthy subjects which showed an increase in cerebral blood flow in treated subjects but no short term effect in cognitive performances [19]. Resveratrol is under investigation versus placebo in a RCT using liquid resveratrol with glucose and malate as a dietary supplement delivered in grape juice in patients with AD with the primary endpoint of assessing ADAS-cog scores at regular intervals up to 1 year after study commencement [clinicaltrials.gov NCT00678431]. Another multi-interventional clinical trial in MCI proposed resveratrol in one arm of intervention to proof any modification of ADAS-Cog scores [clinicaltrials.gov NCT01219244].

Vitamin E belongs to a group of eight structurally related lipophilic chromanol congeners. Vitamin E found in natural food includes both four tocopherols and four tocotrienols all of which have saturated and three double bonds in their phytyl tails respectively. The tocopherols and tocotrienols are further subdivided into α-, β-, γ-, and δ- based on the hydroxyl and methyl substitution in their phenolic rings. Low plasma tocopherols and tocotrienols levels are associated with increased odds of MCI and AD [28].

The most studied congener of vitamin E molecule is α-tocopherol, which shows the most potent biological activity and corrects human vitamin E deficiency symptoms [44]. The most abundant sources of vitamin E are vegetable oils, which typically contain all four tocopherol congeners in varying proportions. Other important sources are nuts and seeds such as sunflower seeds. Vitamin E has been shown to cross the blood–brain barrier and to accumulate at therapeutic levels in the central nervous system, where it is able to lower lipid peroxidation and β-amyloid deposition [46]. It also rescues neuronal damage and β-amyloid deposition in the brain by reducing isoprostane levels [36].

Observational studies on dietary or supplemented vitamin E such as the Cache County Study, the Washington-Heights-Inwood Columbia Aging study, the Age-Related Eye Disease Study and the Women's Health Study led to conflicting results [24, 50, 51]. In a prospective study of 633 persons 65 years and older, and after an average follow-up period of 4.3 years, 91 of the sample participants were diagnosed with AD, but none of the 27 vitamin E supplement users had AD (3.9 predicted) and none of the 23 vitamin C supplement users had AD (3.3 predicted), while there was no relation between Alzheimer's disease and use of multivitamins [32]. Other positive results were obtained in the Honolulu–Asia Aging Study [30], the Chicago Health and Aging Project [30], the Nurses' Health Study [12], and the Rotterdam Study [10]. Vitamin E was the first antioxidant to have been tested in a large placebo-controlled trial, performed on subjects with moderate to severe AD [39]. Those treated with a daily dose of 2,000 IU of vitamin E for 2 years showed a significant delay in AD progression and in nursing home placement but no effect was observed on Mini-Mental State Examination score at the end of the study.

In a recent study, 769 subjects with MCI were randomly assigned to receive 2,000 IU of vitamin E daily or 10 mg of donepezil or placebo for 3 years. At the end of the study 212 developed AD, with 16 % conversion rate per year. The hazard ratio indicating the probability of progression to AD in the vitamin E group was 1.02, not significantly different from that of donepezil (0.80). There were no significant differences in the rate of progression to AD between vitamin E and placebo groups at any point, either considering all patients or the apolipoprotein E ε4 carriers. The secondary outcome measures of the study included changes of cognition as assessed by MMSE, CDR sum of boxes, Global Deterioration Scale, and modified ADAS-Cog which showed few weak significant differences in the vitamin E group compared with the placebo group [35].

The PREADVISE is a phase III RCT expected to enroll 10,400 subjects among men previously involved in the prostate cancer prevention SELECT study. Subjects will be either assigned to α-tocopherol plus selenium, α-tocopherol plus placebo, selenium plus placebo or placebo only, with the primary endpoint of assessing the effects of the treatment on AD incidence within 7–12 years after study commencement [clinicaltrials.gov NCT00040378].

The TEAM-AD clinical trial [clinicaltrials.gov NCT00235716] on patients with mild to moderate AD under stable treatment with acetylcholinesterase inhibitors, assigned to either α-tocopherol, memantine, α-tocopherol plus memantine or matching placebo, has been recently completed but results have been not published yet. Patients have been treated for 3 years with a follow-up period ranging from a minimum of 1 year to a maximum of 4 years. This phase III trial has the ADCS-ADL inventory as primary endpoint and, as secondary outcome measures, the ADAS-cog, the MMSE, and the NPI (Neuropsychiatric inventory).

*Curcumin* is the principal curcuminoid in turmeric, a yellow spice derived from *Curcuma longa*, member of the ginger family indigenous to South Asia. It has several beneficial properties—including anti-inflammatory, antioxidant, and chemotherapeutic—and it was shown to dose-dependently inhibit the formation of β-amyloid fibrils from Aβ40 and Aβ42 [33]. Due to its antioxidant properties, its positive safety results as well as its benefits on cognitive performances shown in animal models of AD [15] the effects of curcumin have been tested in a RCT versus placebo on 34 patients with AD treated with two doses of curcumin per day or placebo, though no differences in MMSE scores or in β-amyloid plasma levels were shown after 6 months [3]. In a preliminary analysis of the results of a RCT versus placebo of phase II type using curcumin in mild and moderate AD for 24 weeks, no significant changes in cognitive performance or in plasma CSF biomarkers among groups were shown [38]. A clinical trial in subjects with MCI treated or not with curcumin for 18 months with the aim of evaluating amyloid deposition in the brain with FDDNP-PET scan is ongoing [clinicaltrials.gov NCT01383161].

The development of *latrepirdine* (Dimebon) as an anti-AD strategy was based on its safety data as anti-histaminic and on his beneficial effects on mitochondria [52] whose damage increases a status of oxidative stress. Latrepirdine has been shown to modulate mitochondrial function in cells at nanomolar concentration, to stabilize mitochondria membrane potential and to improve neuron survival under oxidative stress condition, supporting its role as a mitochondrion-targeted drug [45]. In the

Connection phase III RCT conducted in mild to moderate AD patients treated with up to 20 mg latrepirdine versus placebo, no effect of latrepirdine was shown after 26 weeks on the ADAS-cog, on the MMSE scores or on the ADCS-ADL [18]. A phase III multicenter, randomized, placebo-controlled, double-blind, 12-month safety and efficacy study with latrepirdine versus placebo is ongoing in patients with mild-to-moderate AD on donepezil (Concert study) with ADAS-cog and ADCS-ADL as primary endpoints [clinicaltrials.gov NCT00829374]. More recently, studies showing pro-autophagic properties of this molecule have opened new scenarios for therapeutical proposals as well as for prevention of AD [43].

Another mitochondrial targeted antioxidant is *coenzyme Q10* or *ubiquinone*, a critical subunit of the mitochondrial respiratory chain that transports electrons from complex I (NADH-ubiquinone reductase) and II (succinate-ubiquinone reductase) and from the oxidation of fatty acid and branched-chain amino acids to complex III (ubiquinol-cytochrome c reductase). Ubiquinone is the only lipophilic antioxidant endogenously synthesized and is able to efficiently prevent protein oxidation, lipid peroxidation and the oxidation of DNA, especially of mitochondrial DNA. Also, the reduced form of coenzyme Q10 regenerates α-tocopherol from the α-tocopheroxyl radical by interfering with the initiation and propagation step of lipid peroxidation. It preserves mitochondrial membrane potential during oxidative stress and protects neuronal cells against β-amyloid toxicity through inhibition of the opening of the mitochondrial membrane transition pore. The Alzheimer's Disease Cooperative Study started in 2006 a phase I investigation to evaluate the safety, tolerability and impact on biomarkers of antioxidant treatment of mild to moderate AD. Seventy-five patients with AD underwent treatment with either coenzyme Q or vitamin E plus vitamin C plus α-lipoic acid or placebo three times a day [clinicaltrials.gov NCT00117403]. The primary endpoint were changes in CSF biomarkers after 4 months therapy and secondary outcome measures were the CSF levels of β-amyloid 42, tau and P-tau 181. At the end of the study the authors reported that changes in the CSF biomarkers did not differ among the three groups enrolled. However CSF F2-isoprostane levels decreased on average 19 % from baseline to week 16 in the vitamin E, vitamin C and α-lipoic acid group, but were unchanged in the other groups [11].

*Lipoic acid* is the coenzyme of mitochondrial pyruvate dehydrogenase and α-ketoglutarate dehydrogenase. It exerts powerful antioxidant activity in mitochondria by recycling other antioxidants such as vitamin E, vitamin C and glutathione, scavenging lipid peroxidation products and ROS and by chelating redox-active transition metals [25]. As lipoic acid supplementation has been shown to reduce memory loss and to stabilize cognitive function [14], a phase I/II RCT is ongoing with the aim of treating patients with AD with lipoic acid plus fish oil concentrate versus placebo for 18 months [clinicaltrials.gov NCT01058941].

*Idebenone* is a synthetic analog of coenzyme Q10 with neuronal protective activity against β-amyloid toxicity [5]. Treatment with idebenone in AD in RCTs showed significant improvement in behavior, memory and attention and dose-dependent slowing of disease progression for up to 2 years with a statistical improvement of ADAS-cog scores after 6 months [13]. The last published results on idebenone treatment in AD are available from a 1-year multicentre, double-blind, placebo-controlled RCT conducted on subjects with mild to moderate AD. Patients were randomized to receive either 120 mg, 240 mg, 360 mg idebenone or placebo t.i.d., with the primary endpoint of changes on the ADAS-cog and the Clinical Global Impression of Change (CGIC) scales. No changes were observed among groups in the primary endpoints or in the secondary outcome measures that included activities of daily living and MMSE [47].

*Pramipexole* and other dopamine agonists have been demonstrated neuroprotective effects in vitro and in vivo possibly through antioxidant effects, increasing glutathione (GSH) levels as well as glutathione peroxidase (GSH-Px) and catalase activities [21]. A phase II trial is now ongoing in patients with AD with R-pramipexole twice a day for a total dose of 100 mg per day. After 4 weeks, the dose will be doubled and after another 4 weeks patients will be treated with 300 mg R-pramipexole per day that will be maintained for the subsequent 16 weeks. There is no placebo arm and the primary endpoints are safety issues and effects on cognitive performance, while secondary outcome measures are lowering of oxidative stress as assessed by CSF isoprostanes levels before and after treatment as well as changes in cerebral glucose metabolism as assessed by PET scan before and after treatment [clinicaltrials.gov NCT01388478].

*N-acetyl cysteine* (NAC) is a thiol-containing compound which modulates redox signaling, interferes with redox transition of thiols, inhibits lipid peroxidation, and acts as a substrate for the synthesis of GSH [34]. The NAC-003 P.L.U.S. Program (Progress through Learning Understanding and Support) planned to enroll 800 subjects with early memory impairment who have been prescribed daily CerefolinNAC, an orally administered capsule containing NAC plus methycobalamin and L-METHYLFOLATE calcium. The primary objective of the study is to evaluate the effect of CerefolinNAC on quality of life as assessed by the Quality of Life Alzheimer's disease scale (QOL-AD) after 12 weeks of treatment and should end in May 2012 [clinicaltrials.gov NCT01370954]. A study with a nutriceutical formulation (Memory XL) containing NAC plus to folic acid, vitamin B12, α-tocopherol, *S*-adenosylmethionine, and acetyl-L-carnitine versus placebo in subjects with MCI treated for 1 year [clinicaltrials.gov NCT00903695] did not find any differences in all cognitive and functional tests between groups. Another large multi-site placebo-controlled clinical trial with Memory XL is recruiting patients with AD or subjects with MCI to determine whether or not this formulation can be protective and delay MCI conversion to AD [clinicaltrials.gov NCT01320527].

Extracts of the leaves of the maidenhair tree, *Ginkgo biloba*, have long been used in China as a traditional medicine for various disorders. Due to its effect on vasodilation, improvement of neurotransmission and antioxidant activity [1], its standardized extract EGb 761® has been widely used in Europe to improve memory and concentration problems. A recent Cochrane review of 36 randomized clinical trials on the effects of *Ginkgo biloba* in dementia and cognitive impairment judged most of them not valuable because of low quality, conducted in too small population or too short in time of intervention. Nine trials of 6 months duration on a total of 2,016 patients were selected which tested EGb 761 at different doses and showed inconsistent results for cognition, activities of daily living, mood, depression, and carer burden [4]. The Ginkgo Evaluation of Memory Study (GEMS), a randomized controlled trial of *Ginkgo biloba* in 3,069 elderly persons aged >75 years designed to determine whether *Ginkgo biloba* slows the rates of global or domain-specific cognitive decline in older adults [9] showed no effects on all-cause dementia, on AD or on the rate of progression to dementia in elderly persons with mild cognitive impairment [42]. 4,066 subjects aged 70 years or older and reporting memory complaints were screened for participation in the GuidAge study, 2,854 of which fulfilled the eligibility criteria to be treated with 120 mg EGb twice a day or placebo for 5 years with the primary endpoint of determining the rate of conversion from memory complaints to AD using survival analysis [2]. Results showed that a long-term use of standardized ginkgo biloba extract did not reduce the risk of progression to Alzheimer's disease compared with placebo [48].

## 16.1 Concluding Remarks

Abundant human data consistently support the idea that oxidative stress occurs and is a constant feature of the AD brain pathology. Some recent evidence even indicates that this phenomenon is an early event and might have a functional role in the pathogenesis of the disease. Human observational epidemiological studies are, in general, consistent with the hypothesis that there is an inverse relationship between antioxidant levels and intake, cognitive function, and ultimately the risk to develop AD. However, randomized clinical trials with antioxidants do not fulfill the promises of those studies. Does this fact mean that the oxidative stress hypothesis of AD is not valid anymore? Does it mean that oxidative stress does not play a functional part and is a simple secondary event in the pathogenesis of this complex disease? Based on the current knowledge, we do not have enough information to clearly answer those important questions. Below we will try to summarize some important aspects that need to be taken into account.

If we look at the negative clinical trials it is curious that they all lack some important information required when an antioxidant therapy is administered, such as drug-levels monitoring and/or a surrogate marker for an in vivo therapeutic effect of the drug of interest. Furthermore, the human studies with antioxidants have used different preparations (natural versus synthetic), a wide range of dosages and for a variable length of time. Taken together, these aspects make a comparison of the obtained results at least difficult, if not impossible. Therefore, the negative trials should be considered inconclusive and these examples unbalanced, and in consequence should not be used to rule out the oxidative stress hypothesis of AD.

Finally, no data are available about the antioxidant status of subjects before and after treatment, so it is still unclear if its efficacy is or is not related to this aspect, with a population more or less sensitive according to the redox status.

In conclusion, the oxidative stress hypothesis of AD is still very much alive, but a great deal of work needs to be done to design future studies and appropriate clinical trials that will conclusively establish the role of oxidative stress in AD.

## References


1. Abad MJ, Bedoya LM, Bermejo P. An update on drug interactions with the herbal medicine Ginkgo biloba. Curr Drug Metab. 2010;11:171–81.
2. Andrieu S, Ousset PJ, Coley N, Ouzid M, Mathiex-Fortunet H, Vellas B, GuidAge study Group. GuidAge study: a 5-year double-blind, randomized trial of EGb 761 for the prevention of Alzheimer's disease in elderly subjects with memory complaints. i. rationale, design and baseline data. Curr Alzheimer Res. 2008;5:406–15.
3. Baum L, Lam CW, Cheung SK, Kwok T, Lui V, Tsoh J, Lam L, Leung V, Hui E, Ng C, Woo J, Chiu HF, Goggins WB, Zee BC, Cheng KF, Fong CY, Wong A, Mok H, Chow MS, Ho PC, Ip SP, Ho CS, Yu XW, Lai CY, Chan MH, Szeto S, Chan IH, Mok V. Six-month randomized, placebo-controlled, double-blind, pilot clinical trial of curcumin in patients with Alzheimer's disease. J Clin Psychopharmacol. 2008;28:110–3.
4. Birks J, Grimley Evans J. Ginkgo biloba for cognitive impairment and dementia. Cochrane Database Syst Rev. 2009;1:CD003120.
5. Chaturvedi RK, Beal MF. Mitochondrial approach for neuroprotection. Ann N Y Acad Sci. 2008;1147:395–412.
6. Commenges D, Scotet V, Renaud S, Jacqmin-Gadda H, Barberger-Gateau P, Dartigues JF. Intake of flavonoids and risk of dementia. Eur J Epidemiol. 2000;16:357–63.
7. Dai Q, Borenstein AR, Wu Y, Jackson JC, Larson EB. Fruit and vegetable juices and Alzheimer's disease: the Kame Project. Am J Med. 2006;119:751–9.
8. Das DK, Mukherjee S, Ray D. Resveratrol and red wine, healthy heart and longevity. Heart Fail Rev. 2010;15:467–77.
9. De Kosky ST, Fitzpatrick A, Ives DG, Saxton J, Williamson J, Lopez OL, Burke G, Fried L, Kuller LH, Robbins J, Tracey R, Woolard N, Dunn L, Kronmal R, Nahin R, Furberg C, GEMS investigators. The Ginkgo Evaluation of Memory (GEM) study: design and baseline data of a randomized trial of Ginkgo biloba extract in prevention of dementia. Contemp Clin Trials. 2006;27:238–53.
10. Engelhart MJ, Geerlings MI, Ruitenberg A, van Swieten JC, Hofman A, Witteman JC, Breteler MM. Dietary intake of antioxidants and risk of Alzheimer's disease. JAMA. 2002;287:3223–9.
11. Galasko DR, Peskind E, Clark CM, Quinn JF, Ringman JM, Jicha GA, Cotman C, Montine T, Thomas RG, Aisen P, Alzheimer's Disease Cooperative Study. Antioxidants for AD: a randomized clinical trials with cerebrospinal fluid biomarker measures. Arch Neurol. 2012;69:836–41.
12. Grodstein F, Chen J, Willett WC. High-dose antioxidant supplements and cognitive function in community-dwelling elderly women. Am J Clin Nutr. 2003;77:975–84.
13. Gutzmann H, Hadler D. Sustained efficacy and safety of idebenone in the treatment of Alzheimer's disease: update on a two-year double-blind multicentre study. J Neural Transm Suppl. 1998;54:301–10.
14. Hager K, Kenklies M, McAfoose J, Engel J, Munch GG. Alpha-lipoic acid as a new treatment option for Alzheimer's disease – a 48 months follow-up analysis. J Neural Transm Suppl. 2007;72:189–93.
15. Hamaguchi T, Ono K, Yamada M. Curcumin and Alzheimer's disease. CNS Neurosci Ther. 2010;16:285–97.
16. Harman D. Aging: a theory based on free radical and radiation chemistry. J Gerontol. 1956;11:298–300.
17. Ho RC, Cheung MW, Fu E, Win HH, Zaw MH, Ng A, Mak A. Is high homocysteine level a risk factor for cognitive decline in elderly? A systematic review, meta-analysis, and meta-regression. Am J Geriatr Psychiatry. 2011;19:607–17.
18. Jones RW. Dimebon disappointment. Alzheimers Res Ther. 2010;2:25.
19. Kennedy DO, Wightman EL, Reay JL, Lietz G, Okello EJ, Wilde A, Haskell CF. Effects of resveratrol on cerebral blood flow variables and cognitive performance in humans: a double-blind, placebo-controlled, crossover investigation. Am J Clin Nutr. 2010;91:1590–7.

20. Laurin D, Masaki KH, Foley DJ, White LR, Launer LJ. Midlife dietary intake of antioxidants and risk of late-life incident dementia: the Honolulu-Asia Aging Study. Am J Epidemiol. 2004;159:959–67.
21. Le WD, Jankovic J, Xie W, Appel SH. Antioxidant property of pramipexole independent of dopamine receptor activation in neuroprotection. J Neural Transm. 2000;107:1165–73.
22. Letenneur L, Proust-Lima C, Le Gouge A, Dartigues JF, Barberger-Gateau P. Flavonoid intake and cognitive decline over a 10-year period. Am J Epidemiol. 2007;165:1364–71.
23. Lotito SB, Frei B. Consumption of flavonoid-rich foods and increased plasma antioxidant capacity in humans: cause, consequence, or epiphenomenon? Free Radic Biol Med. 2006;41: 1727–46.
24. Luchsinger JA, Wang MX, Shea S, Mayeux R. Antioxidant vitamin intake and risk of Alzheimer disease. Arch Neurol. 2003;60:203–8.
25. Maczurek A, Hager K, Kenklies M, Sharman M, Martins R, Engel J, Carlson DA, Munch G. Lipoic acid as an anti-inflammatory and neuroprotective treatment for Alzheimer's disease. Adv Drug Deliv Rev. 2008;60:1463–70.
26. Mangialasche F, Polidori MC, Monastero R, Ercolani S, Camarda C, Cecchetti R, Mecocci P. Biomarkers of oxidative and nitrosative damage in Alzheimer's disease and mild cognitive impairment. Ageing Res Rev. 2009;8:285–305.
27. Mangialasche F, Solomon A, Winblad B, Mecocci P, Kivipelto M. Alzheimer's disease: clinical trials and drug development. Lancet Neurol. 2010;9:702–16.
28. Mangialasche F, Xu W, Kivipelto M, Costanzi E, Ercolani S, Pigliautile M, Cecchetti R, Baglioni M, Simmons A, Soininen H, Tsolaki M, Kloszewska I, Vellas B, Lovestone S, Mecocci P, AddNeuroMed Consortium. Tocopherols and tocotrienols plasma levels are associated with cognitive impairment. Neurobiol Aging. 2012;33:2282–90.
29. Mariani E, Polidori MC, Cherubini A, Mecocci P. Oxidative stress in brain aging, neurodegenerative and vascular diseases: an overview. J Chromatogr B Analyt Technol Biomed Life Sci. 2005;827:65–75.
30. Masaki HK, Losonczy KG, Izmirlian G, Foley DJ, Ross GW, Petrovitch H, Havlik R, White LR. Association of vitamin E and C supplement use with cognitive function and dementia in elderly men. Neurology. 2000;54:1265–72.
31. Morillas-Ruiz JM, Rubio-Perez JM, Albaladejo MD, Zafrilla P, Parra S, Vidal-Guevara ML. Effect of an antioxidant drink on homocysteine levels in Alzheimer's patients. J Neurol Sci. 2010;299:175–8.
32. Morris MC, Beckett LA, Scherr PA, Hebert LE, Bennett DA, Field TS, Evans DA. Vitamin E and vitamin C supplement use and risk of incident Alzheimer disease. Alzheimer Dis Assoc Disord. 1998;12:121–6.
33. Ono K, Hasegawa K, Yoshiike Y, Takashima A, Yamada M, Naiki H. Nordihydroguaiaretic acid potently breaks down preformed Alzheimer's β-amyloid fibrils in vitro. J Neurochem. 2002;81:434–40.
34. Parasassi T, Brunelli R, Costa G, De Spirito M, Krasnowska E, Lundeberg T, Pittaluga E, Ursini F. Thiol redox transitions in cell signaling: a lesson from N-acetylcysteine. Scientific World Journal. 2010;10:1192–202.
35. Petersen RC, Thomas RG, Grundman M, Bennett D, Doody R, Ferris S, Galasko D, Jin S, Kaye J, Levey A, Pfeiffer E, Sano M, van Dyck CH, Thal LJ. Vitamin E and donepezil for the treatment of mild cognitive impairment. N Engl J Med. 2005;352:2379–88.
36. Praticò D, Clark CM, Liun F, Rokach J, Lee VY, Trojanowski JQ. Increase of brain oxidative stress in mild cognitive impairment: a possible predictor of Alzheimer disease. Arch Neurol. 2002;59:972–6.
37. Rinaldi P, Polidori MC, Metastasio A, Mariani E, Mattioli P, Cherubini A, Catani M, Cecchetti R, Senin U, Mecocci P. Plasma antioxidants are similarly depleted in mild cognitive impairment and in Alzheimer's disease. Neurobiol Aging. 2003;24:915–9.
38. Ringman JM, Cole GM, Tend E. Oral curcumin for the treatment of mild-to-moderate Alzheimer's disease: tolerability and clinical and biomarker efficacy results of a placebo-controlled 24-week study. Proceedings of the International Conference of Alzheimer's disease 2008, Chicago, abstract.

39. Sano M, Ernesto C, Thomas RG, Klauber MR, Schafer K, Grundman M, Woodbury P, Growdon J, Cotman CW, Pfeiffer E, Schneider LS, Thal LJ. A controlled trial of selegiline, alpha-tocopherol, or both as treatment for Alzheimer's disease. The Alzheimer's Disease Cooperative Study. N Engl J Med. 1997;336:1216–22.
40. Sies H. Oxidative stress: introductory remarks. In: Sies H, editor. Oxidative Stress. Orlando: Academic; 1985.
41. Smith M, Nunomura A, Lee HG, Zhu X, Moreira PI, Avila J, Perry G. Chronological primacy of oxidative stress in Alzheimer disease. Neurobiol Aging. 2005;26:579–80.
42. Snitz BE, O'Meara ES, Carlson MC, Arnold AM, Ives DG, Rapp SR, Saxton J, Lopez OL, Dunn LO, Sink KM, DeKosky ST, Ginkgo evaluation of memory (GEM) study investigators. Ginkgo biloba for preventing cognitive decline in older adults: a randomized trial. JAMA. 2009;302:2663–70.
43. Steele JW, Lachenmayer ML, Ju S, Stock A, Liken J, Kim SH, Delgado LM, Alfaro IE, Bernales S, Verdile G, Bharadwaj P, Gupta V, Barr R, Friss A, Dolios A, Wang R, Ringe D, Fraser P, Westaway D, St George-Hyslop PH, Szabo P, Relkin NR, Buxbaum JD, Glabe CG, Protter AA, Martins RR, Ehrlich ME, Petsko GA, Yue Z, Gandy S. Latrepirdine improves cognition and arrests progression of neuropathology in an Alzheimer's mouse model, Mol Psychiatry, 2012; [Epub ahead of print].
44. Stocker R. Vitamin E. Novartis Found Symp. 2007;282:77–87.
45. Su B, Wang X, Bonda D, Perry G, Smith M, Zhu X. Abnormal mitochondrial dynamics – a novel therapeutic target for Alzheimer's disease? Mol Neurobiol. 2010;41:87–96.
46. Sung S, Yao Y, Uryu K, Yang H, Lee VM, Trojanowski JQ, Pratico D. Early vitamin E supplementation in young but not aged mice reduces Abeta levels and amyloid deposition in a transgenic model of Alzheimer's disease. FASEB J. 2004;18:323–5.
47. Thal LJ, Grundman M, Berg J, Ernstrom K, Margolin R, Pfeiffer E, FF M, Weiner E, Zamrini R, Thomas G. Idebenone treatment fails to slow cognitive decline in Alzheimer's disease. Neurology. 2003;61:1498–502.
48. Vellas B, Coley N, Ousset PJ, Berrut G, Dartigues JF, Dubois B, Grandjean H, Pasquier F, Piette F, Robert P, Touchon J, Garnier P, Mathiex-Fortunet H, Andrieu S, GuidAge Study Group. Long-term use of standardised Ginkgo biloba extract for the prevention of Alzheimer's disease (GuidAge): a randomised placebo-controlled trial. Lancet Neurol. 2012;11:851–9.
49. Williams RJ, Spencer JP, Rice-Evans C. Flavonoids: antioxidants or signaling molecules? Free Radic Biol Med. 2004;36:838–49.
50. Yaffe K, Clemons TE, McBee WL, Lindblad AS. Impact of antioxidants, zinc and copper on cognition in the elderly: a randomized, controlled trial. Neurology. 2004;63:1705–7.
51. Zandi PP, Anthony JC, Khachaturian AS, Stone SV, Gustafson D, Tschanz JT, Norton MC, Welsh-Bohmer KA, Breitner JC. Reduced risk of Alzheimer disease in users of antioxidant vitamin supplements: the Cache County Study. Arch Neurol. 2004;61:82–8.
52. Zhang S, Hedskog L, Petersen CA, Winblad B, Ankarcrona M. Dimebon (latrepirdine) enhances mitochondrial function and protects neuronal cells from death. J Alzheimers Dis. 2010;21:389–402.
53. Zhu N, Huang TC, Yu Y, LaVoie EJ, Yang CS, Ho CT. Identification of oxidation products of (-)-epigallocatechin gallate and (-)-epigallocatechin with H(2)O(2). J Agric Food Chem. 2000;48:979–81.

# About the Editors

**Dr. Domenico Praticò** received his M.D. from the University of Rome "La Sapienza," School of Medicine where he completed a residency program in Internal Medicine. After receiving a European Research Award fellowship, he spent 2 years of postdoctoral training at the University of Dublin as a fellow in the Center for Cardiovascular Science. Next, he joined the University of Pennsylvania where he rose to the the rank of Associate Professor. He is currently Professor of Pharmacology, Microbiology, and Immunology and member of the Center for Translational Medicine at Temple University. His main research interest is oxidative biology with a particular focus on bioactive lipids, their metabolic pathways, and their role as biological mediators of cellular and molecular events in the pathogenesis of human diseases, including neurodegenerative and cardiovascular diseases. Dr. Pratico has published more than 200 papers in peer-reviewed scientific journals and has received many awards for his research accomplishments including the Irvine H. Page Award, Neurosciences Education and Research Award, The Zenith Award from the Alzheimer's Association, the Dorothy Dillon Eweson Lectureship, and the Alzheimer's Art Quilt Initiative Award.

**Dr. Patrizia Mecocci** received her M.D. from the University of Perugia, School of Medicine where she completed a residency program in Geriatrics. She also earned a Ph.D. in Biology and Physiopathology of Aging at the University of Modena (Italy). After receiving a grant from the Italian National Research Council (CNR), she spent 3 years as research fellow first at the Department of Psychiatry and Neurochemistry at the University of Lund (Sweden) and then at the Department of Neurology, MGH-Harvard Medical School in Boston. She then started the academic career at the University of Perugia where she is now Full Professor in Gerontology and Geriatrics and Director of the Geriatric Clinic. Dr. Mecocci's research is mainly devoted to clinical geriatrics, with a specific interest in brain aging and dementia and in age-related biological modifications, with a particular focus on the role of oxidants and antioxidants in neurodegenerative and cardiovascular diseases. She has been PI or coinvestigator in research projects funded by the European Commission and by the Italian Ministry of Research. She is the author of more than 180 papers in peer-reviewed scientific journals and 30 books and book chapters.

D. Praticò and P. Mecocci (eds.), *Studies on Alzheimer's Disease*, Oxidative Stress in Applied Basic Research and Clinical Practice, DOI 10.1007/978-1-62703-598-9,

# Index



D. Praticò and P. Mecocci (eds.), *Studies on Alzheimer's Disease*, Oxidative Stress in Applied Basic Research and Clinical Practice, DOI 10.1007/978-1-62703-598-9,

MIX
Papier aus verantwortungsvollen Quellen
Paper from responsible sources
FSC® C105338

If you have any concerns about our products,
you can contact us on
**ProductSafety@springernature.com**

In case Publisher is established outside the EU,
the EU authorized representative is:
**Springer Nature Customer Service Center GmbH**
**Europaplatz 3, 69115 Heidelberg, Germany**

Printed by Libri Plureos GmbH
in Hamburg, Germany